Lab Manual to Accompany Introduction to Digital Electronics

Kenneth J. Reid
Indiana University Purdue University Indianapolis,
Indianapolis, Indiana

Robert K. Dueck
Red River College, Winnipeg, Manitoba

Australia • Brazil • Japan • Korea • Mexico • Singapore • Spain • United Kingdom • United States

Lab Manual to Accompany Introduction to Digital Electronics
Kenneth J. Reid, Robert K. Dueck

Vice President, Technology and Trades ABU: David Garza

Director of Learning Solutions: Sandy Clark

Managing Editor: Larry Main

Executive Editor: Stephen Helba

Senior Product Manager: Michelle Ruelos Cannistraci

Marketing Director: Deborah S. Yarnell

Marketing Manager: Guy Baskaran

Marketing Coordinator: Shanna Gibbs

Director of Production: Patty Stephan

Production Manager: Andrew Crouth

Associate Content Project Manager: Niamh Matthews-Schweitzer

Art Director: Benjamin Gleeksman

Production Technology Analyst: Thomas Stover

Senior Editorial Assistant: Dawn Daugherty

For product information and technology assistance, contact us at
Cengage Learning Customer & Sales Support, 1-800-354-9706

For permission to use material from this text or product,
submit all requests online at **www.cengage.com/permissions**
Further permissions questions can be emailed to
permissionrequest@cengage.com

ISBN-13: 978-1-4180-4104-5

ISBN-10: 1-4180-4104-1

Delmar
Executive Woods
5 Maxwell Drive
Clifton Park, NY 12065
USA

Cengage Learning is a leading provider of customized learning solutions with office locations around the globe, including Singapore, the United Kingdom, Australia, Mexico, Brazil, and Japan. Locate your local office at **www.cengage.com/global**

Cengage Learning products are represented in Canada by Nelson Education, Ltd.

To learn more about Delmar, visit **www.cengage.com/delmar**

Purchase any of our products at your local bookstore or at our preferred online store **www.cengagebrain.com**

Printed in the United States of America
2 3 4 5 6 16 15 14 13 12

FD128

Contents

Preface

Intended Audience

This series of labs is intended for use in a digital systems or digital design sequence, as part of a Computer/Electronics Engineering Technology (CET/EET) program. The labs are based on material in *Introduction to Digital Electronics* (Ken Reid and Robert Dueck, Delmar, Cengage Learning, © 2008).

Programmable Logic as a Vehicle for Teaching Digital Design

Historically, digital logic or digital design courses at the EET level have focussed on using fixed-function TTL and CMOS integrated circuits as the vehicle for teaching principles of logic design. However, the digital design field has changed; nearly all digital designs are being implemented in Programmable Logic Devices (PLDs), rendering popular fixed-function devices virtually obsolete.

This lab manual, and the textbook it accompanies, address this trend by focusing primarily on PLDs as a vehicle for teaching digital logic. The labs use the Student Edition of Altera's Quartus II PLD design and programming software. The exercises in this manual are configured for use with either of two hardware platforms: the Altera UP-2 University Program Laboratory Design Package or the RSR PLDT-2 Programmable Logic Trainer. The DeVry eSOC board is another acceptable platform.

Both boards contain an Altera EPM7128S Complex Programmable Logic Device (CPLD), a Sum-of-Products (SOP) device with 128 macrocells, based on EEPROM technology. This chip is in-system programmable and thus can be programmed and erased multiple times, via a cable from a PC parallel port, without removing it from the board. In addition to the CPLD, both boards include a number of standard input and output devices, such as switches, LEDs, and seven-segment numerical displays.

The Quartus II software is bundled with the *Introduction to Digital Electronics* textbook. Each installed copy must be activated by a license file available, free of charge, by e-mail from the Altera website (**www.altera.com**). More detailed information on software installation and licensing can be found in Labs 8 and 9 in this lab manual.

The Altera UP-2 circuit board is available from Altera's University Program for sale to students ($99.00, as of September 2007) or can be requested by educational institutions, either for purchase or on a donation basis for those institutions that are members of the Altera University Program. Institutions can also request donations of full-version software and CPLD chips.

The RSR PLDT-2 board, available from RSR Electronics/Electronix Express (**http://www.elexp.com**), can be purchased for $79.00 (as of September 2007). This board has some features, such as jumpered inputs and outputs, debounced switches, clearly numbered pin connections, active-HIGH LEDs, and more wiring room around the chip, that make it more user-friendly than the Altera UP-2 board.

Traditional SSI Labs

Many instructors find it useful to start the digital lab sequence with a few exercises based on more traditional Small Scale Integrated (SSI) Circuit components. Hands-on wiring of devices on a breadboard teaches skills in circuit construction that aren't found in software-based design entry. Labs 2 through 7 give students opportunity for circuit breadboarding; students will get to practice finding devices, pin numbers, wiring discrete components, and connecting (or forgetting to connect) power supplies to circuits. These labs will usefully reinforce basic digital principles for the student in the first few weeks of the course before the CPLD-based material is introduced.

CPLD Labs

Labs 8, 9, and 10 cover the installation and licensing of Altera Quartus II software. If students are familiar with the design entry features of Quartus II, it may be used in labs 6 and 7, although it is not required. Labs 11 and beyond make extensive use of the design entry and simulation features of Quartus II, including schematic capture and use of components from the Altera Library of Parameterized Modules (LPM). Labs are stepped in difficulty from initial exercises, where students are asked only to follow given procedures without making too many decisions, to later labs with broader directions, requiring some independent decision making. The principles taught are brought together in design projects: combinational design projects are found in labs 21 and 22; sequential design projects are found in labs 27, 30, and 32. More advanced labs include lab 31 (state machines) and labs 36, 37, and 38 (A/D and D/A conversion).

Class Time Requirements

Most of the labs in this manual are suitable for traditional two-hour lab periods assuming that students are required to prepare in advance. Labs begin with Experimental Notes, which should be studied prior to lab, and certainly prior to beginning the procedure.

Early labs involving a more traditional focus on TTL devices and breadboarding may require additional time. However, when programming CPLDs with Quartus II, the majority of design setup is done in software. This lends a flexibility and portability for students who may now work on their labs on their own time, at home or in a computer lab. Hardware is then required only for final design test and demonstration.

Labs written as projects (labs 21, 22, 27, 30, and 32) may be extended over multiple lab sessions and/or assigned as team projects as the instructor wishes. Other labs that cover related topics may be combined: for example, labs 14 (Multiplexers) and 15 (Demulitplexers) may be assigned in a single lab period; this would, of course, require additional work on the part of the student to be adequately prepared.

CPLD pin assignments for labs 11 through 34 (with possible exceptions for project labs) are standard; this allows students to wire the switch and LED connections on the Altera UP-2 or RSR PLDT-2 boards once and leave the wires in place for the next lab period without the need to rewire. This wiring procedure is found in lab 10.

Due to differences between the Altera UP-2 board and RSR PLDT-2 board (active-HIGH vs. active-LOW inputs and outputs and different oscillator speeds), students will need to make adjustments if moving from one type of board to another. These differences are outlined in individual lab exercises.

Hardware and Software Requirements

Lab 1:
This lab involves searching the Internet for information on digital devices and assembling a library for the student's use. Other than Internet access, no equipment is required.

Labs 2–7:
These labs involve SSI devices and breadboarding. Specific hardware requirements are listed in each lab, but generally consist of a few selected TTL devices, breadboard, power supply, and wires.

Labs 8–10:
These labs involve the downloading, installation, and licensing of Altera Quartus II software. All modern computer systems should be able to run Quartus II. Lab 10 covers a standard wiring procedure for a selection of hardware development boards. Most universities choose a single development board platform: either the Altera UP-2, RSR Electronics PLDT-2, or DeVry eSOC board. Students and instructors choosing the UP-2 or PLDT-2 board should be aware that these program through a connection to the computer's parallel port: newer laptop systems without a parallel port may not be able to support these boards.

Labs 11–35
These labs use the selected development board and Altera Quartus II software.

Labs 36–38
These labs use the same development board and Quartus II, but require additional hardware as specified in each lab.

Quartus II Design Files

Two sets of design and programming files are available:

1. **For Instructors:** a full set of required Graphic Design Files (gdf), VHDL files (vhd), and Programmer Object Files (pof) for all labs, as well as the related project files for each design. These files are available from Delmar, Cengage Learning as part of the e.resource package for *Introduction to Digital Electronics* (ISBN: 1-4180-4103-3).
2. **For Students:** a limited number of design entry files required in some CPLD labs. These are generally for components that are needed to make a particular design work properly, but which the student is not expected to create at that point in the lab sequence. These files are included on the CD-ROM that accompanies the textbook *Introduction to Digital Electronics,* ISBN: 1-4180-4102-5, Ken Reid & Robert Dueck.

Acknowledgments

The author and Delmar Publishers would like to thank the following reviewers:

Pravin Patel, Durham College, Oshawa, Ontario

Billy Graham, Northwest Technical Institute, Springdale, Arizona

William Routt, Wake Technical Community College, Raleigh, North Carolina

Norm Grossman, DeVry University, Long Beach, California

Lab 1

Creating a Library of Electronic Component Datasheets

Name ______________________ Class ____________ Date ____________

"A technical degree is good for about five years. After that, everything you know is from vendors' literature."—Attributed to Dilbert

It's sad, but true. After all your hard work in acquiring a college diploma or university degree, your training will be obsolete within a few short years. However, this also means that you are becoming part of a dynamic, ever-changing field that holds new challenges for the foreseeable future. In the context of that climate, it is essential that you know how to search out information about electronic components and devices that are new to you. In this assignment, you will use some of the available resources to create a simple data library of electronic components, use the library to label a digital circuit, and use a component catalog to calculate the component cost for the circuit. You will be able to add to this library as you require new information.

Objectives

1. Using part numbers or part descriptions, find datasheets from:
   - Search engines (Google)
   - Datasheet archives (alldatasheet.com)
   - Component catalogs (DigiKey)
   - Manufacturer's websites (e.g., Texas Instruments, Freescale Semiconductor, Fairchild, or Philips)
2. Create a portfolio of datasheets in an electronic form on one of the following media:
   - USB flash drive, or;
   - a separate folder on a department network drive (if available), or;
   - CD-RW (not CD-R; you want to be able to add to it in the future).
3. Index the datasheets in a separate document (e.g., a Word document or Excel spreadsheet) so that they can easily be referenced. ***The datasheet library can be added to as new components are required or discovered. Your index system should accommodate this.***
4. Use the datasheets to label a digital circuit with component designators, part numbers, and pin numbers.
5. Calculate the cost of components required for the circuit.

Reference Ken Reid and Robert Dueck, *Introduction to Digital Electronics*

- Search engine: www.google.com
- Datasheet archive: www.alldatasheet.com
- Component catalog: www.digikey.com
- Manufacturers' websites: www.fairchildsemi.com
  www.nxp.com (Philips)
  www.onsemi.com/PowerSolutions/home.do
  www.ti.com

Experimental Notes

Electronic components, such as digital logic gates, are described by datasheets that specify component connection and function.

We can use information on a datasheet to show how to connect two or more gates, as shown in Figure 1.1. In this figure, pin 3 of gate *A* is connected to pin 4 of gate *B*.

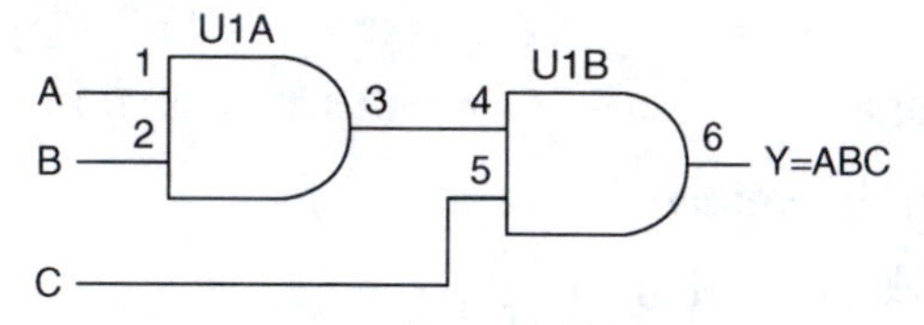

Figure 1.1 Showing Pin Numbers Derived from a Single Device Datasheet

The device can be wired as shown in Figure 1.2 to make the circuit shown in Figure 1.1. This figure shows the placement of the four AND gates in the device, relative to its input and output pins. Inputs *A*, *B*, and *C* (from logic switches) are connected to pins 1, 2, and 5. Output *Y* (to an LED) is connected to pin 6. In this circuit, an active-HIGH LED comes ON when *A*, *B*, and *C* are all HIGH.

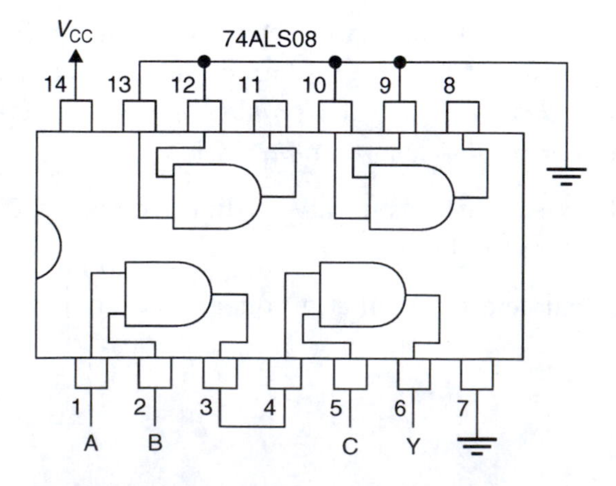

Figure 1.2 Wiring Diagram for the Circuit in Figure 1.1

A circuit having two or more devices would have separate names for each part, with pin numbers selected from a relevant datasheet. In Figure 1.3, the two parts are designated as *U1* and *U2*. Pin 3 of gate *A* in *U1* is connected to pin 1 of gate *A* in *U2*. Inputs *A* and *B* (from logic switches) are connected to pins 1 and 2 of gate *U1A*. Input *C* (logic switch) is connected to pin 2 of gate *U2A*. Output *Y* (to an LED) is connected to pin 3 of gate *U2A*. In this circuit, an active-LOW LED comes ON when *A*, *B*, and *C* are all HIGH.

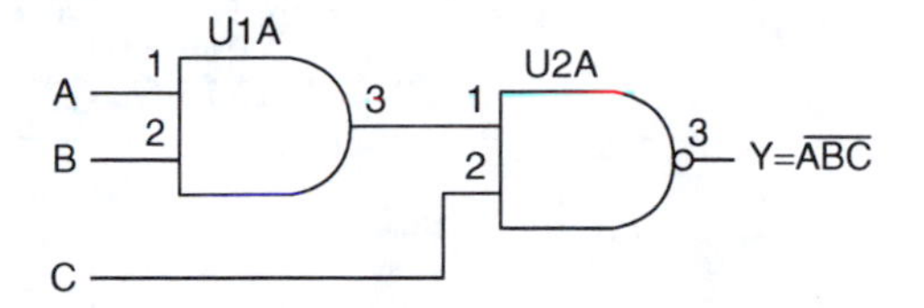

U1: 74ALS08 Quadruple 2-input AND gate
U2: 74ALS00 Quadruple 2-input NAND gate

Figure 1.3 Pin Numbers of Gates from Two Devices

The circuit in Figure 1.3 can be wired as shown in Figure 1.4.

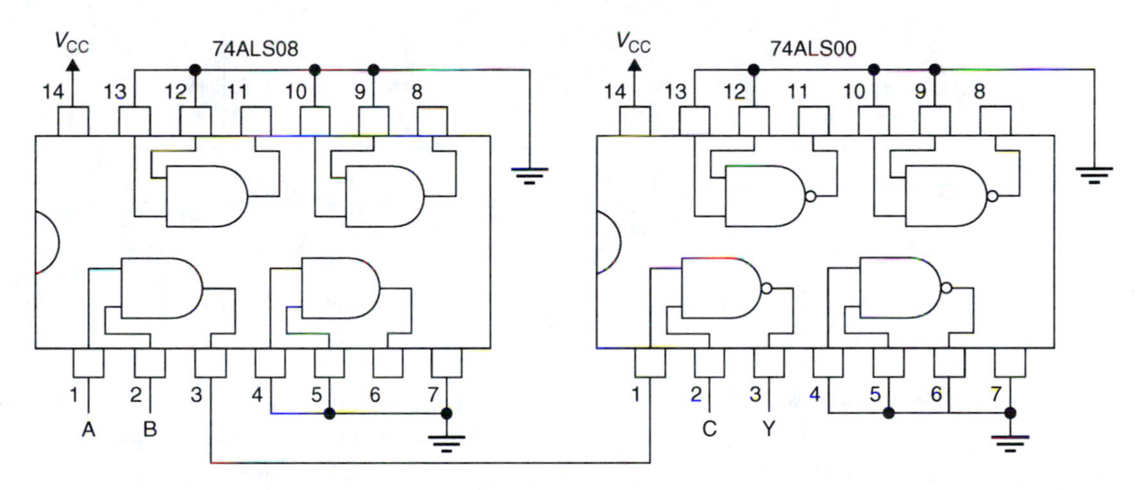

Figure 1.4 Wiring Diagram for the Circuit in Figure 1.3

- We can determine the cost of a circuit by looking up its parts in a catalog, such as the one published online by DigiKey; use the website www.digikey.com and search for 74ALS.

Figure 1.5 shows a partial result from the search for a 74ALS08 device. Figure 1.6, on the following page, shows a detailed view of the table in Figure 1.5, showing the part number and price more clearly.

Digi-Key Part Number	Manufacturer Part Number	Description	Vendor	Number of Circuits	Series	Package / Case	Logic Type	Features	Packaging	Quantity Available	Unit Price Min Qty
296-1123-1-ND	SN74ALS08DR	IC QUAD 2IN POS-AND GATE 14-SOIC	Texas Instruments	4 Quad	74ALS	14-SOIC	2 Input AND	-	Cut Tape (CT)	0	0.72000 1
296-1123-2-ND	SN74ALS08DR	IC QUAD 2IN POS-AND GATE 14-SOIC	Texas Instruments	4 Quad	74ALS	14-SOIC	2 Input AND	-	Tape & Reel (TR)	2500	0.23400 2500
296-1123-5-ND	SN74ALS08D	IC QUAD 2-INPUT AND GATE 14-SOIC	Texas Instruments	4 Quad	74ALS	14-SOIC	2 Input AND	-	Tube	972	0.72000 1
296-1123-6-ND	SN74ALS08DR	IC QUAD 2IN POS-AND GATE 14-SOIC	Texas Instruments	4 Quad	74ALS	14-SOIC	2 Input AND	-	Digi-Reel	0	0.72000 1
296-1684-5-ND	SN74ALS08N	IC QUAD 2-INPUT AND GATE 14-DIP	Texas Instruments	4 Quad	74ALS	14-DIP	2 Input AND	-	Tube	944	0.76000 1

Figure 1.5 Sample Component Cost from DigiKey

Suppose we want to find this device in a 14-pin Dual In-Line Package (14-DIP). We have one choice from the table: the fifth entry. The fifth entry is available from Texas Instruments, minimum quantity 1, at $0.76, with 944 units available. This is a good choice if you require immediate availability.

Digi-Key Part Number	Manufacturer Part Number	Description	Vendor	Number of Circuits	Series	Package / Case	Logic Type	Packaging	Quantity Available	Unit Price Min Qty
296-1123-1-ND	SN74ALS08DR	IC QUAD 2IN POS-AND GATE 14-SOIC	Texas Instruments	4 Quad	74ALS	14-SOIC	2 Input AND	Cut Tape (CT)	0	0.72000 1
296-1123-2-ND	SN74ALS08DR	IC QUAD 2IN POS-AND GATE 14-SOIC	Texas Instruments	4 Quad	74ALS	14-SOIC	2 Input AND	Tape & Reel (TR)	2500	0.23400 2500
296-1123-5-ND	SN74ALS08D	IC QUAD 2-INPUT AND GATE 14-SOIC	Texas Instruments	4 Quad	74ALS	14-SOIC	2 Input AND	Tube	972	0.72000 1
296-1123-6-ND	SN74ALS08DR	IC QUAD 2IN POS-AND GATE 14-SOIC	Texas Instruments	4 Quad	74ALS	14-SOIC	2 Input AND	Digi-Reel	0	0.72000 1
296-1484-5-ND	SN74ALS08N	IC QUAD 2-INPUT AND GATE 14-DIP	Texas Instruments	4 Quad	74ALS	14-DIP	2 Input AND	Tube	944	0.76000 1

Figure 1.6 Detail of DigiKey Search Result

Note that the second table entry is not a good choice for two reasons: wrong type of package (14-pin Small Outline Integrated Circuit; it won't fit on a prototyping breadboard) and large minimum quantity (2500 units). However, if these conditions suit your purpose, the part is only $0.234 per unit or $585 for 2500 units.

We can perform a similar search for the 74ALS00 NAND gate. The cost of the circuit in Figure 1.3 can be calculated as shown in Table 1.1.

Table 1.1 Component Cost for Figure 1.3

Quantity	Component Number	Device	Part Number	Manufacturer	Package	Unit Price	Extended Price
1	U1	Quad 2-input AND	SN74ALS08N	Texas Instruments	14-DIP	$0.76	$0.76
1	U2	Quad 2-input NAND	SN74ALS00N	Texas Instruments	14-DIP	$0.80	$0.80
Total							$1.56

The cost of the circuit in Figure 1.1 would be $0.76, since it uses only one 74ALS08 device, which contains four gates. (It may not be a good idea to order this small a quantity of goods, since most mail-order sites charge for shipping.)

Procedure

1. **Find datasheets** for the following devices, shown in Figure 1.7:
    - 74ALS04 Hex inverter
    - 74ALS11 Triple 3-input AND
    - 74ALS21 Dual 4-input AND
    - 74ALS27 Triple 3-input NOR

    Use two or more sources for locating the datasheets. In your index document (see below), list the Web link that indicates the source of each datasheet. **Save the datasheets electronically,** on one of the following storage media: a USB flash drive, your network drive, or a CD-RW (not a CD-R).

Figure 1.7 Sample Digital Circuit

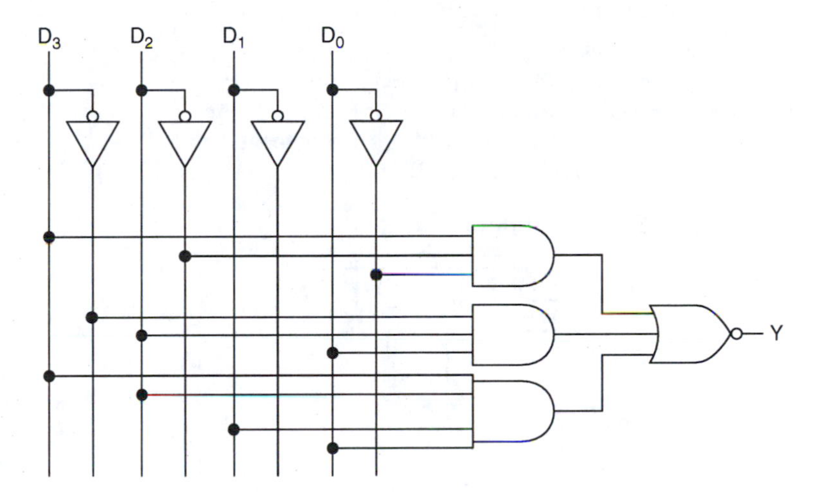

2. **Create an index document** that allows easy location of the datasheets within the data library. It should contain the following information about each datasheet: part number, part description, file name, manufacturer, and Internet source, including date accessed. A sample is shown in Table 1.2, below. A Word document, Excel spreadsheet, database file, or pdf document would be suitable examples of an index document.

Table 1.2 Sample Index Document

Part Number	Description	File Name	Manu-facturer	Internet Source
DM74ALS00A	Quad 2-in NAND	DM74ALS00.pdf	Fairchild Semi-conductor	http://www.fairchildsemi.com/ds/DM/DM74ALS00A.pdf Accessed May 9, 2006
SN74ALS08N	Quad 2-in AND	SN74ALS08.pdf	Texas Instruments	http://pdf1.alldatasheet.com/datasheet-pdf/view/27693/TI/SN74ALS08N.html Accessed May 9, 2006

3. **Redraw the circuit** in Figure 1.7 on a separate sheet. **Label the gates** shown in Figure 1.7 with part and pin numbers, such as those shown in Figure 1.3, assuming each device is an N-package DIP. Include a list that associates each component designation, (e.g., U1A) with a part number (e.g., 74ALSxx).

4. **Calculate a cost** for the circuit, using the DigiKey catalog. The information should be similar to that shown in Table 1.1.

Appendix 1A: Sample Datasheet (first page)

SN54ALS08, SN54AS08, SN74ALS08, SN74AS08
QUADRUPLE 2-INPUT POSITIVE-AND GATES

SDAS191A – APRIL 1982 – REVISED DECEMBER 1994

- **Package Options Include Plastic Small-Outline (D) Packages, Ceramic Chip Carriers (FK), and Standard Plastic (N) and Ceramic (J) 300-mil DIPs**

description

These devices contain four independent 2-input positive-AND gates. They perform the Boolean functions $Y = A \bullet B$ or $Y = \overline{\overline{A} + \overline{B}}$ in positive logic.

The SN54ALS08 and SN54AS08 are characterized for operation over the full military temperature range of −55°C to 125°C. The SN74ALS08 and SN74AS08 are characterized for operation from 0°C to 70°C.

FUNCTION TABLE
(each gate)

INPUTS		OUTPUT
A	B	Y
H	H	H
L	X	L
X	L	L

SN54ALS08, SN54AS08 . . . J PACKAGE
SN74ALS08, SN74AS08 . . . D OR N PACKAGE
(TOP VIEW)

Signal	Pin	Pin	Signal
1A	1	14	V_{CC}
1B	2	13	4B
1Y	3	12	4A
2A	4	11	4Y
2B	5	10	3B
2Y	6	9	3A
GND	7	8	3Y

SN54ALS08, SN54AS08 . . . FK PACKAGE
(TOP VIEW)

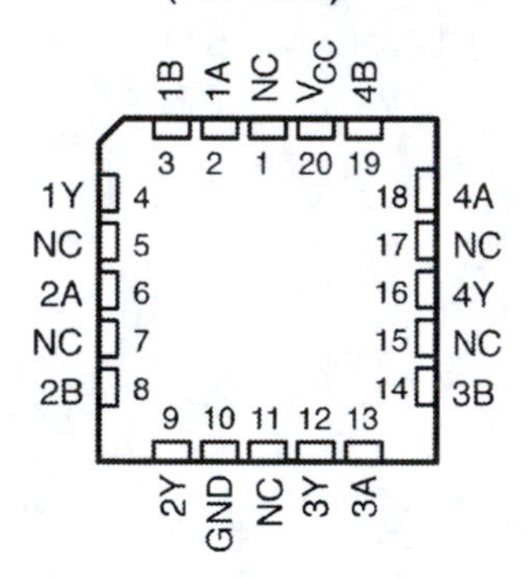

NC – No internal connection

logic symbol†

Input	Pin		Pin	Output
1A	1	&	3	1Y
1B	2			
2A	4		6	2Y
2B	5			
3A	9		8	3Y
3B	10			
4A	12		11	4Y
4B	13			

† This symbol is in accordance with ANSI/IEEE Std 91-1984 and IEC Publication 617-12.
Pin numbers shown are for the D, J, and N packages.

logic diagram (positive logic)

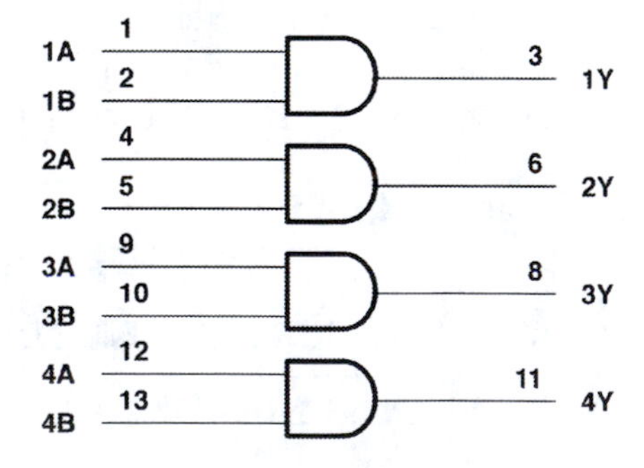

Lab 2

Logic Switches and LED Indicators

Name ______________________ Class __________ Date __________

Objectives

- To create a circuit for a logic switch using DIP switches and SIP resistor networks.
- To demonstrate the operation of light-emitting diodes (LEDs).

Reference

Ken Reid and Robert Dueck, *Introduction to Digital Electronics*

- Chapter 2, Logic Function and Gates
- Section 2.5, Logic Switches and LED Indicators

Equipment Required

Qty.	Description
1	Digital trainer or 5-volt power supply
1	10 kΩ SIP resistor network
1	330Ω resistor
1	DIP switch
1	LED
1	74LS04 Hex Inverter Integrated Circuit

Experimental Notes

Logic Switches

Figure 2.1 shows a single-pole single-throw (SPST) switch connected as a logic switch.

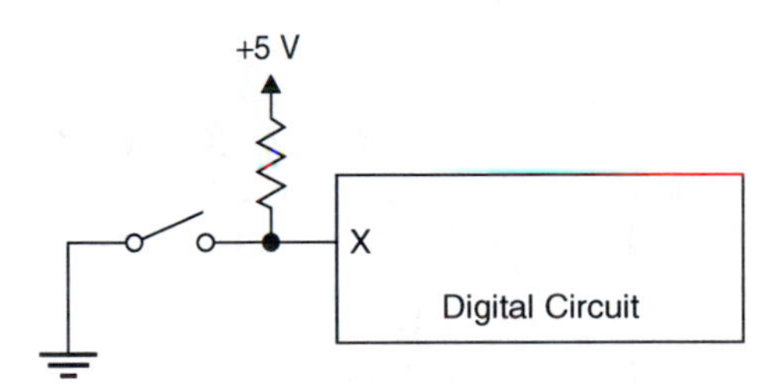

Figure 2.1 SPST Logic Switch

An important premise of the logic switch circuit is that the digital circuit input to which it is connected has a very high resistance to current. When the switch is open, the current flowing through the **pull-up resistor** from +5 volts to the digital circuit is very small. Since the current is small, Ohm's Law states that very little voltage drops across the pull-up resistor; the voltage is about the same at one end as at the other. Therefore, an open switch generates a logic HIGH at point X.

When the switch is closed, the majority of current flows to ground, limited only by the value of the pull-up resistor (typically between 1 kΩ and 10 kΩ). Point X is approximately at ground potential, or logic LOW. Thus, the switch generates a HIGH when open and a LOW when closed. The pull-up resistor provides a connection to +5 V in the HIGH state and limits power supply current in the LOW state.

LED Indicators

A device used to indicate the status of a digital output is the **light-emitting diode** or **LED**. (This is sometimes pronounced as a word ["led"] and sometimes said as separate initials ["ell ee dee"].) The circuit symbol, shown in Figure 2.2, has two terminals, called the anode (positive) and cathode (negative). The arrow coming from the symbol indicates emitted light.

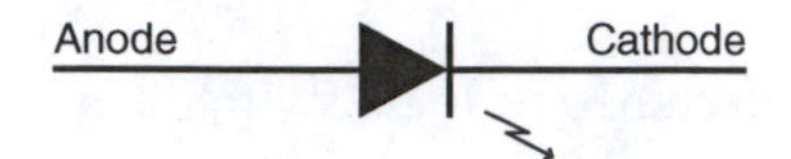

Figure 2.2 LED Showing Anode and Cathode Terminals

The electrical requirements for the LED are simple: current flows through the LED if the anode is more positive than the cathode by more than a specified value (about 1.5 volts). If enough current flows, the LED illuminates. If more current flows, the illumination is brighter. (If too much flows, the LED burns out, so a series resistor is used to keep the current in the required range.) Figure 2.3 shows a circuit in which an LED illuminates when a switch is closed.

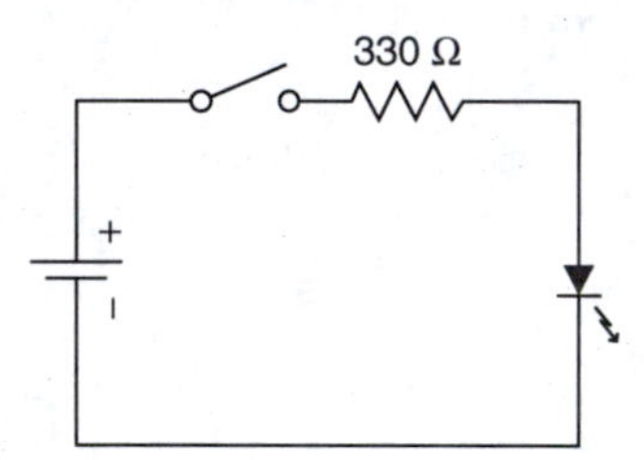

Figure 2.3 LED Turns ON when Switch Closes

The left side of Figure 2.4 shows a NOT gate (inverter) driving an LED. The LED is on when the inverter output is HIGH (~5 volts), since the anode of the LED is more positive than the cathode. This happens when the inverter input A is LOW (~0 volts).

In the diagram on the right side of Figure 2.4, the LED turns on when the inverter output is LOW (~0 volts), again since the anode is more positive than the cathode. This happens when input the inverter input A is HIGH.

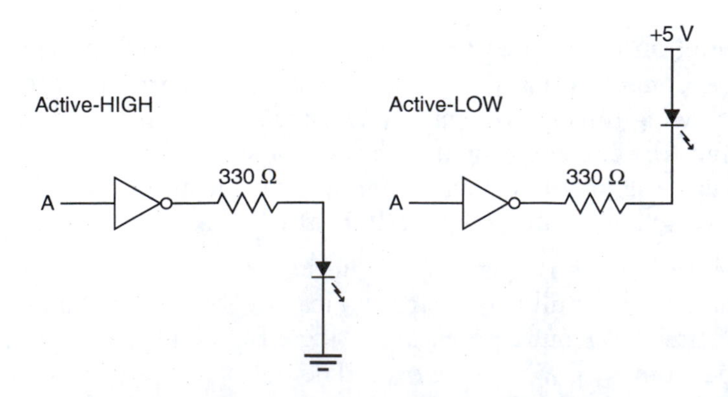

Figure 2.4 Driving an LED with a Logic Gate

Procedure

DIP Switches as Logic Switches

1. DIP (Dual In-Line Package) switches can be combined with resistors packaged as a SIP (Single In-Line Package) to make multiple logic switches on a breadboard. A SIP resistor package is shown in Figure 2.5. If you measure the resistance between pin 1 and any other pin, you should get the value of 1 resistor. Measuring between any two other pins will give the value of two resistors added together.

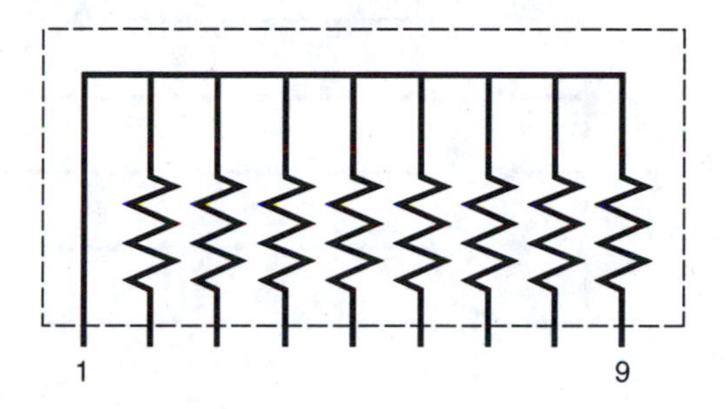

Figure 2.5 SIP Resistor Network (Bussed)

2. Measure the resistance between pins 1 and 2 of the SIP resistor. Resistance: ________

   Measure the resistance between pins 1 and 3. Resistance: ________

   Measure the resistance between pins 2 and 3. Resistance: ________

   Measure the resistance between pins 2 and 4. Resistance: ________

   What do you conclude from these resistance measurements?

   ____

   ____

   ____

3. Connect the components shown in Figure 2.6 to make eight logic switches. Test each one with a logic probe to ensure its correct operation. Demonstrate the circuit to your instructor. ***Do not disconnect the circuit yet. It will be used in the next part of this lab.***

   Instructor's Initials ________________

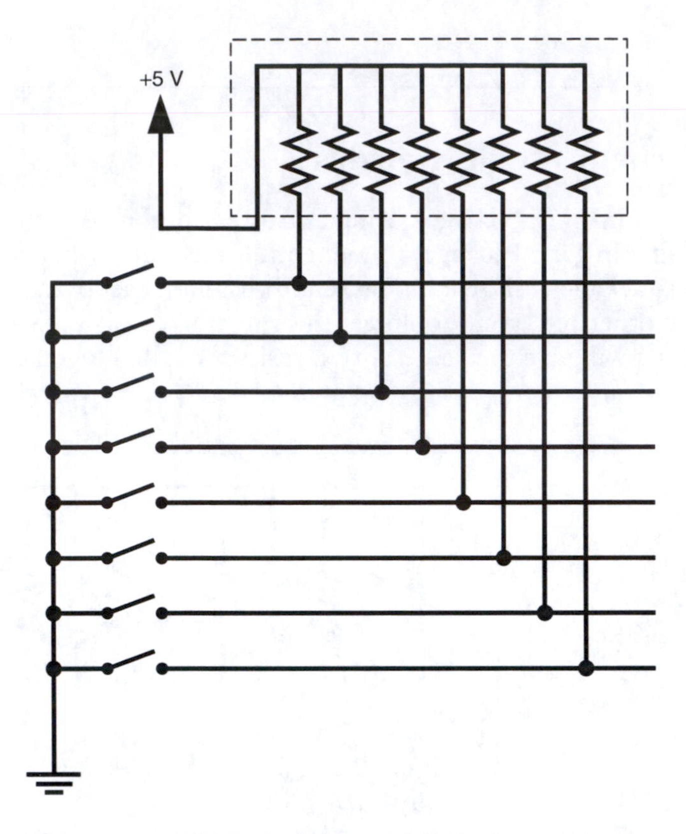

Figure 2.6 DIP Switches and SIP Resistors Used as Logic Switches

LED Operation

Connect the circuit shown in Figure 2.7. This circuit is similar to Figure 2.4. The device with six inverters is labeled 74LS04 or similar.

- Pin 14 of the inverter chip should be connected to the +5 V power supply.
- Pins 3, 5, 7, 9, 11, and 13 should be connected to ground.
- Connect one logic switch from the previous section of the lab to pin 1 of the inverter.
- Connect pin 2 of the inverter to one end of a 330 Ω resistor (identified by three colored bands that are in the order orange, orange, and brown).
- Connect the other end of the 330 Ω resistor to the anode of the LED.
- Connect the cathode of the LED to ground. (The LED case has a flat spot that marks the cathode of the LED. Have your instructor check the LED direction if you are unsure.)

Figure 2.7 LED Driver Circuit 1

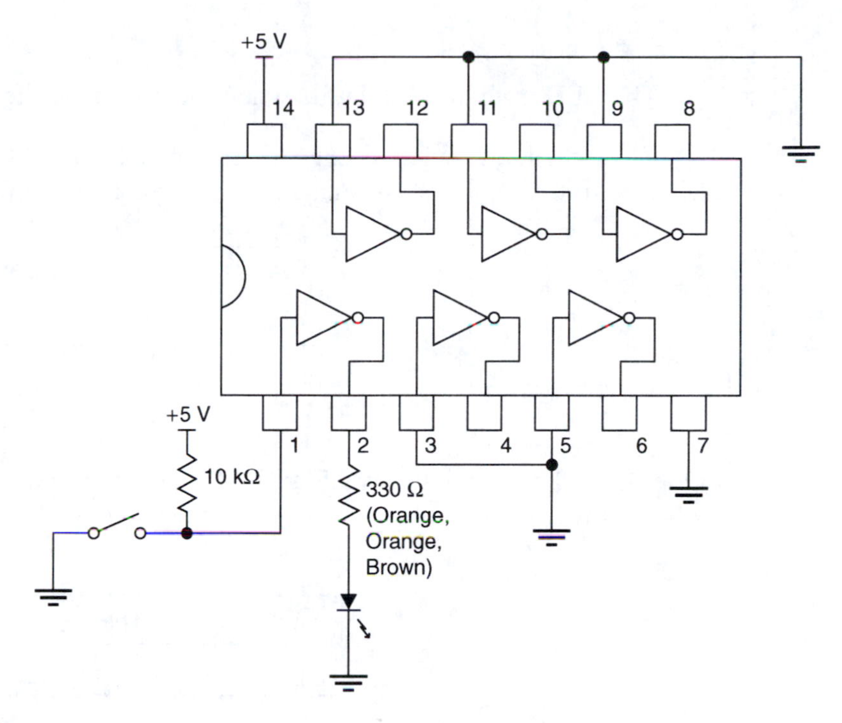

Circle the correct answer:

The LED in Figure 2.7 lights when the output of the inverter is HIGH/LOW.

Close the logic switch connected to pin 1. Is the LED ON or OFF? (Circle one.) Measure the logic levels at pins 1 and 2 of the inverter with a logic probe or a voltmeter.

Pin Number	Logic Level
1	
2	

Instructor's Initials ____________________

Reverse the direction of the LED, connecting the anode to the 5 volt supply, as shown in Figure 2.8.

Figure 2.8 LED Driver Circuit 2

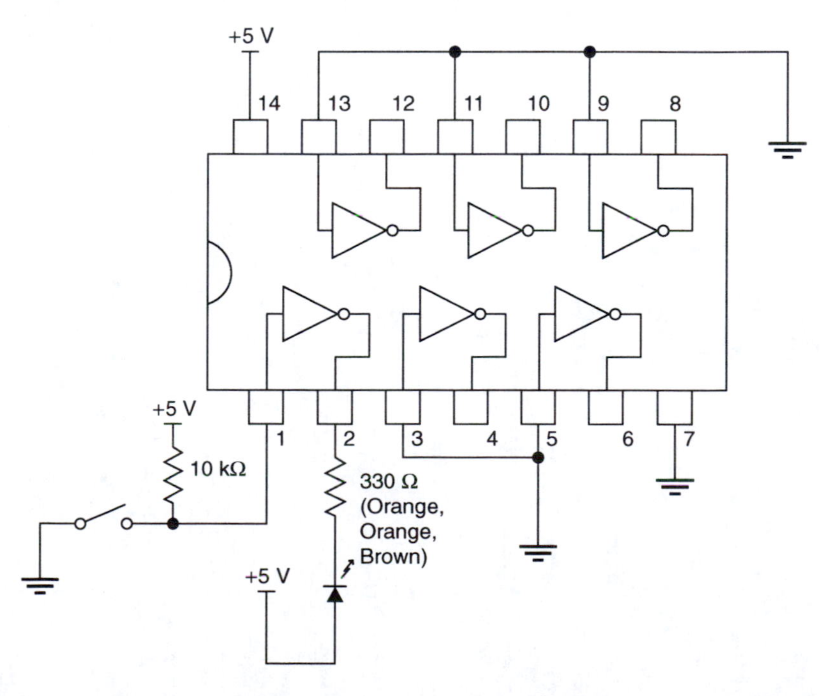

Circle the correct answer:

The LED in Figure 2.8 lights when the output of the inverter is HIGH/LOW.

Close the logic switch connected to pin 1. Is the LED ON or OFF? (Circle one.) Measure the logic levels at pins 1 and 2 of the inverter with a logic probe or a voltmeter.

Pin Number	Logic Level
1	
2	

One of the LEDs in Figures 2.7 and 2.8 is active-LOW and one is active-HIGH. State which is which.

Figure 2.7	active- _______
Figure 2.8	active- _______

Instructor's Initials ________________

Lab 3

DIP Integrated Circuits

Name ______________________ Class __________ Date __________

Objectives Upon completion of this laboratory exercise, you should be able to:

- Describe the configuration of several basic logic gates in dual in-line packages.
- Wire a logic gate integrated circuit (IC) on a prototyping breadboard.
- Obtain the truth tables of each gate to be tested.

Reference Ken Reid and Robert Dueck, *Introduction to Digital Electronics*
Chapter 2: Logic Functions and Gates

Equipment Required
+5-volt power supply or digital trainer
Breadboard
Wire strippers
#22 solid-core wire, as required
Components as follows:

Part No.	Qty.	Description
74LS00 or 74HC00	1	Quad 2-input NAND gate
74LS02 or 74HC02	1	Quad 2-input NOR gate
74LS04 or 74HC04	1	Hex Inverter
74LS08 or 74HC08	1	Quad 2-input AND gate
74LS32 or 74HC32	1	Quad 2-input OR gate
74LS86 or 74HC86	1	Quad 2-input XOR gate
		If not using a digital trainer:
	1	4PST or 8PST DIP switch
	1	LED
	1	Resistor, 330 Ω, ¼ W
	2	Resistor, 10 kΩ, ¼ W

Experimental Notes

A common way to package logic gates is in a plastic or ceramic **dual in-line package,** or **DIP,** which has two parallel rows of pins. The standard spacing between pins in one row is 0.1″ (or 100 mil). For packages having less than 28 pins, the spacing between rows is 0.3″ (or 300 mil). For some larger packages, the rows are spaced by 0.6″ (or 600 mil).

The outline of a 14-pin DIP is shown in Figure 3.1. There is a notch on one end to show the orientation of the pins. When the IC is oriented as shown, pin 1 is at the bottom left corner and the pins number counterclockwise from that point.

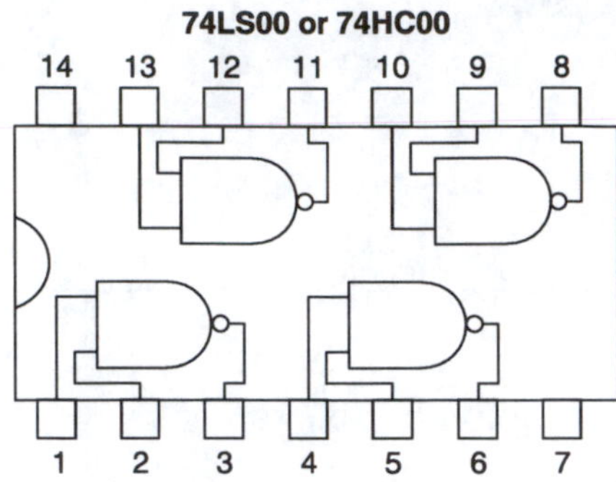

Figure 3.1 DIP IC (Quad 2-Input NAND)

Figure 3.2 shows the internal diagrams of gates listed in the Equipment Required section. These can be found in manufacturer's datasheets describing the devices. In addition to the gate inputs and outputs, there are two more connections to be made on each chip: the power and ground connections. V_{CC} (pin 14) must be connected to +5 volts and GND (pin 7) to ground to provide power supply connections. ***The gates won't work without these connections.*** Logic levels at the other pin inputs are derived from these power supply voltages by connecting them to +5 volts for logic 1 and ground for logic 0.

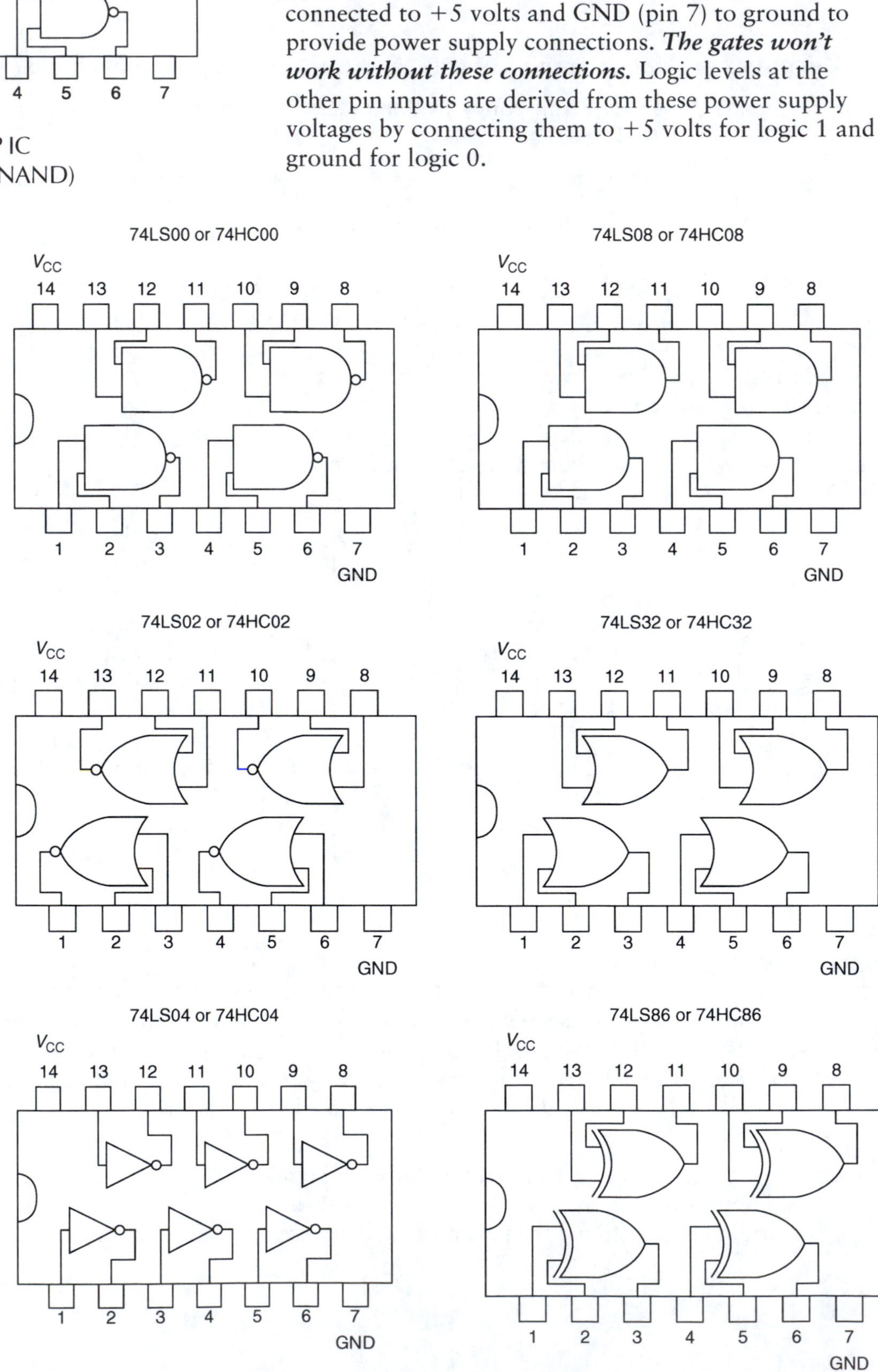

Figure 3.2 Pinouts of Some Basic Logic ICs

A **truth table** of a digital logic gate or circuit is a table showing the gate or circuit output for all possible combinations of inputs. These ***must*** be shown in standard binary order. Table 3.1 shows an example of a truth table for one of the NAND gates from Figure 3.1.

Note that the input combinations count up from 00 to 11 in binary. This is the standard order for all truth tables. Truth tables for other gates would have the same input pattern and a unique output pattern.

Table 3.1 Truth Table

A	*B*	*Y*
0	0	1
0	1	1
1	0	1
1	1	0

Procedure

1. Insert a 74LS00 or 74HC00 quad 2-input NAND gate into the circuit breadboard. Connect pin 14 to V_{CC} (the + terminal of the +5-volt power supply) and pin 7 to ground (the – terminal of the power supply), as shown in Figure 3.3.

Figure 3.3 Testing a Gate in a DIP IC (Without a Digital Trainer)

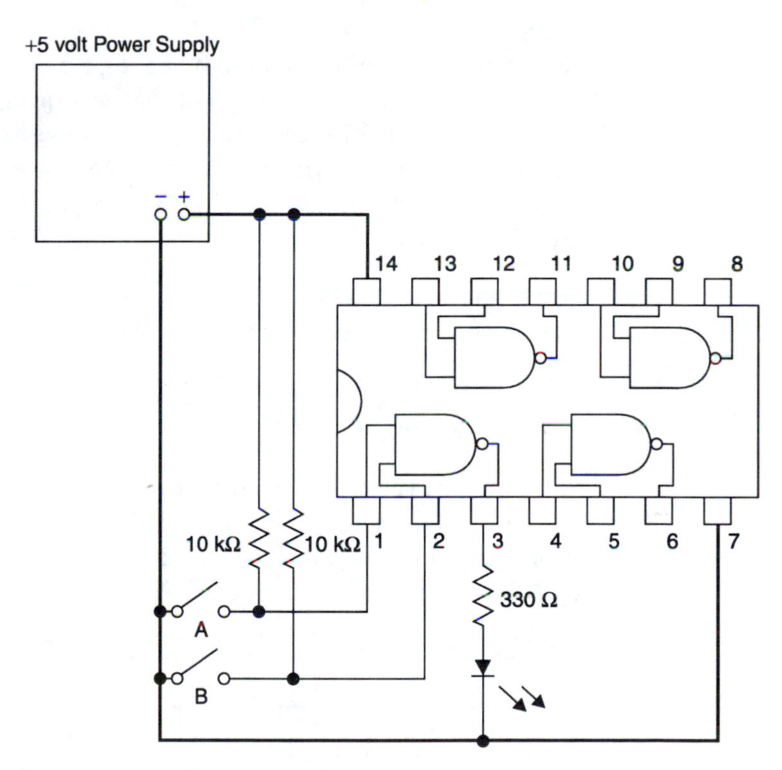

2. If you are not using a digital trainer with logic switches and LED indicators, connect one end of a single-pole single-throw (SPST) switch to ground and the other end to pin 1 of the NAND IC. Connect the junction of the switch and pin 1 to V_{CC} through a 10-kΩ resistor. This represents input A of the gate. Make a similar connection to pin 2 of the IC for input B.

   Note If you are using a digital logic trainer with logic switches and LED indicators as shown in Figure 3.4, you may use those instead of the switches and LED shown in Figure 3.3. In this case, the resistors are not required. Simply connect pins 1 and 2 to the pins associated with the logic switches. If in doubt, ask your instructor.

3. If not using a digital trainer, connect pin 3 of the NAND IC to an LED through a 330-Ω series resistor or to the pin for an LED indicator on a digital trainer. The cathode (indicated by a short lead or by a flat spot on the LED case) should go to ground. If you are using a logic trainer, the series resistor is not required. Simply connect pin 3 of the NAND IC to the LED on the trainer.

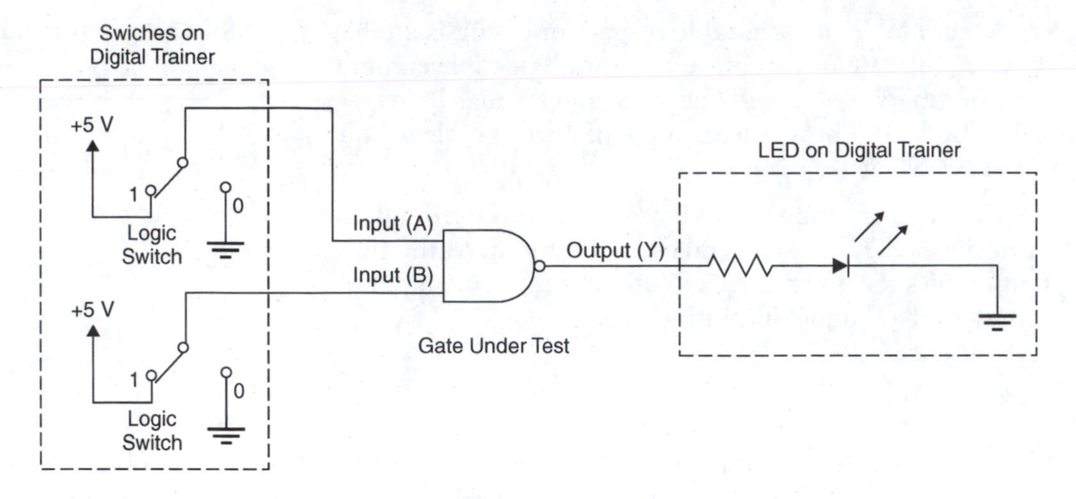

Figure 3.4 Testing a Gate in a DIP IC (with a Digital Trainer)

4. Take the truth table of the gate by changing inputs *A* and *B* to make all possible combinations of input (AB = 00, 01, 10, 11) and writing down whether the LED at output 3 indicates logic 1 (ON) or logic 0 (OFF) for each input combination. Fill in Table 3.2 with your results.

Table 3.2 NAND Truth Table

A	*B*	*Y*
0	0	
0	1	
1	0	
1	1	

5. Move the logic input connections to pins 4 and 5 and the lamp to pin 6 and repeat Step 4. Also repeat with:

   logic switches at pins 9 and 10 and lamp at pin 8, and;

   logic switches at pins 12 and 13 and lamp at pin 11.

6. Repeat the above steps with the other gates listed in the parts list. (Note that the 74LS02/74HC02 and 74LS04/74HC04 gates require different connections than the rest of the gates. Consult Figure 3.2 for details.) Make a truth table for one gate in each IC. See Tables 3.3 through 3.7.

Table 3.3 AND Truth Table

A	*B*	*Y*
0	0	
0	1	
1	0	
1	1	

Table 3.4 NOR Truth Table

A	*B*	*Y*
0	0	
0	1	
1	0	
1	1	

Table 3.5 OR Truth Table

A	*B*	*Y*
0	0	
0	1	
1	0	
1	1	

Table 3.6 Inverter Truth Table

A	*B*
0	
1	

Table 3.7 XOR Truth Table

A	*B*	*Y*
0	0	
0	1	
1	0	
1	1	

Instructor's Initials ____________

Lab 4

Expanding Logic Gates

Name ______________________ Class __________ Date __________

Objectives Upon completion of this laboratory exercise, you should be able to:

- Configure a NAND or NOR gate to act as an inverter.
- Expand an AND, OR, NAND, or NOR gate from a 2-input to a 3-input gate.
- Draw a logic diagram that conforms to standard digital design practice.
- Obtain the truth tables of each gate to be tested.

Reference Ken Reid and Robert Dueck, *Introduction to Digital Electronics*

Chapter 2: Logic Functions and Gates

Equipment Required

Digital trainer
Wire strippers
#22 solid-core wire, as required
Components as follows:

Part No.	Qty.	Description
74LS00 or 74HC00	1	Quad 2-input NAND gate
74LS02 or 74HC02	1	Quad 2-input NOR gate
74LS32 or 74HC32	1	Quad 2-input OR gate

Experimental Notes

When using more than one logic gate in a circuit, it is a good practice to draw a **logic diagram,** such as the one shown in Figure 4.1. Such a diagram simplifies the process of troubleshooting and allows for neat documentation of the circuit.

Figure 4.1 shows several characteristics of a good logic diagram. A logic diagram shows all connections between logic elements (the two AND gates), but may omit symbols for input devices, such as logic switches; output devices, such as LEDs; or power and ground connections of the chip.

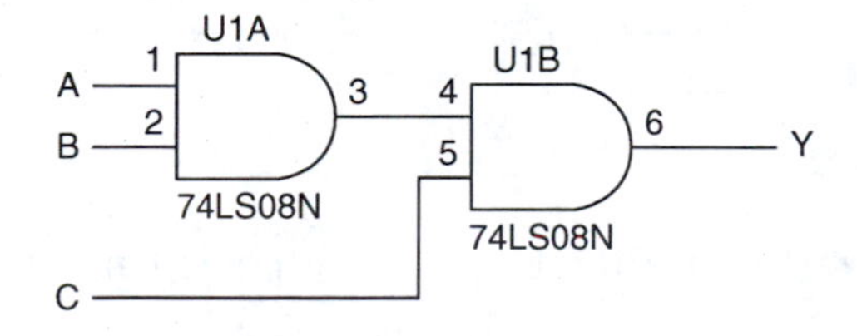

Figure 4.1 Logic Diagram Showing Two Gates from the Same Chip

Each gate is separately numbered (U1A, U1B) and the pin numbers are shown for each connection. The gate numbers each indicate a chip (U1) and a gate designator (A or B). The part number is shown for each logic element in Figure 4.1, although this does not need to be shown for each gate in a particular chip.

Figure 4.2 shows how to wire the AND chip for the circuit. Note that ***unused inputs are grounded and unused outputs are left open.*** This is very important for the reliable operation of a device, especially a CMOS (74HC) device.

If two or more gates from different chips are used, the gates have different numbers, (U1, U2) showing that they come from different chips. In this case, the gate designators (A) are the same on both gates, indicating that each is the first gate on each chip. Figure 4.3 indicates connections between the chips. In this case, the AND output, pin 3 of chip U1, is connected to one of the NAND inputs, pin 1 of chip U2. The connections are shown in the wiring diagram of Figure 4.4. Again, note that ***unused inputs are grounded and unused outputs are left open.***

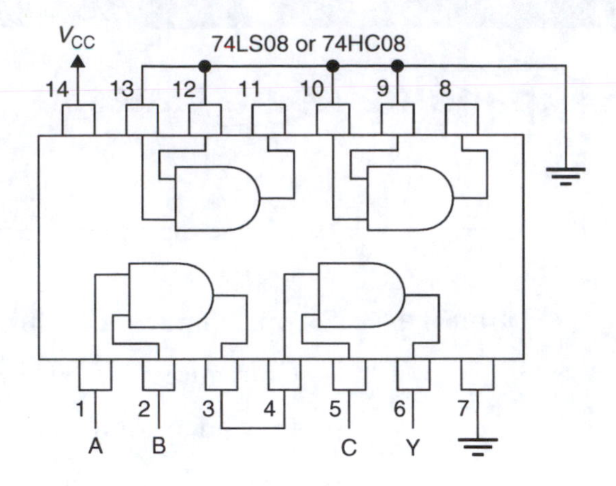

Figure 4.2 Wiring Diagram for Logic Circuit in Figure 4.1

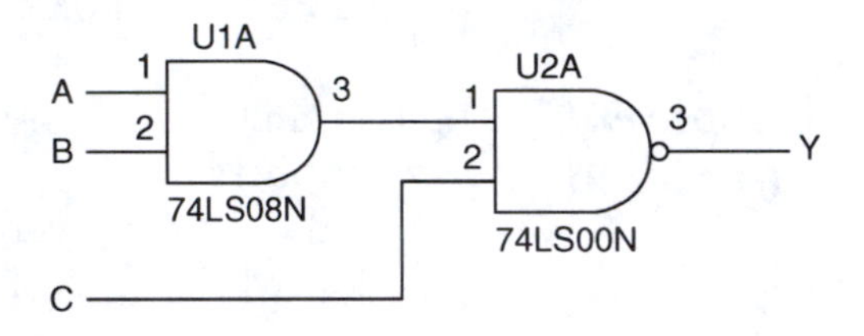

Figure 4.3 Logic Diagram Showing Gates from Different Chips

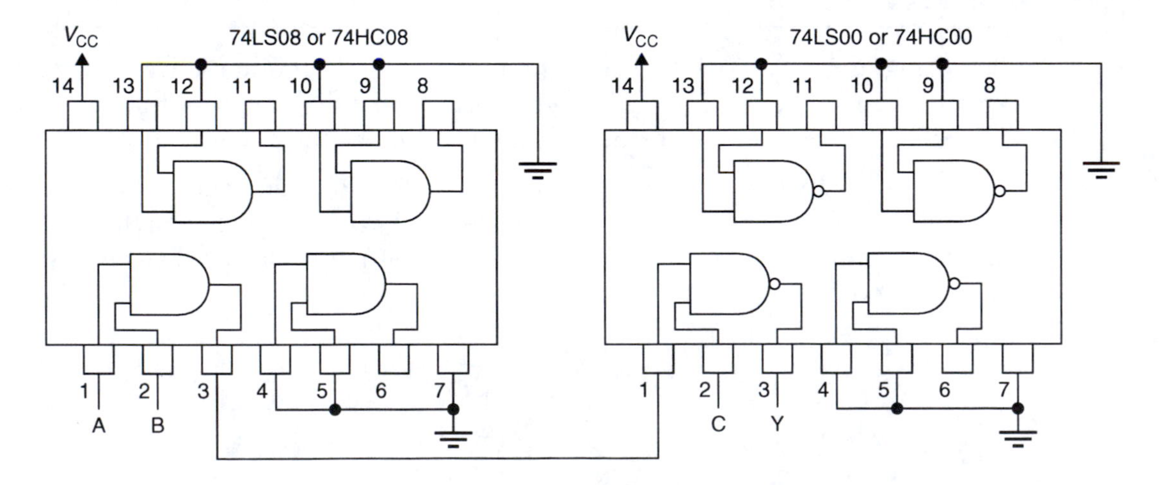

Figure 4.4 Wiring Diagram for Logic Circuit of Figure 4.3

Note Logic diagrams are the usual way of indicating the connections in a digital circuit. ***Wiring diagrams are seldom used,*** as they quickly become complicated and hard to follow as more components are added to the circuit. Also, wiring diagrams contribute little information regarding the operation and function of the circuit.

Procedure

1. Draw the logic diagram of a NAND gate connected as an inverter. Include chip designators (U1A, etc.), pin numbers, and part numbers.

2. Connect the NAND-gate inverter drawn in Step 1 and verify its operation. Complete the following truth table.

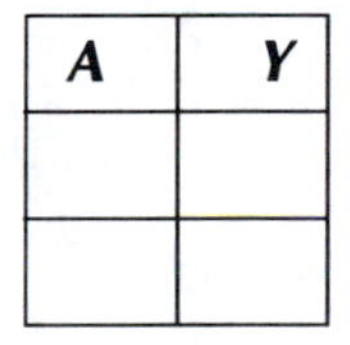

A	*Y*

Instructor's Initials ______________________

3. Draw the logic diagram of a NOR gate connected as an inverter. Include pin numbers, part numbers, and chip designators (U1A, etc.).

4. Connect the NOR-gate inverter drawn in Step 3 and verify its operation. Complete the following truth table.

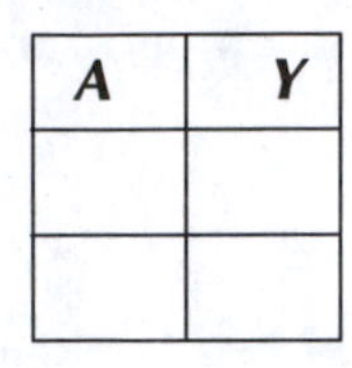

A	*Y*

Instructor's Initials ____________________

5. Make a logic diagram, including pin numbers, part numbers, and chip designators (U1A, etc.), that creates a 3-input OR function, using only 2-input OR gates.

6. Connect the logic diagram from Step 5 and verify its operation. Complete the following truth table.

A	*B*	*C*	*Y*

Instructor's Initials ____________________

7. Make a logic diagram, including pin numbers, part numbers, and chip designators (U1A, etc.), that creates a 3-input NOR function, using only 2-input OR and NOR gates.

8. Connect the logic diagram from Step 7 and verify its operation. Complete the following truth table.

A	*B*	*C*	*Y*

Instructor's Initials ____________

9. Draw the logic diagram, including pin numbers, part numbers, and chip designators (U1A, etc.), of a circuit that makes a 3-input NOR function, using as many 2-input NOR gates from a single chip as required.

   Hint *Refer to the logic diagrams you drew for the NOR gate as inverter and the 3-input NOR that used 2-input OR and NOR gates. How can you replace the OR with a NOR-only circuit?*

10. Connect the logic diagram from Step 9 and verify its operation. Complete the following truth table.

A	*B*	*C*	*Y*

Instructor's Initials ________________

Lab 5

Pulsed Operation of Logic Gates

Name ______________________________ Class ____________ Date ____________

Objectives Upon completion of this laboratory exercise, you should be able to:

- Determine how basic logic gates can be used to pass (enable) or block (inhibit) time-varying digital signals by examining the gate truth tables.
- Monitor the pulsed behavior of logic gates with LEDs and with an oscilloscope.

Reference Ken Reid and Robert Dueck, *Introduction to Digital Electronics*
Chapter 2: Logic Functions and Gates

Equipment Required
Digital trainer
Wire strippers
#22 solid-core wire, as required
Components as follows:

Part No.	Qty.	Description
74LS00 or 74HC00	1	Quad 2-input NAND gate
74LS02 or 74HC02	1	Quad 2-input NOR gate
74LS08 or 74HC08	1	Quad 2-input AND gate
74LS32 or 74HC32	1	Quad 2-input OR gate
74LS86 or 74HC86	1	Quad 2-input XOR gate

Experimental Notes

Any logic gate can be used as a switch to pass or block a time-varying waveform. When the waveform is passed by the gate, we say that the gate is ***enabled;*** when the waveform is blocked, the gate is ***inhibited.*** Each type of logic gate has a particular set of enable/inhibit characteristics: the gate may be enabled by a logic 1 or a logic 0; the time-varying input might pass through the gate in ***true*** (noninverted) or ***complement*** (inverted) form.

A pulse waveform is just a sequence of logic 0s and 1s. We can use the truth table of any gate predict how the gate will respond to a waveform input. For example, the truth table of an AND gate is shown in Table 5.1.

Table 5.1 AND Gate Truth Table Showing Enable/Inhibit Properties

A	*B*	*Y*	
0	0	0	Y = 0
0	1	0	(Inhibit)
1	0	0	Y = 1
1	1	1	(Enable)

If we hold input A at a constant logic 0 level and vary input B between logic 0 and logic 1, we will get a waveform similar to the first half of the timing diagram in Figure 5.1. If we then change A to its opposite level (logic 1), we will then get a waveform like the one in the second half of Figure 5.1.

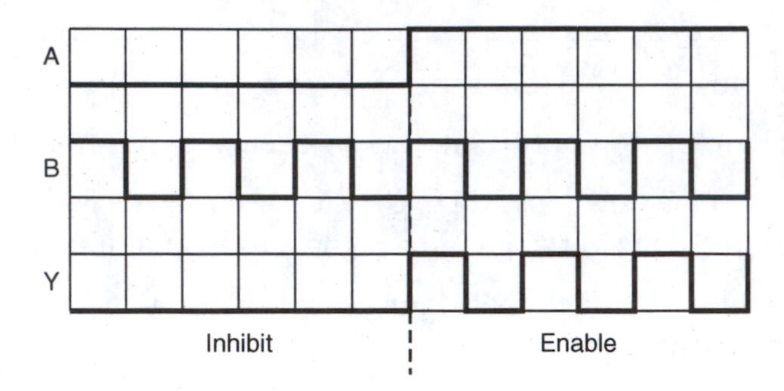

Figure 5.1 Enable/Inhibit Waveforms for an AND Gate

If input A is 0 and B is pulsing, the gate output is always 0, since an AND gate requires all inputs HIGH to make the output HIGH. This never happens when $A = 0$. The first two lines of the truth table in Table 5.1 show this condition. If A is 1 and B is pulsing, the gate output is the same as B, as shown by the last two lines of the truth table. Thus, when $A = 0$, $Y = 0$ and when $A = 1$, $Y = B$.

We can designate the A input as the "control" input because it controls whether the waveform at B will pass through the gate or not. Input B is designated as the "signal" input. Our examination of the AND gate truth table tells us that an AND gate is enabled by a 1 at its control input and that the signal is passed through in true (noninverted) form when the gate is enabled. Other logic gates have similar characteristics, as predicted by examination of their truth tables.

Procedure

1. Connect the circuit shown in Figure 5.2, using a 74LS08 or 74HC08 quad 2-input AND gate. The pulse source (the clock source on the digital trainer) can be run at various frequencies. The 10 kHz setting is suitable for viewing on an oscilloscope, and the 1 Hz setting is slow enough to observe directly on the LEDs. The logic level at the gate's control input is also selected by a logic switch on the digital trainer.

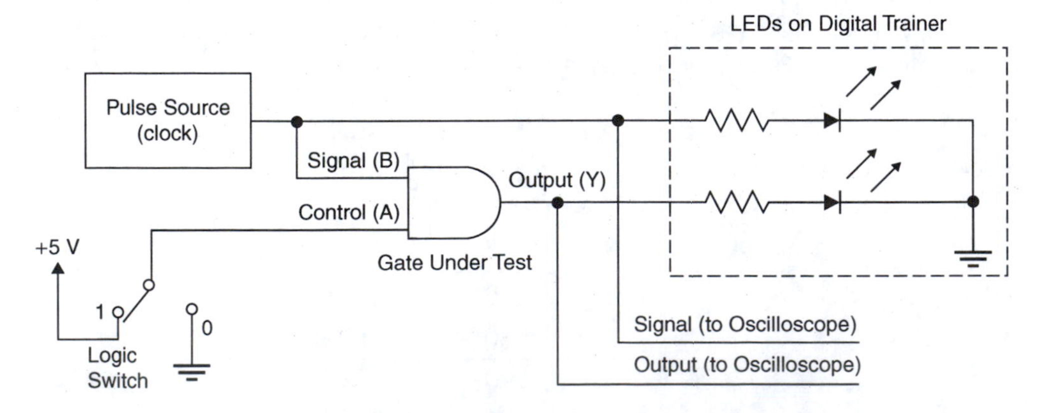

Figure 5.2 Testing a Gate with a Pulse Waveform

2. Set the Control input to 0 and the pulse source to **1 Hz.** Monitor the Signal input and the gate output on the two LEDs. There are four possibilities for the behavior of the output LED, relative to the *B* input LED. The output LED could be:

   a. always ON

   b. always OFF

   c. flashing the same as *B* (in phase)

   d. flashing opposite to *B* (out of phase)

   Given one of the four choices above, how do the *B* and *Y* LEDs relate to each other when:

   a) $A = 0$? ______________________________

   b) $A = 1$? ______________________________

3. Set the pulse source frequency to 10 kHz. Monitor the signal input to the gate on channel *B* of the oscilloscope and the gate output on channel *A*. Trigger the oscilloscope on channel *B*. Draw waveforms similar to Figure 5.1 for the cases when $A = 0$ and $A = 1$.

Instructor's Initials ______________

4. Repeat Steps 1 through 3 for the following logic gates using Figure 5.2:

   a. 74LS00 or 74HC00 quad 2-input NAND gate

   b. 74LS02 or 74HC02 quad 2-input NOR gate

   c. 74LS32 or 74HC32 quad 2-input OR gate

   d. 74LS86 or 74HC86 quad 2-input XOR gate

Instructor's Initials ______________

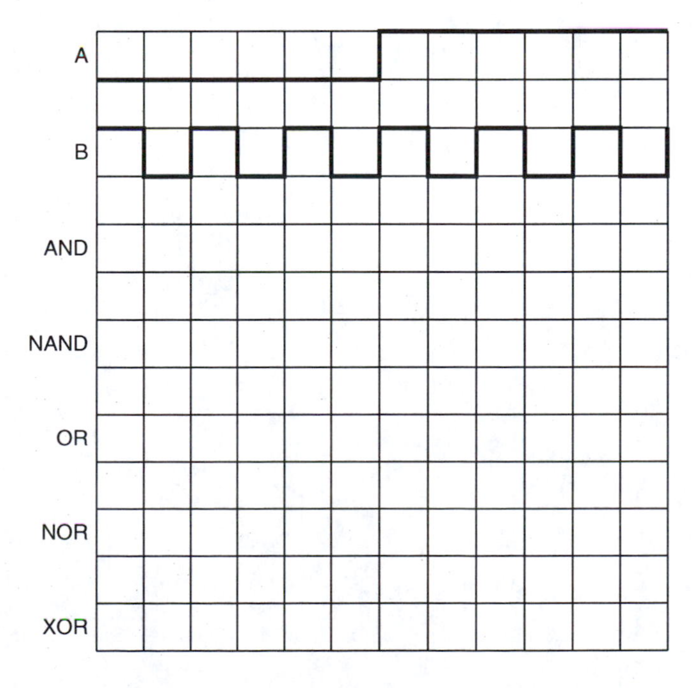

Figure 5.3 Gate Output Waveforms

Assignment Questions

1. Make a table relating control input, signal input, and output for all gates (e.g., for an AND gate, $Y = 0$ when $A = 0$; $Y = B$ when $A = 1$).

2. Hand in the results for this lab on a separate sheet. Draw a timing diagram for each gate, showing the relationship between control input, signal input, and gate output. When drawing the waveforms, use a ruler or drafting software.

Lab 6

Boolean Algebra: SOP Forms

Name ______________________________ Class ____________ Date ____________

Objectives Upon completion of this laboratory exercise, you should be able to:

- Draw the logic diagram of a combinational circuit from a Boolean expression.
- Take a truth table from a logic gate network and use it to derive a sum-of-products (SOP) Boolean expression for the network.
- Use Boolean algebra to simplify a logic gate network and to prove that two gate networks are equivalent.
- Use DeMorgan equivalent forms of logic gates to simplify the Boolean expression of a logic gate network.

Reference Ken Reid and Robert Dueck, *Introduction to Digital Electronics*

Chapter 2: Logic Functions and Gates

Chapter 3: Boolean Algebra and Combinational Logic

Equipment Required

Digital trainer
Wire strippers
#22 solid-core wire, as required

Components as follows:

Part No.	Qty.	Description
74LS00 or 74HC00	1	Quad 2-input NAND gate
74LS02 or 74HC02	1	Quad 2-input NOR gate
74LS04 or 74HC04	1	Hex inverter
74LS08 or 74HC08	1	Quad 2-input AND gate
74LS10 or 74HC10	2	Triple 3-input NAND gate

Experimental Notes

Any combinational logic circuit can be described by a Boolean expression written in sum-of-products (SOP) form. This form can be derived from a truth table by noting the lines on the truth table where the output has a value of 1 and writing a **product term** for each such line. Each product term consists of all input variables in either true or complement form. If an input is 0, it is written in complement form (with a bar); if the input is 1, it is written in true form (no bar).

For example, in Table 6.1 there are three product terms. These three terms are combined in an OR function to make a sum-of-products Boolean expression:

$$Y = \bar{A}\bar{B}\bar{C} + \bar{A}BC + AB\bar{C}$$

Table 6.1 Product Terms from a Truth Table

A	*B*	*C*	*Y*	
0	0	0	1	$\bar{A}\,\bar{B}\,\bar{C}$
0	0	1	0	
0	1	0	0	
0	1	1	1	$\bar{A}\,B\,C$
1	0	0	0	
1	0	1	0	
1	1	0	1	$A\,B\,\bar{C}$
1	1	1	0	

Figure 6.1 shows the logic diagram derived from this Boolean expression. The circuit can be implemented by a combination of AND, OR, and NOT gates.

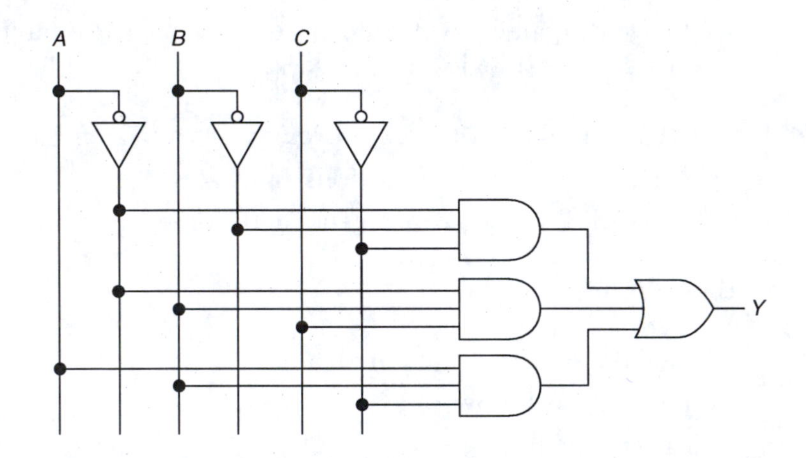

Figure 6.1 Sum-of-Products Network (AND-OR Construction)

An alternative configuration, shown in Figure 6.2, can be constructed using only NAND gates and inverters. The output gate is a NAND gate, shown in its DeMorgan equivalent form. The circuit can be derived from an AND-OR configuration by inverting all AND outputs and all OR inputs, as shown. In some cases, this can result in a more efficient circuit implementation (i.e., with fewer logic gate packages) than the AND-OR circuit. In this case, both circuits are equally efficient to build.

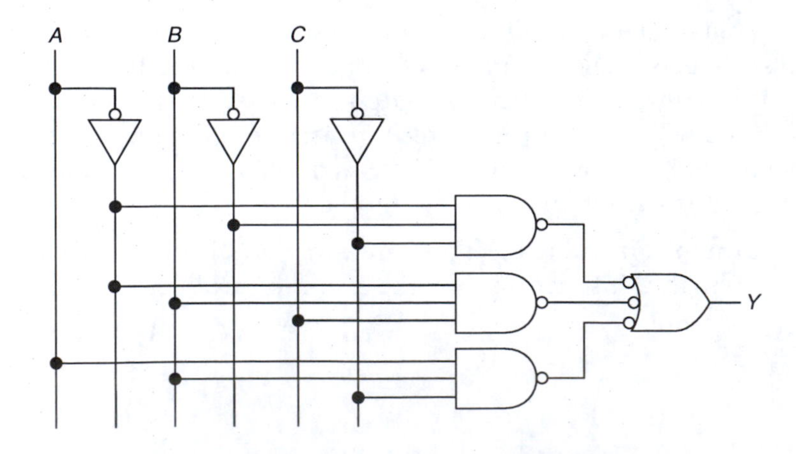

Figure 6.2 Sum-of-Products Network (NAND-NAND Construction)

Procedure

1. Draw the unsimplified logic diagram represented by the following Boolean expression:

$$Y = (\overline{AD + B\overline{D}})\,C$$

2. Connect the circuit drawn in Step 1, using 74LS08 or 74HC08 AND gates, 74LS02 or 74HC02 NOR gates, and 74LS04 or 74HC04 inverters. Refer to Figure 6.3 at the end of this lab for the IC pinouts. Connect a logic switch to each input and an LED monitor to the output.

   Refer to previous labs for configuration of logic switches and LED monitors. If you are using a digital trainer, use the logic switches and LED monitor on the trainer.

3. Construct the truth table of the circuit in Step 2 by setting the input switches to all possible input combinations and noting the output value for each combination. Write the truth table in the space provided below.

4. Write the SOP expression derived from the truth table. Use Boolean algebra to simplify the expression as much as possible.

5. Draw the circuit described by the simplified Boolean expression in Step 4. Connect the circuit using only 74LS10 or 74HC10 triple 3-input NAND gates and 74LS04 or 74HC04 inverters. Refer to Figures 6.2 and 6.3 for guidance.

6. Take the truth table of the circuit drawn in Step 5. Verify that it is the same as the table constructed in Step 3.

Instructor's Initials ____________________

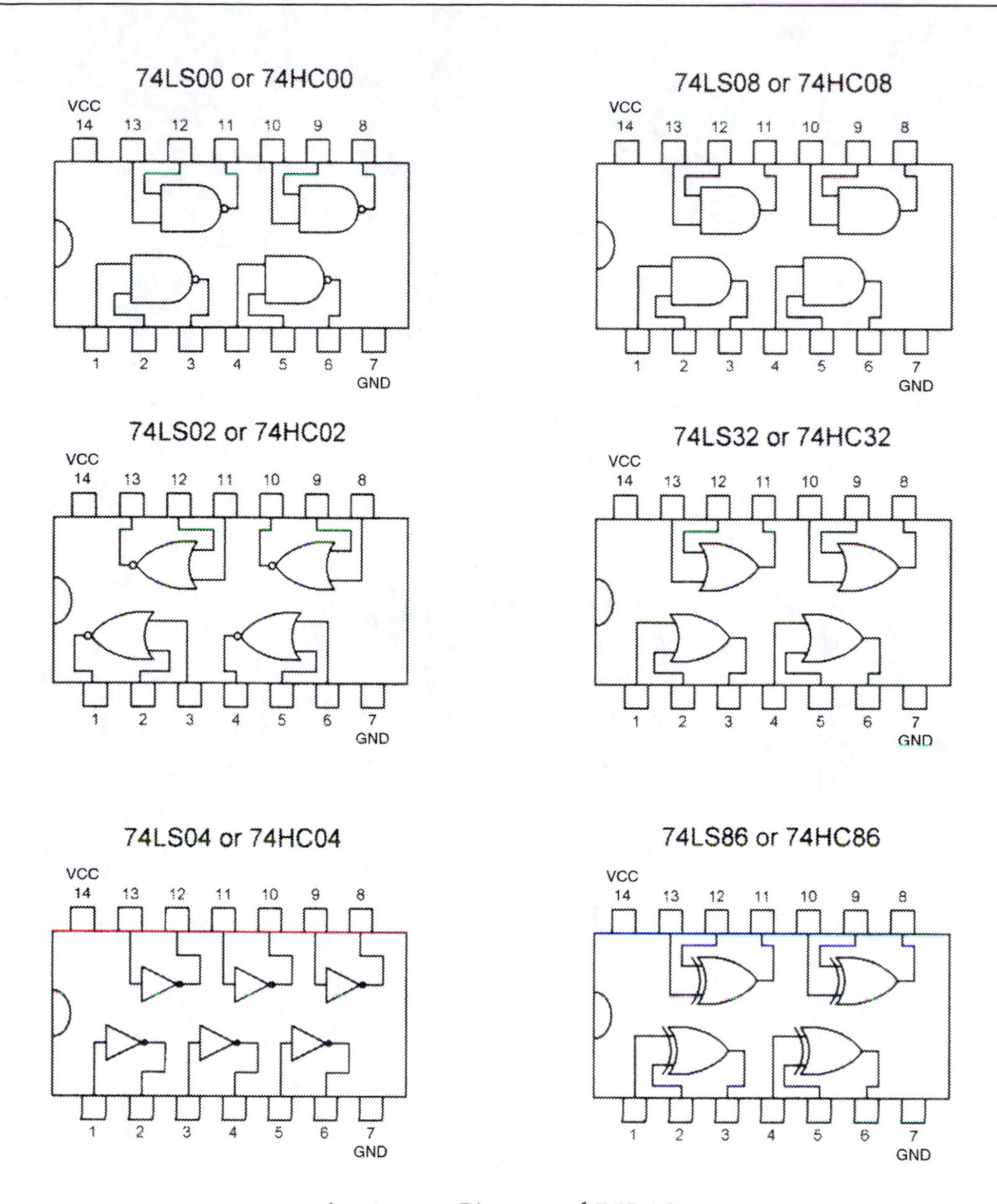

Figure 6.3 Pinouts of DIP ICs

Lab 7

Boolean Algebra: DeMorgan Equivalent Forms

Name ______________________ Class __________ Date __________

Objectives Upon completion of this laboratory exercise, you should be able to:

- Use Boolean algebra to simplify a logic gate network and to prove that two gate networks are equivalent.
- Use DeMorgan equivalent forms of logic gates to simplify the Boolean expression of a logic gate network.

Reference Ken Reid and Robert Dueck, *Introduction to Digital Electronics*
Chapter 3: Boolean Algebra and Combinational Logic

Equipment Required
Digital Trainer
Wire strippers
#22 solid-core wire, as required
Components as follows:

Part No.	Qty.	Description
74LS00 or 74HC00	1	Quad 2-input NAND gate
74LS02 or 74HC02	1	Quad 2-input NOR gate
74LS04 or 74HC04	1	Hex inverter

Procedure

1. Write the unsimplified Boolean expression for the logic diagram shown in Figure 7.1.

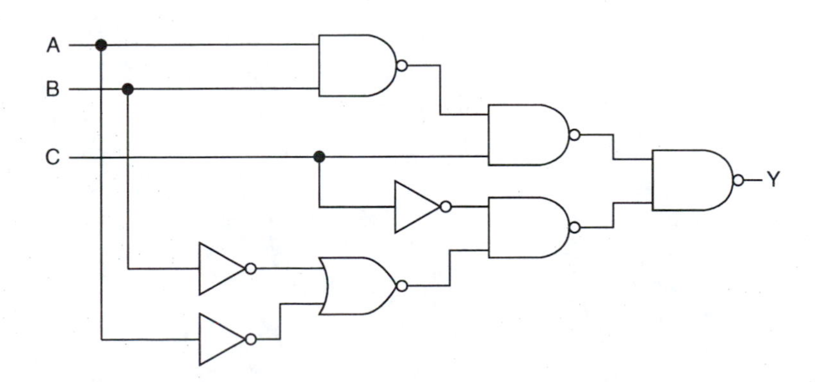

Figure 7.1 Logic Gate Network

2. Redraw the logic diagram in Figure 7.1, replacing some of the gates with their DeMorgan equivalent forms, so that each gate output with a bubble drives a gate input with a bubble and each gate output with no bubble drives a gate input with no bubble. Write the Boolean expression of the redrawn circuit.

3. Construct the circuit of step 2 using the minimum number of IC packages.

   Note Recall that an inverter can be constructed from a NAND or NOR gate by connecting the gate inputs together.

   Take the truth table of the circuit. Write the equivalent SOP expression of this circuit on the next page.

Instructor's Initials ____________________

4. Use Boolean algebra to simplify the expression derived in step 2 to the simplest possible SOP form. Draw the logic diagram of the simplified expression. Use Boolean algebra to prove that this circuit is equivalent to the circuit constructed in step 3.

How many logic gate ICs are required to construct this circuit if AND-OR construction is used? ______________

Which ones? ____

5. Redraw the simplified SOP circuit of step 4 to use NAND-NAND construction, as in lab 6. How many logic gate ICs are required to make this circuit?

____

Which ones? ____

Construct the circuit as a NAND-NAND circuit and take its truth table. Verify that this is the same as the truth table taken in step 3. Refer to Figure 7.2.

Instructor's Initials ________________

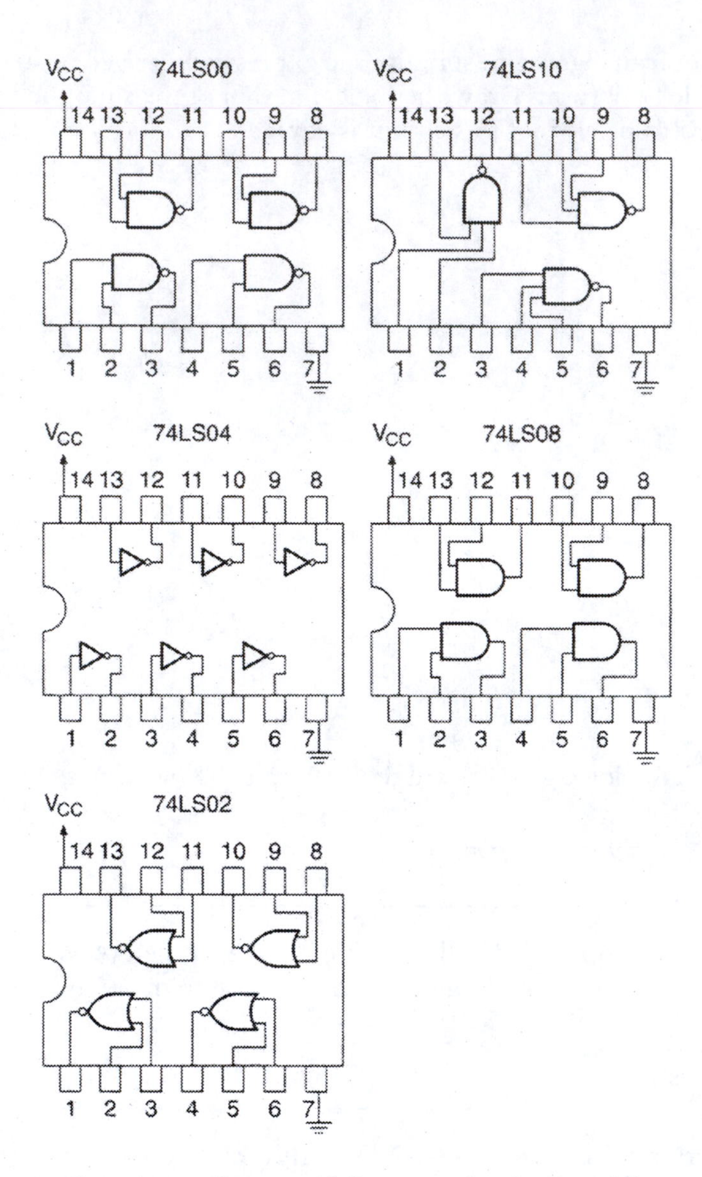

Figure 7.2 Pinouts of Common Logic Gate ICs

Lab 8

Installing and Licensing Quartus II on Your PC (Tutorial)

Name ______________________ Class ____________ Date ____________

Objectives Upon completion of this laboratory exercise, you should be able to:

- Download updated versions of the Quartus II Web Edition software.
- Install and license the software for use on your home PC.
- Install a driver for the Altera ByteBlaster parallel port programmer.

Reference Ken Reid and Robert Dueck, *Introduction to Digital Electronics*
Chapter 4: Introduction to PLDs and Quartus II

Equipment Required Quartus II Web Edition Software
PC running Windows, with Internet access and e-mail

Experimental Notes

This lab tutorial is intended to give instructions for installation of the Quartus II Web Edition software on your computer at home. ***It is not intended for use in a networked computer lab at your school.*** Quartus II should already be installed on your school's network. If it is not, ask your instructor to consult your school's system administrator.

The tutorial consists of four parts:

1. **Downloading the most recent version of the Quartus II Web Edition software.** The textbook *(Introduction to Digital Electronics)* comes with a CD that contains the most recent version of Quartus II, as of the date of publication. However, Altera is continually making upgrades to its software, so a more recent version may be available online. In most cases, this is not a problem and the software may simply be installed from the CD. If you wish to download a more recent version, then you should follow this procedure.

2. **Obtaining a license file from Altera.** Licensing of the Quartus II Web Edition is at no cost to the user. However, a license file is required for every copy of the software that you install on a separate computer. The license may be temporary and then must be renewed, again at no cost.

3. **Installing the license file.** Before all features of the software can be used, Quartus II must know how to locate the license file. Follow this procedure to properly install the license file.

4. **Installing the ByteBlaster driver.** Before you can set up the Quartus II software for programming a CPLD, you must install a driver for the Altera ByteBlaster parallel port programmer. This must done for most CPLD trainer boards.

Procedure

Downloading the Most Recent Version of the Quartus II Web Edition Software (optional)

1. Start your PC and open your web browser. Go to the Altera home page at *http://www.altera.com*. The page should look similar to the screen shot in Figure 8.1.

Figure 8.1 Altera Home Page

2. Click on *University Program* under **Education and Events.** The university program page, shown in Figure 8.2, will open.

3. On the left side of the page, click on *Design Software* under **Educational Materials.** The university program development software page will open (Figure 8.3).

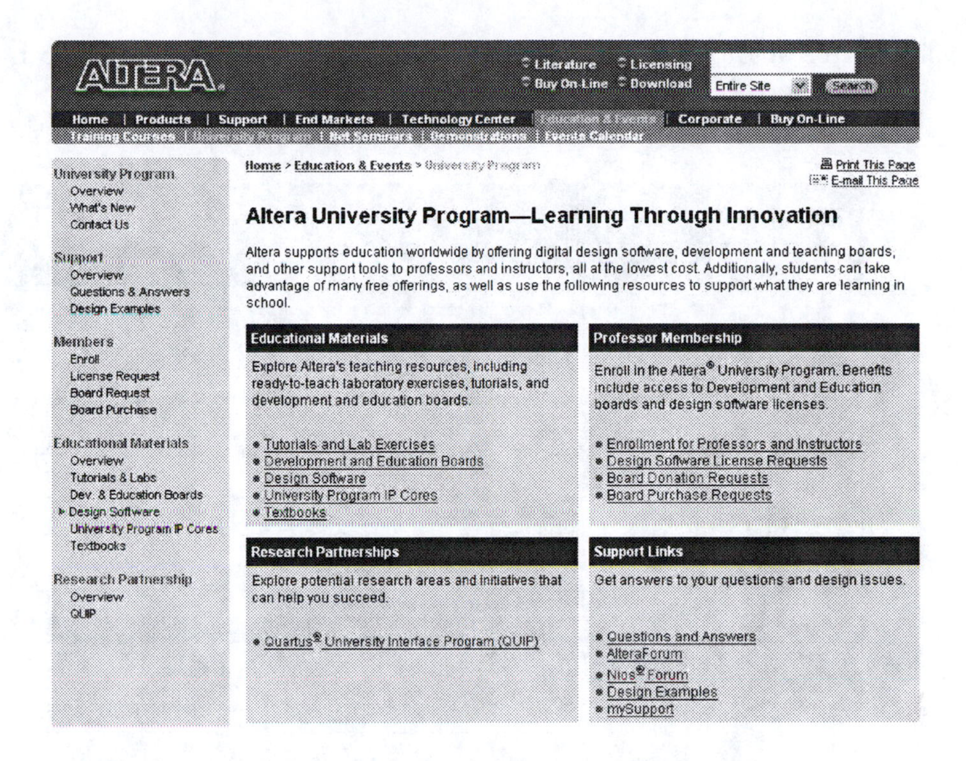

Figure 8.2 Altera University Program Page

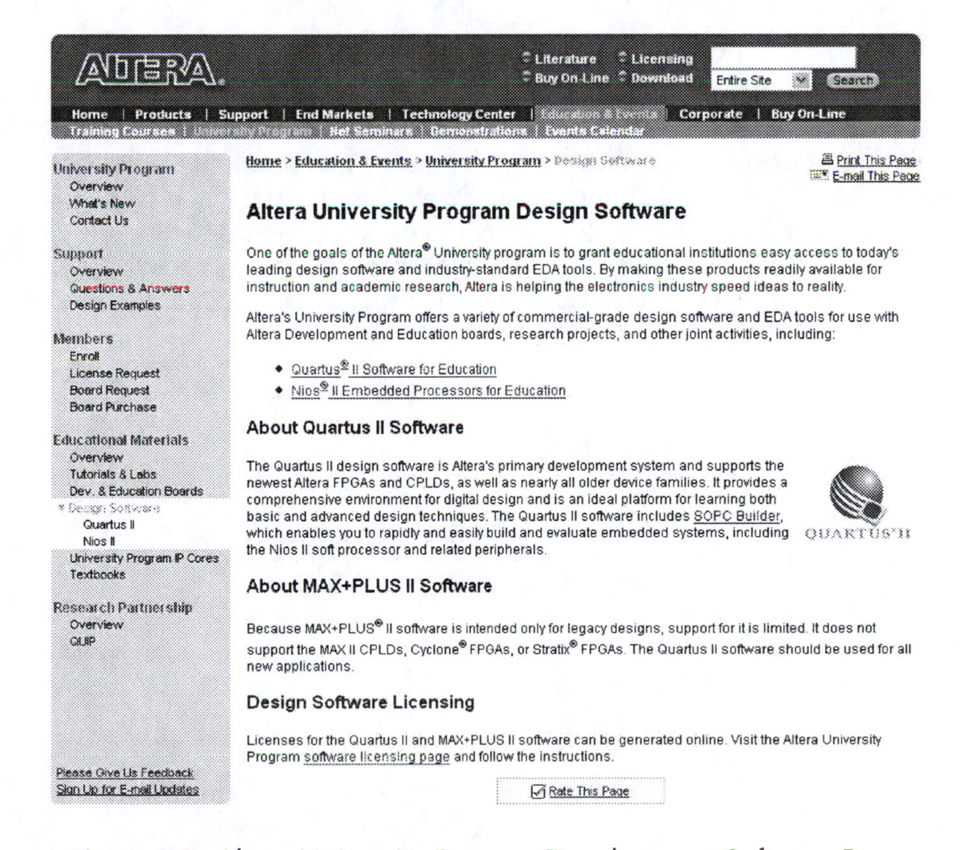

Figure 8.3 Altera University Program Development Software Page

4. Click on the link for the *Quartus II Software for Education.* This will open the page shown in Figure 8.4.

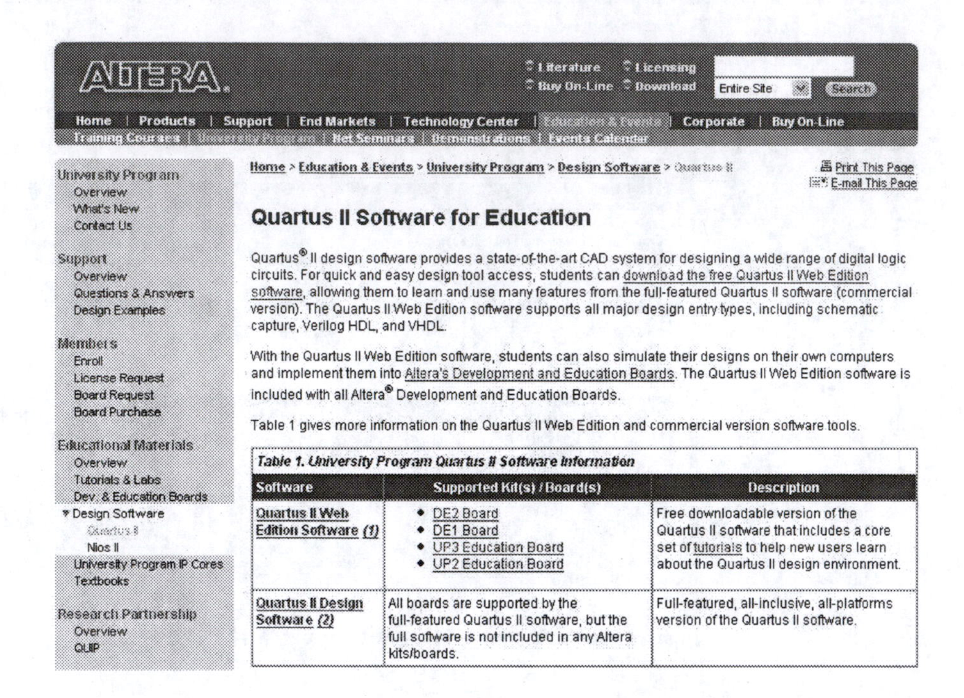

Figure 8.4 Quartus II Web Edition Download Page

5. Click on the link that says *Download the Quartus II Web Edition Software.* This gets you to the download page, shown in Figure 8.5. Click **Download free Quartus II Web Edition software,** taking you to the page where you can access the software and licenses (Figure 8.6). Click the yellow **Download** button to start the Altera Download manager; this will lead you through the steps required to download the software. When downloading, save the file to a convenient folder on your PC hard drive. After downloading is complete, follow the *installation instructions,* available on this web page.

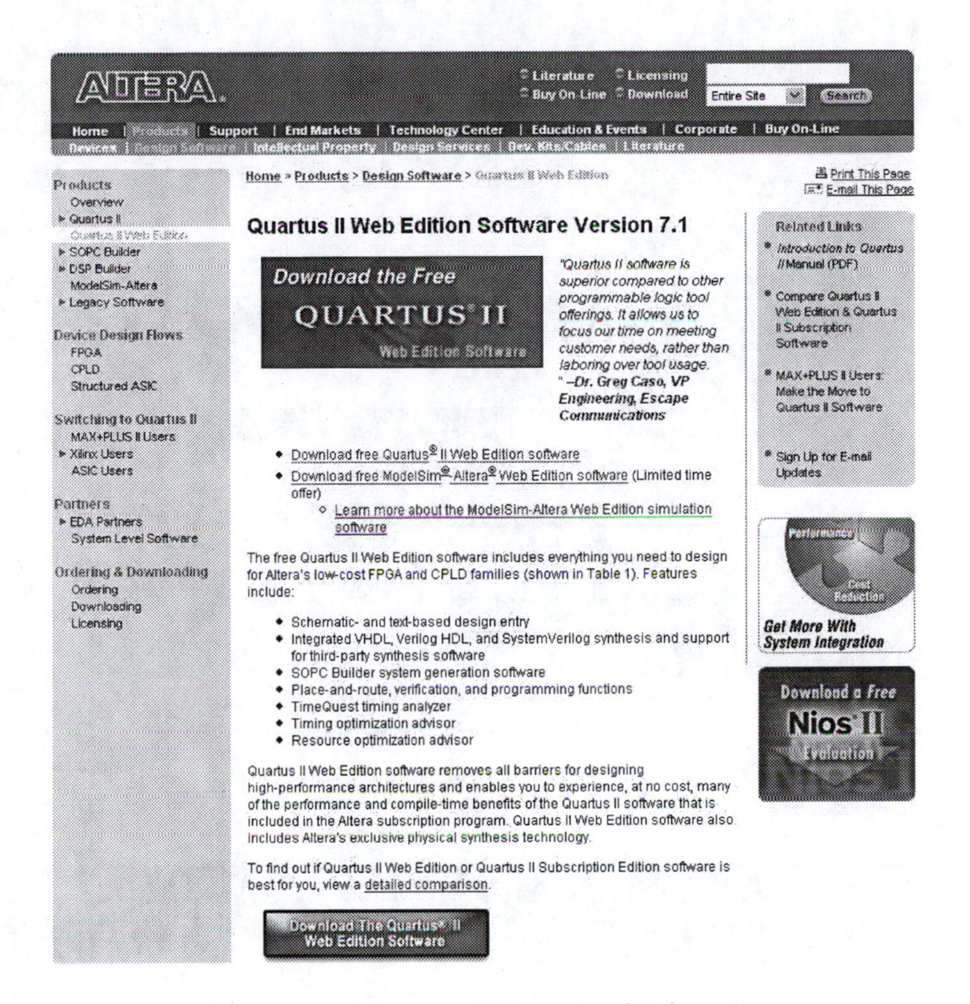

Figure 8.5 Quartus II Download Select

Obtaining a License File from Altera

1. Before requesting a license file, you must know the physical address (MAC address) of the network interface card (NIC) in the computer on which you are installing the software. You can obtain this from the command prompt in Windows. Click the Windows **Start** menu and select **Run.** Type **command** in the box that opens and click **OK.** At the command prompt, type **ipconfig/all.** This will give a list of network settings for your computer. Look for the physical address of your NIC and write it down. It will consist of a string of twelve hexadecimal digits, separated by dashes between pairs of digits (e.g., 00-08-02-6C-E6-F2). This number is unique to your computer. Close the command window by typing **exit** at the command prompt and pressing the **Enter** key.

2. Follow the steps in the previous section to get to the web page shown in Figure 8.5. Click the link for *Download free Quartus. . . .* The page shown in Figure 8.6 will open. Click on "Get licenses".

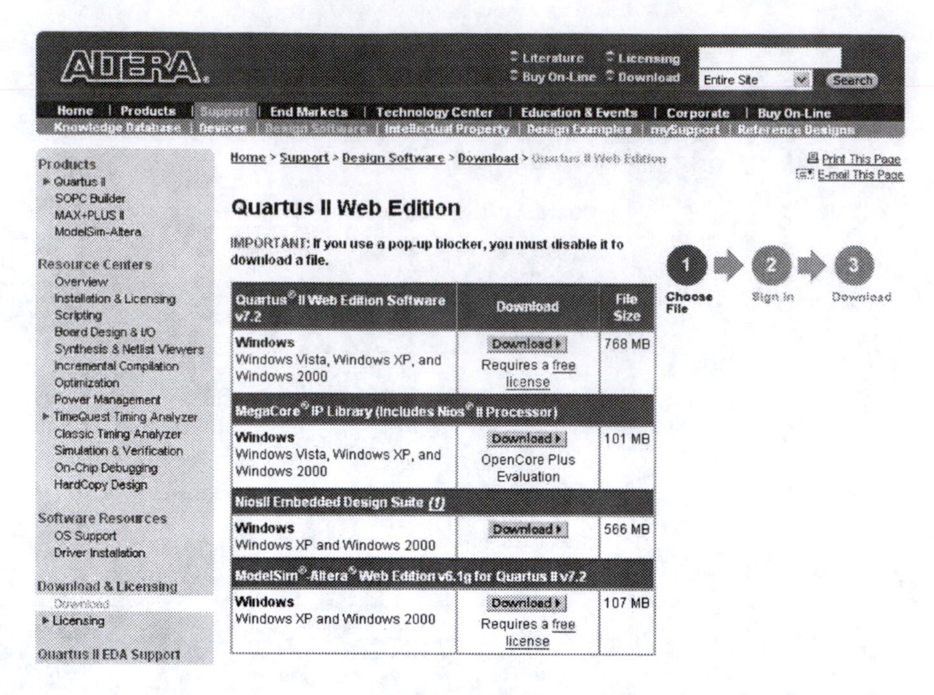

Figure 8.6 Quartus II License and Download Page

3. Click the **free license** link. The page shown in Figure 8.7 will be displayed.

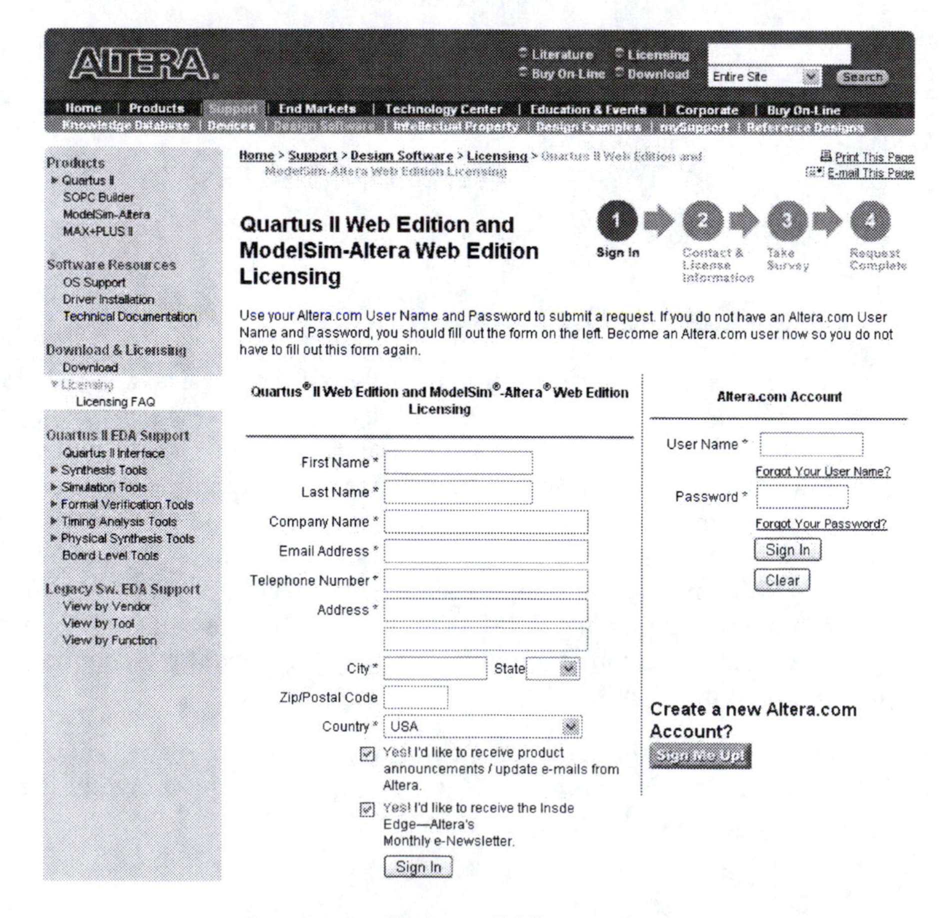

Figure 8.7 Quartus II License Request

4. Unless you have an existing Altera login, you should fill in your information requesting a free license from Altera.

5. A license file will be mailed to the e-mail address you specified on the form. Under most circumstances, this should happen almost immediately, but due to network conditions it may be delayed up to several hours.

Installing the License File

1. When you receive the e-mail from Altera with your license file, the file will be included as an attachment and will also be embedded as text in the e-mail. If at all possible, use the attachment. Save it to a separate folder, such as **c:\qlicense_we\.** If you are unable to save the attachment, copy the embedded text and save it in a text editor as **license.dat.** (Note that if you use Notepad to do this, it will append a **.txt** extension to your file, so that it will be called **license.dat.txt.** When saving the file, use quotes: type "license.dat" instead of license.dat.)

2. Start Quartus II. Open the license setup dialog by clicking **License Setup** from the **Tools** menu, as shown in Figure 8.8.

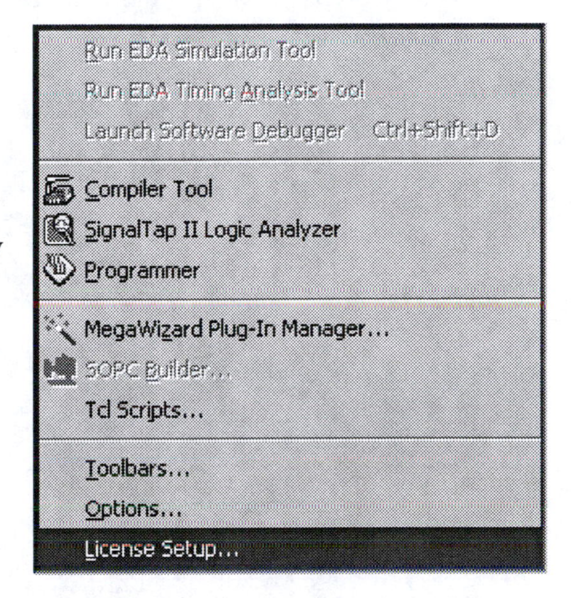

Figure 8.8 License Setup (Tools Menu)

3. In the box shown in Figure 8.9, browse to locate the license file by clicking the button labeled (. . .). For this example, the folder with the license file is assumed to be **c:\qlicense\.** Click **OK** to close the box. The Quartus II software is now licensed.

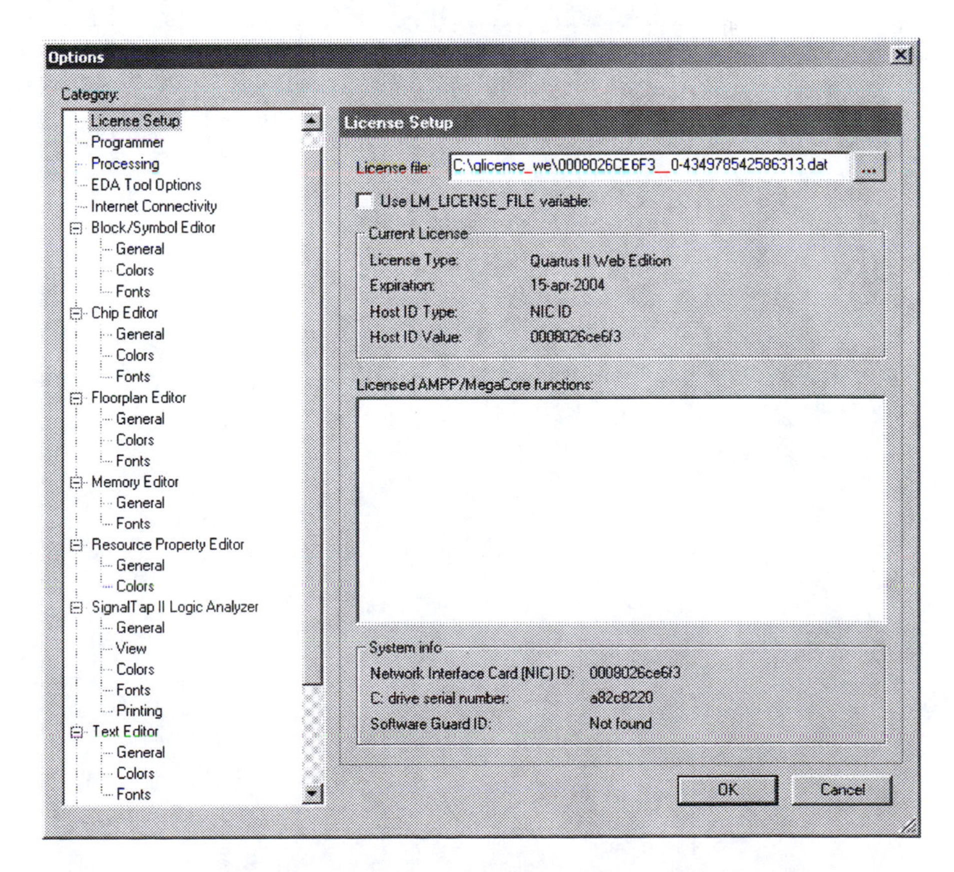

Figure 8.9 License Setup Dialog

Installing the ByteBlaster Driver

1. Browse to *http://www.altera.com/support/software/drivers/dri-index.html* to see a table (Figure 8.10) showing driver requirements for various Altera programming devices and Windows operating systems. Installation instructions are available on this web page under the link labeled *Install driver* for the ByteBlaster MV cable.

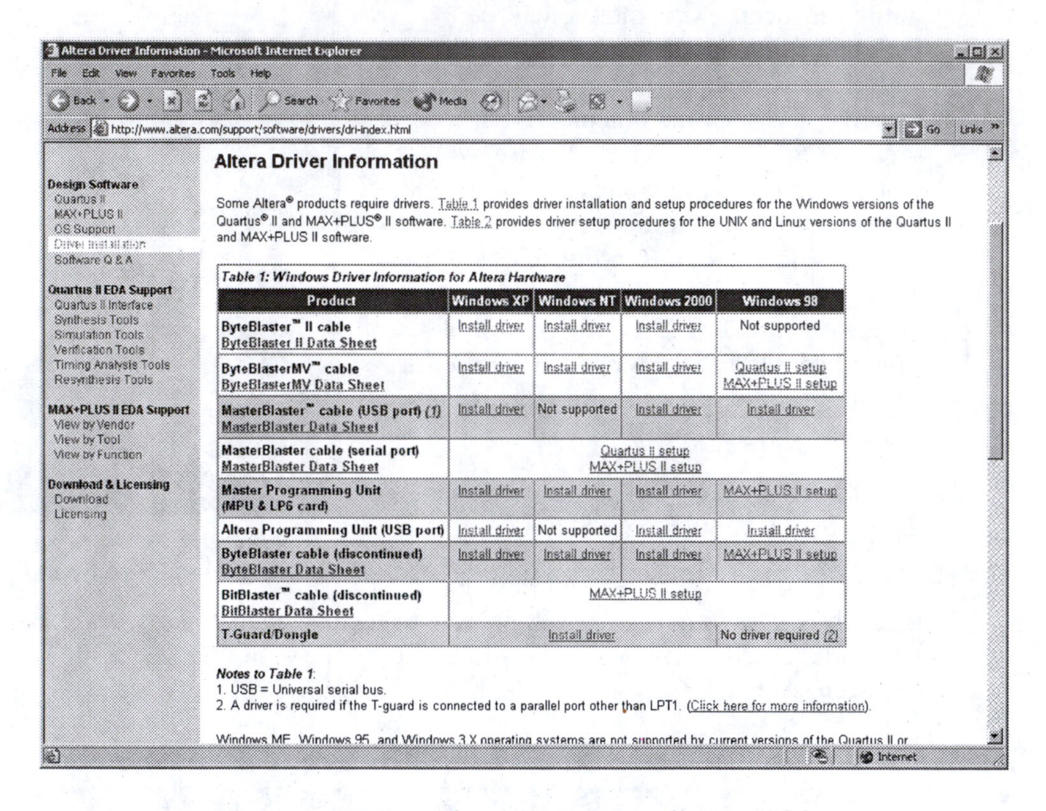

Altera Driver Information

Some Altera® products require drivers. Table 1 provides driver installation and setup procedures for the Windows versions of the Quartus® II and MAX+PLUS® II software. Table 2 provides driver setup procedures for the UNIX and Linux versions of the Quartus II and MAX+PLUS II software.

Table 1: Windows Driver Information for Altera Hardware

Product	Windows XP	Windows NT	Windows 2000	Windows 98
ByteBlaster™ II cable ByteBlaster II Data Sheet	Install driver	Install driver	Install driver	Not supported
ByteBlasterMV™ cable ByteBlasterMV Data Sheet	Install driver	Install driver	Install driver	Quartus II setup MAX+PLUS II setup
MasterBlaster™ cable (USB port) (1) MasterBlaster Data Sheet	Install driver	Not supported	Install driver	Install driver
MasterBlaster cable (serial port) MasterBlaster Data Sheet	Quartus II setup MAX+PLUS II setup			
Master Programming Unit (MPU & LP6 card)	Install driver	Install driver	Install driver	MAX+PLUS II setup
Altera Programming Unit (USB port)	Install driver	Not supported	Install driver	Install driver
ByteBlaster cable (discontinued) ByteBlaster Data Sheet	Install driver	Install driver	Install driver	MAX+PLUS II setup
BitBlaster™ cable (discontinued) BitBlaster Data Sheet	MAX+PLUS II setup			
T-Guard/Dongle	Install driver			No driver required (2)

Notes to Table 1:
1. USB = Universal serial bus.
2. A driver is required if the T-guard is connected to a parallel port other than LPT1. (Click here for more information).

Windows ME, Windows 95, and Windows 3.X operating systems are not supported by current versions of the Quartus II or

Figure 8.10 ByteBlaster Driver Installation Table

Lab 9

Introduction to Quartus II (Tutorial)

Name ______________________ Class ____________ Date ____________

Objectives Upon completion of this laboratory exercise, you should be able to:

- Create a project in Quartus II.
- Use the Quartus Block Editor to enter a graphical design in Quartus II.
- Compile and simulate the design.
- Program an Altera CPLD with the design.
- Test the design on a CPLD test board to determine its truth table.

Reference Ken Reid and Robert Dueck, *Introduction to Digital Electronics*
Chapter 4: Introduction to PLDs and Quartus II

Equipment Required CPLD Trainer:
Altera UP-2 circuit board with ByteBlaster Download Cable, or
DeVry eSOC board with Straight-Through Parallel Port Cable, or
RSR PLDT-2 circuit board with Straight-Through Parallel Port Cable,
or Equivalent CPLD trainer board with Altera EPM7128S CPLD
Quartus II Web Edition software
AC adapter, minimum output: 7 VDC, 250 mA DC
Anti-static wrist strap
#22 solid-core wire
Wire strippers

Experimental Notes

This lab will follow the tutorial steps laid out in Chapter 4 of *Introduction to Digital Electronics*. This example creates a project in Quartus II, enters a design file for a majority vote circuit, and programs an Altera CPLD with the design.

Procedure

Creating a Quartus II Project and Block Diagram File

To create a new Block Diagram File, select **New** from the **File** menu. From the dialog box, shown in Figure 9.1, choose the **Device Design Files** tab and select **Block Diagram/Schematic File.** The Quartus II Block Editor will open, as shown in Figure 9.2.

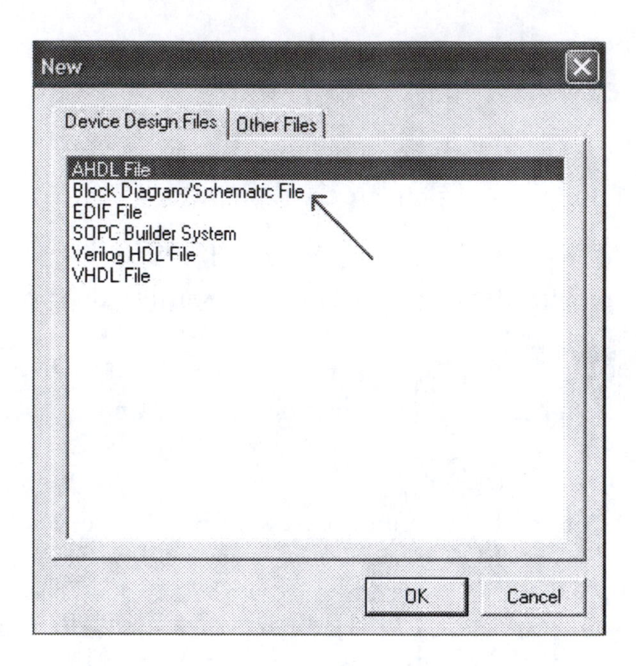

Figure 9.1 New File Dialog Box (Block Diagram File)

Figure 9.2 Block Diagram File (Blank)

Before entering any design information, we will create a new project, with the new Block Diagram File as the top-level file in a design **hierarchy.**

To create the new project, first save the blank Block Diagram File, using the **Save As** dialog, shown in Figure 9.3. Change the file name to **majority_vote.bdf** and save the file in the folder ***drive:*\qdesigns\labs\lab09\majority\.** Make sure the box labeled **Create new project based on this file** is checked. The dialog box in Figure 9.4 will appear, asking you to confirm your choice. Click **Yes.**

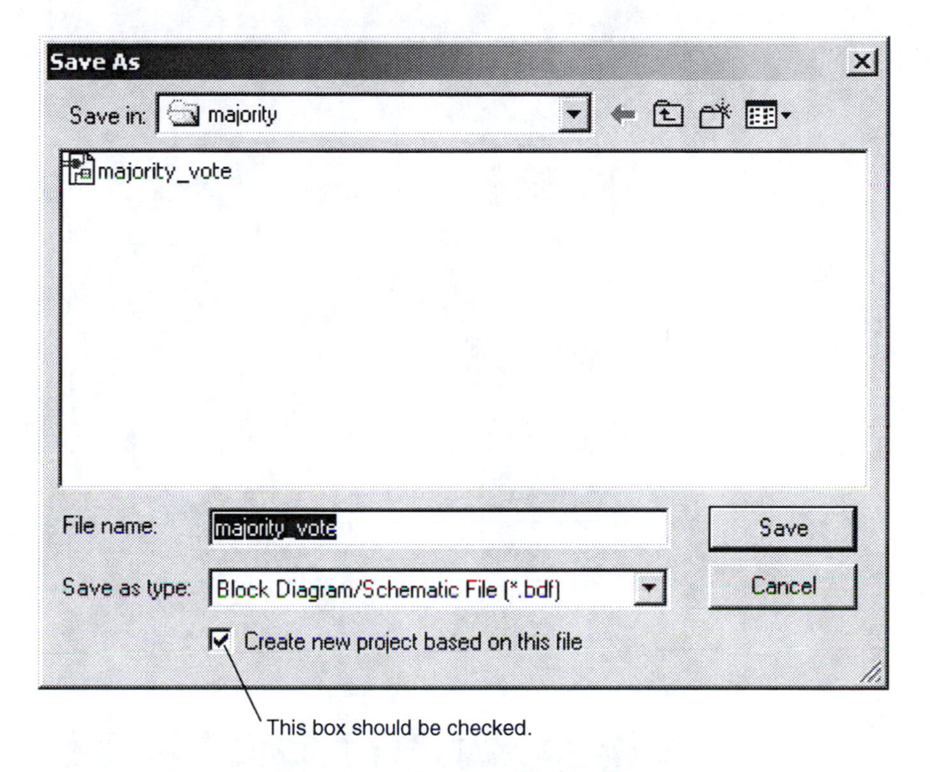

Figure 9.3 Save as Dialog Box

Figure 9.4 New Project Query

After saving your file, the New Project Wizard will automatically appear. The New Project Wizard is a series of setup screens that asks the user for information about a project. Initially, there is an introductory screen, shown in Figure 9.5. The next screen, in Figure 9.6, asks for information about the directory the project will use, project name, and name of the top-level design entity. A directory name should be unique for each project and can be created by typing it in or selecting it from an existing directory list. By default, the screen in Figure 9.6 contains the directory where the new Block Diagram File was stored and its file name. (Figure 9.6 shows the file path for the example shown in Chapter 4 of *Introduction to Digital Electronics*.) For most cases, simply accept the default names and click **Next.**

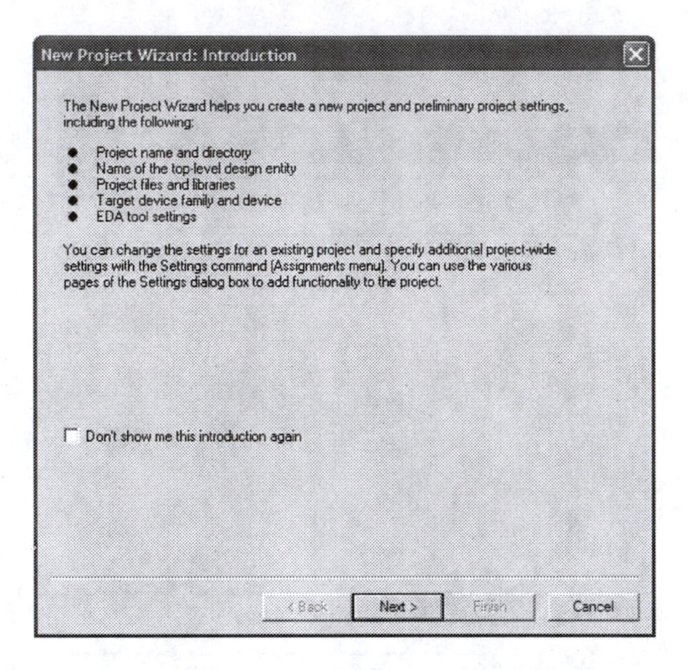

Figure 9.5 New Project Wizard (Intro)

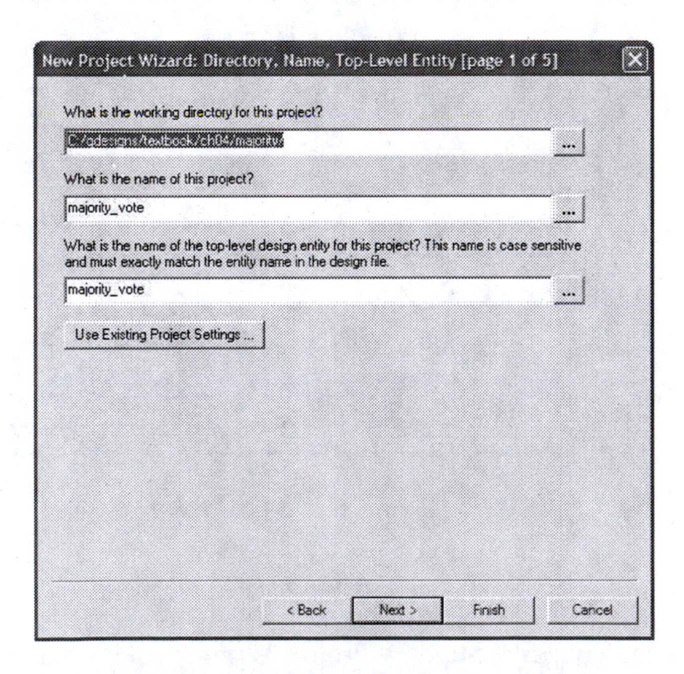

Figure 9.6 New Project Wizard (Directory, Entity)

The next screen, in Figure 9.7, shows the files currently included in the project and asks if any more should be added. Click **Next.**

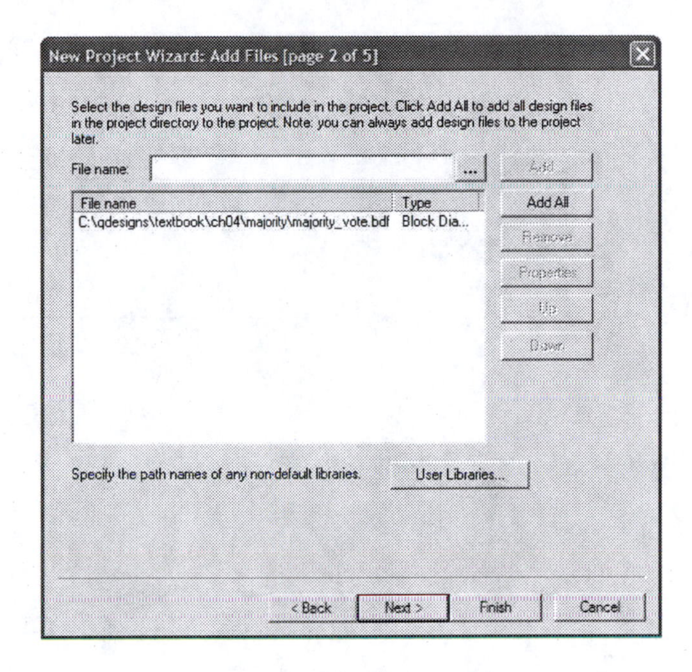

Figure 9.7 New Project Wizard (Files)

The screen in Figure 9.8 select the device family (MAX 7000S) and target device (EPM7128SLC84-7) for the project. Click **Next** after making these selections.

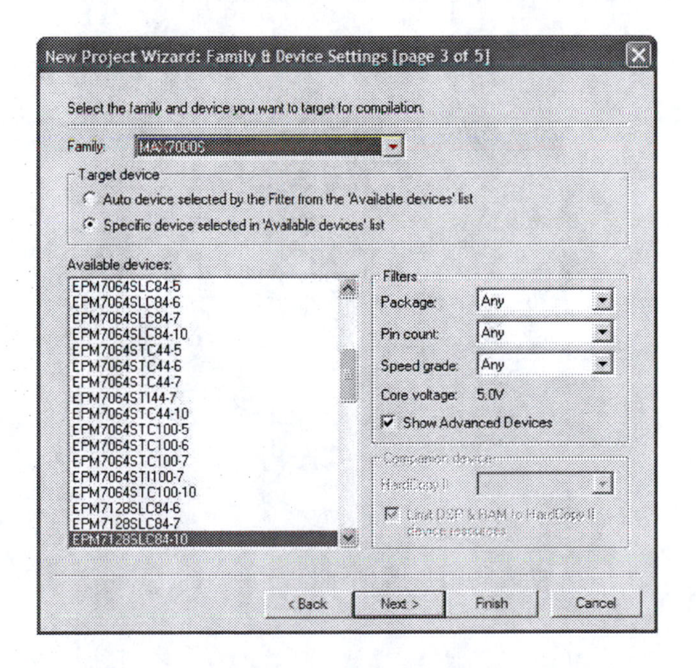

Figure 9.8 New Project Wizard (Device Family)

The screen in Figure 9.9 allows us to specify additional tools. We don't need to, so click **Next.**

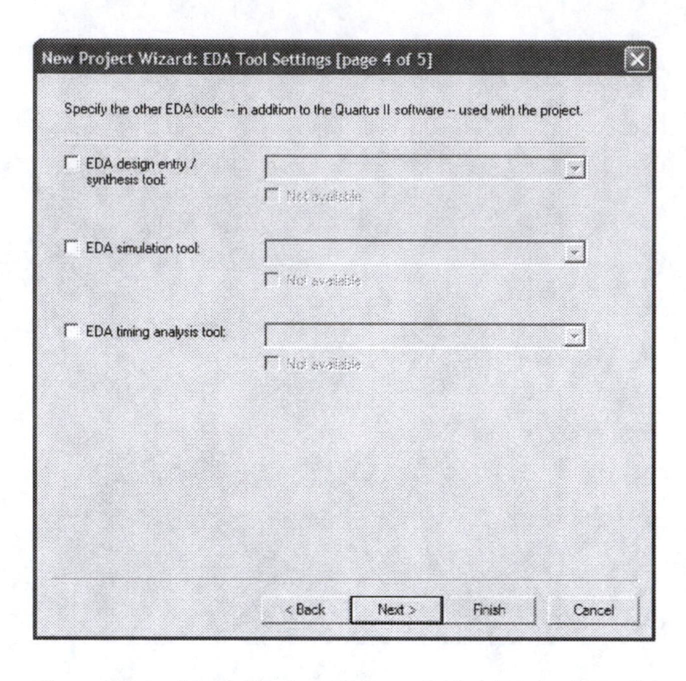

Figure 9.9 New Project Wizard (Additional Tools)

The final screen, shown in Figure 9.10 is a summary of project settings provided by the user. Click **Finish** to exit the wizard. The wizard can be exited at any point if all the relevant information has been entered. Project settings can be altered at any time via the **Settings** dialog box, accessible from the **Assignments** menu.

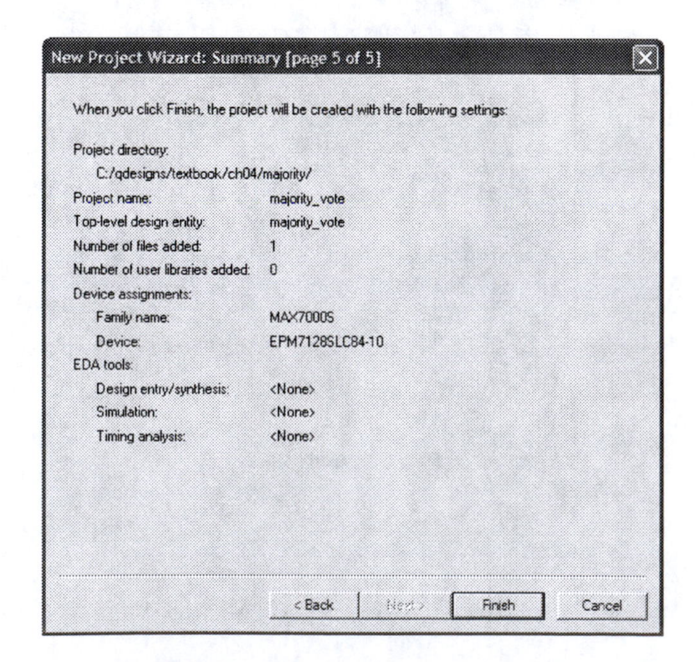

Figure 9.10 New Project Wizard (Summary)

Entering Components

The first step in entering the majority vote circuit in the Quartus II Block Editor is to lay out and align the required components. We require three 2-input AND gates, a 3-input OR gate, three input pins, and one output pin. These basic components are referred to as **primitives.** Let us start by entering three copies of the AND gate primitive, called **and2.**

Open the **Edit** menu, shown in Figure 9.11, and select **Insert Symbol,** or simply double-click on the Block Editor desktop. In the **Symbol** dialog box (Figure 9.12) type **and2** in the box labeled **Name.** The **and2** symbol appears in the desktop area on the right. Since we want to enter three **instances** of the symbol, check the box that is labeled **Repeat-insert mode.** Click **OK.**

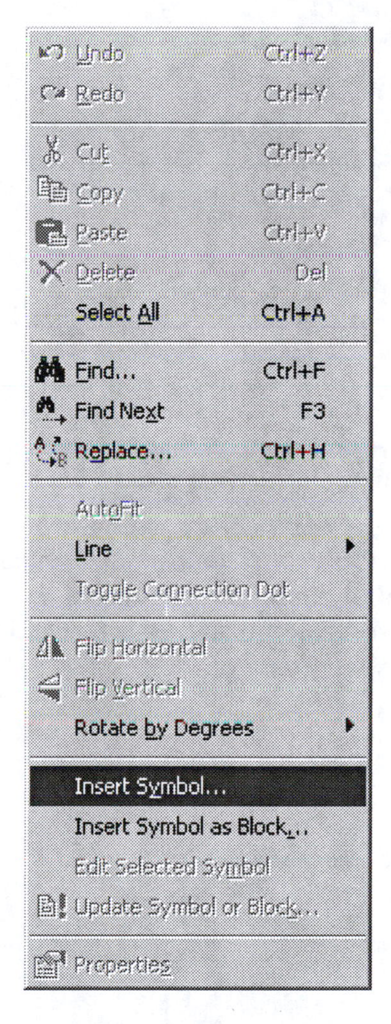

Figure 9.11 Edit Menu (Insert Symbol)

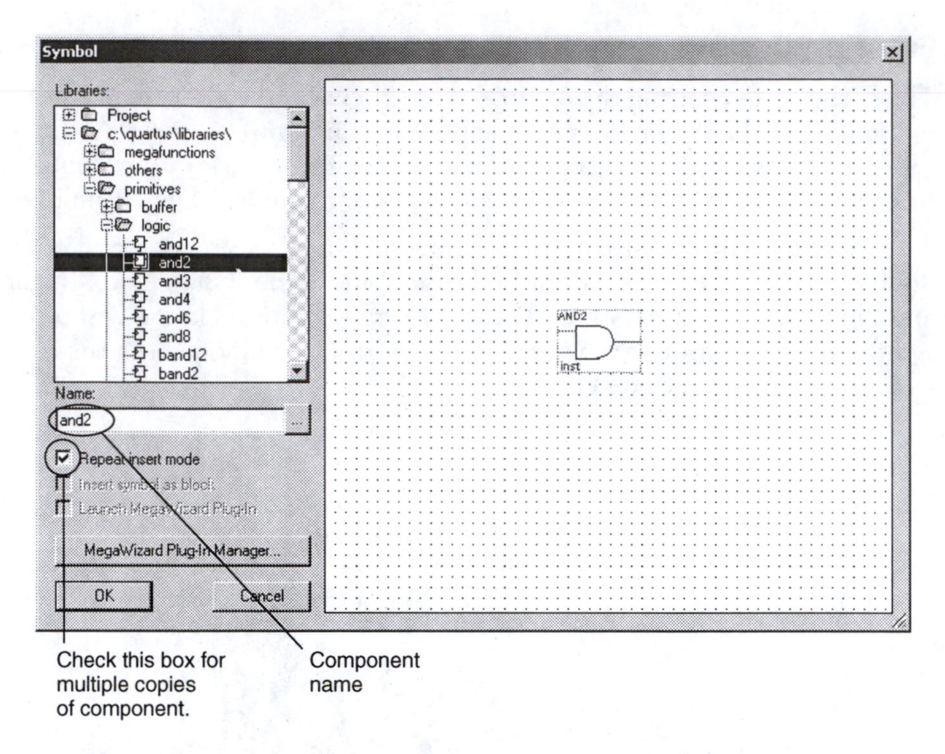

Figure 9.12 Symbol Dialog Box (and2)

Click on the desktop in the Block Editor window to place an instance of the **and2** component. Don't worry about its exact placement for the moment. Click two more times to place two more gates, then use the **ESC** key to exit the insert-repeat mode.

Enter the remaining components by following the **Insert Symbol** procedure outlined above. The primitives are called **or3, input,** and **output.** The insert-repeat mode is not necessary for the **or3** and **output** components, as there is only one of each. When all components are entered we can align them, as in Figure 9.13 by highlighting, then dragging each one to a desired location.

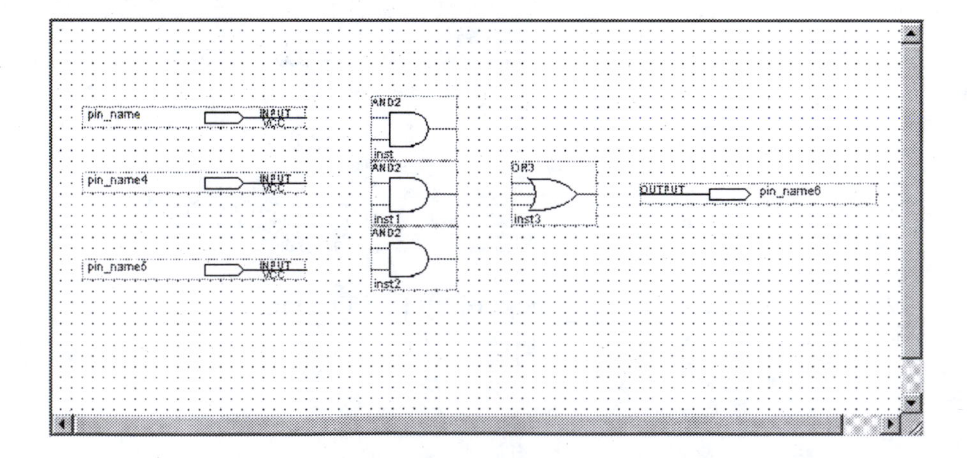

Figure 9.13 Aligned Components

Connecting Components

To connect components, click over one end of one component and drag a line to one end of a second component, as shown in Figure 9.14. When you hover over a line end, the cursor changes from an arrow to a crosshair with a right-angle symbol. When you drag the line, a horizontal and a vertical grid helps you align connections properly.

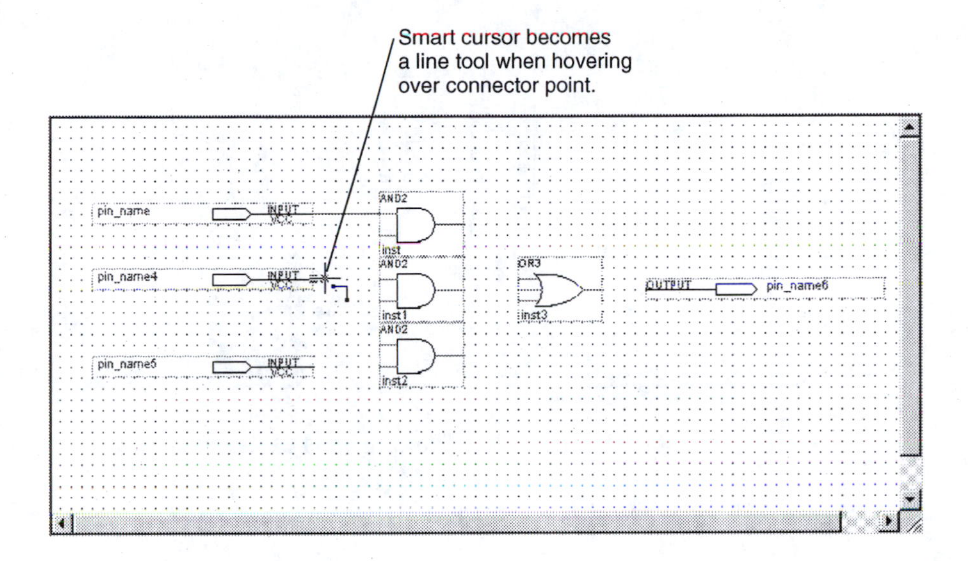

Figure 9.14 Dragging a Line to Connect Components

A line will automatically make a connection to a perpendicular line, as shown in Figure 9.15.

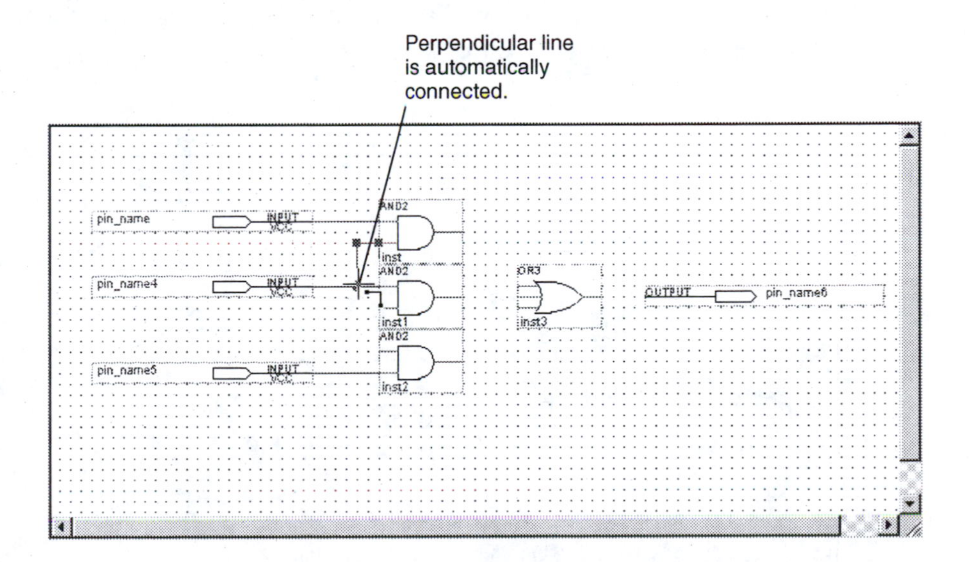

Figure 9.15 Making a 90° Bend and a Connection

A line can have one 90° bend, as at the inputs of the AND gates. If a line requires two bends, such as shown at the OR inputs in Figure 9.16, you must draw two separate lines.

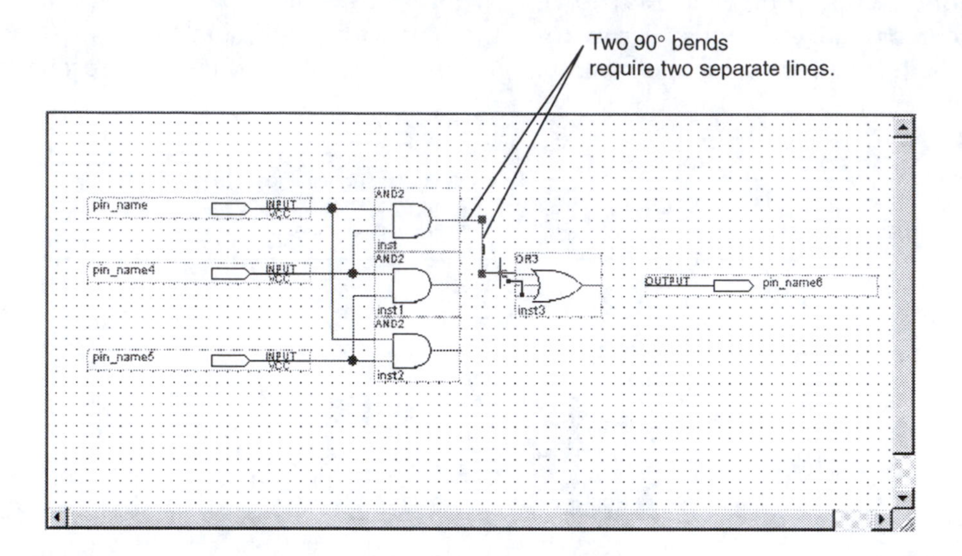

Figure 9.16 Making Two 90° Bends

Assigning Pin Names

Before a design can be compiled, its inputs and outputs must be assigned names. Figure 9.17 shows the naming procedure. Pins **a** and **b** have already been assigned names. Double-click the pin name (not the pin symbol) to highlight the name. Type in the new name.

If there are several pins that are spaced one above the other, such as **a**, **b**, and **c** in Figure 9.17, you can highlight the top pin name, as described above, then highlight successive pin names by using the **Enter** key.

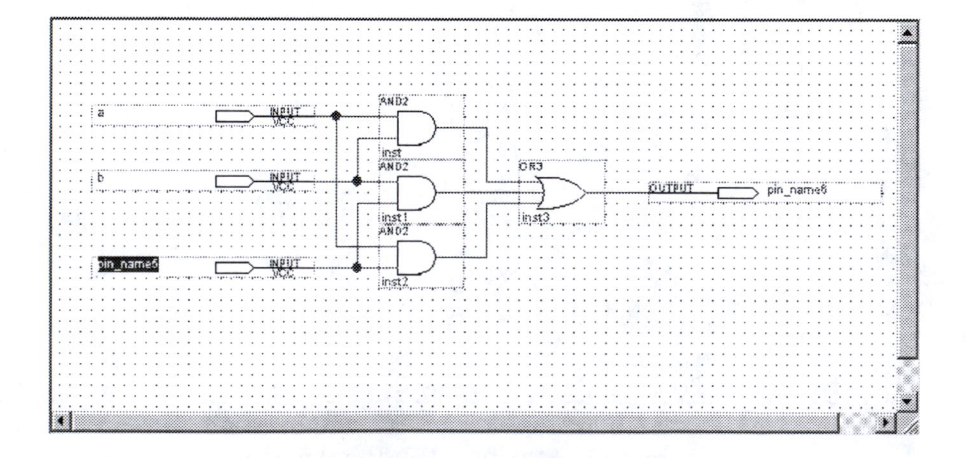

Figure 9.17 Assigning Pin Names

Compiling and Simulating a Design in Quartus II

To start the compilation process, click the **Start Compilation** button on the Quartus II toolbar, as shown in Figure 9.18.

Figure 9.18 Start Compilation (Toolbar Button)

When compilation is complete, Quartus II displays a **Compilation Report,** shown in Figure 9.19.

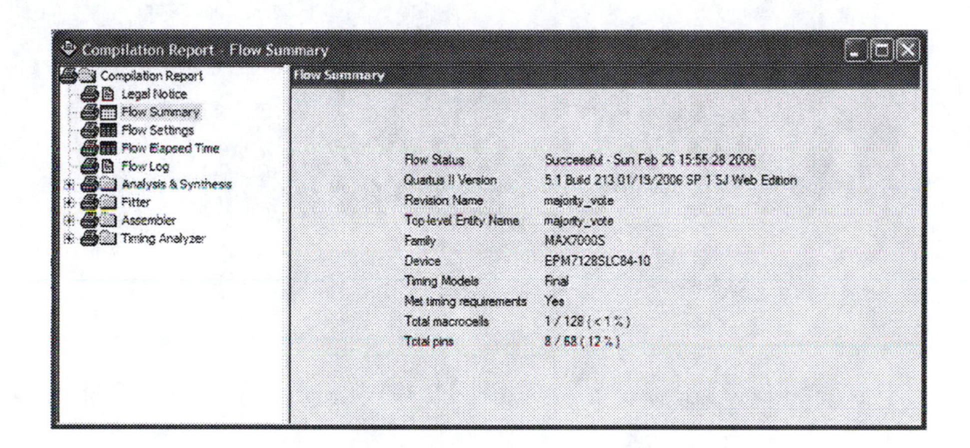

Figure 9.19 Compilation Report Window

Instructor's Initials ____________

Simulation

Before we create a simulation for the majority vote circuit of our tutorial example, we should write a set of simulation criteria for the circuit. The circuit must generate a HIGH output when a majority of inputs is HIGH. Since we have three inputs, this means two or more inputs must be HIGH. The most thorough and systematic way to test the circuit is to apply all possible input combinations to the circuit in an ascending binary sequence, in other words, to take its truth table. We can summarize the criteria, as follows.

Simulation Criteria

- Apply all possible inputs in an ascending binary sequence.
- For any input combination having two or more HIGH inputs (011, 101, 110, 111), the output must be HIGH.
- For any other input combination (000, 001, 010, 100), the output must be LOW.

A simulation is based on a **Vector Waveform File (vwf),** which contains simulation input and output values in the form of graphical waveforms. To create a new **vwf,** select **New** from the **File** menu or click the appropriate toolbar button. From the box in Figure 9.20, click the **Other Files** tab and select **Vector Waveform File.**

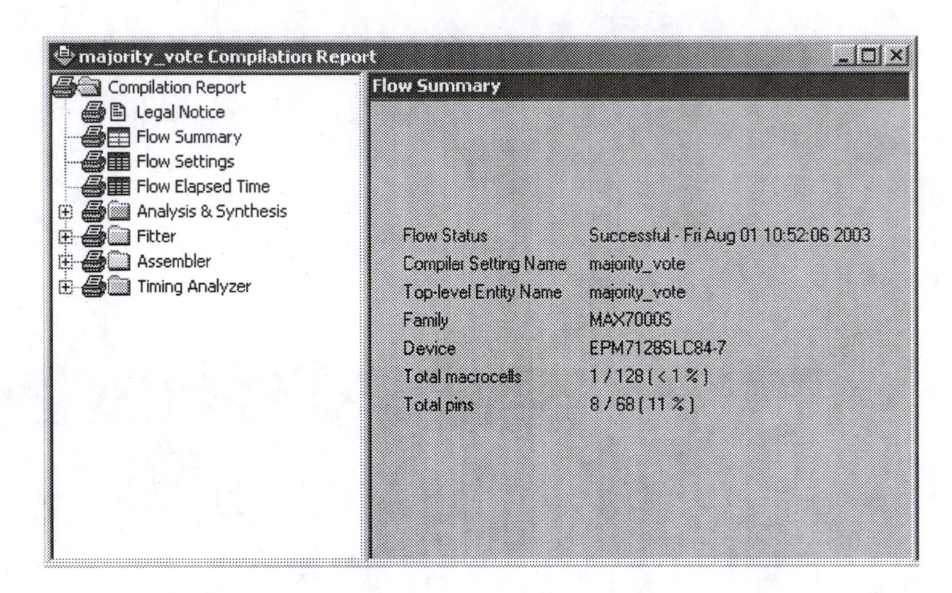

Figure 9.20 New File Dialog Box (Vector Waveform File)

The default window of the Quartus II Waveform Editor will appear, as shown in Figure 9.21.

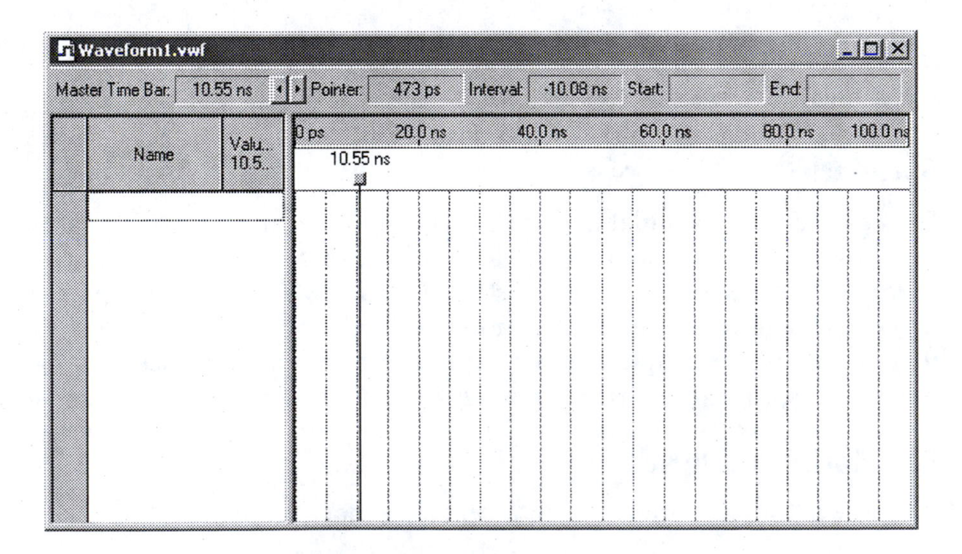

Figure 9.21 Waveform Editor (Default)

To add waveforms to the window, we can use the **Node Finder.** Start the Node Finder by clicking on the **Node Finder** toolbar button or selecting **Utility Windows, Node Finder** from the **View** menu. The default window opens, as shown in Figure 9.22. Click **Start** to display a list of nodes, or signal points, for the project. The window in Figure 9.23 shows a list of available nodes for the project. We are only interested in **a, b, c,** and y; the others represent logic levels internal to the circuit.

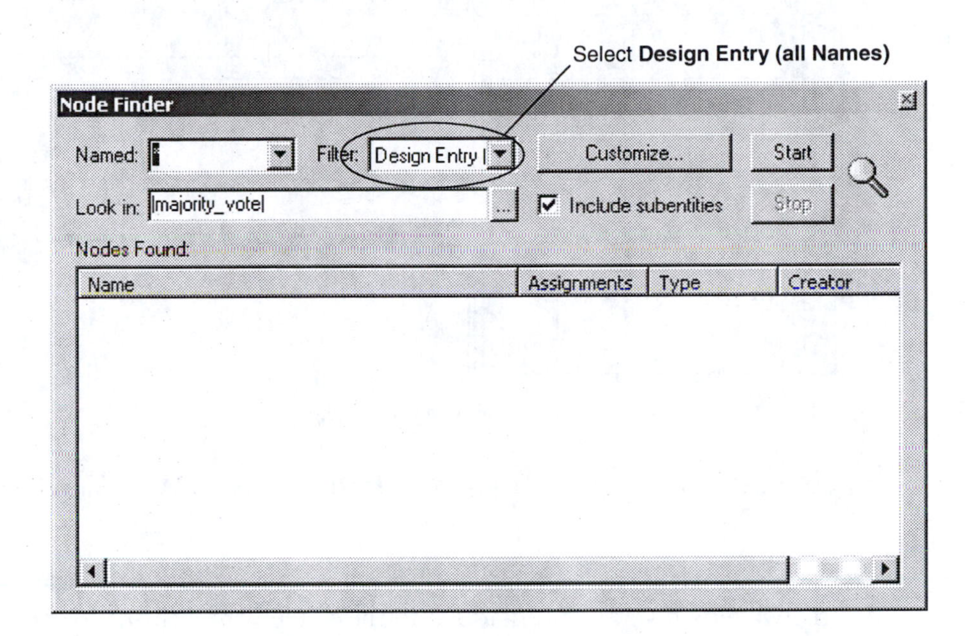

Figure 9.22 Node Finder (Default)

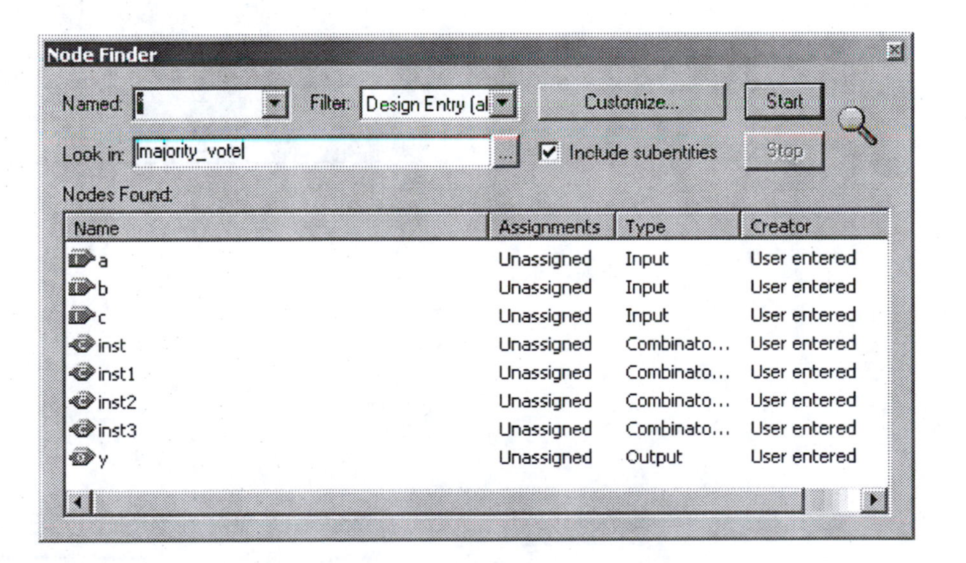

Name	Assignments	Type	Creator
a	Unassigned	Input	User entered
b	Unassigned	Input	User entered
c	Unassigned	Input	User entered
inst	Unassigned	Combinato...	User entered
inst1	Unassigned	Combinato...	User entered
inst2	Unassigned	Combinato...	User entered
inst3	Unassigned	Combinato...	User entered
y	Unassigned	Output	User entered

Figure 9.23 Node Finder (Node Selected)

Add the required nodes to the Vector Waveform File by dragging them from the Node Finder to the Waveform Editor window. Do this by clicking on a waveform to highlight it, then drag and drop to place it in the Waveform Editor, as shown in Figure 9.24.

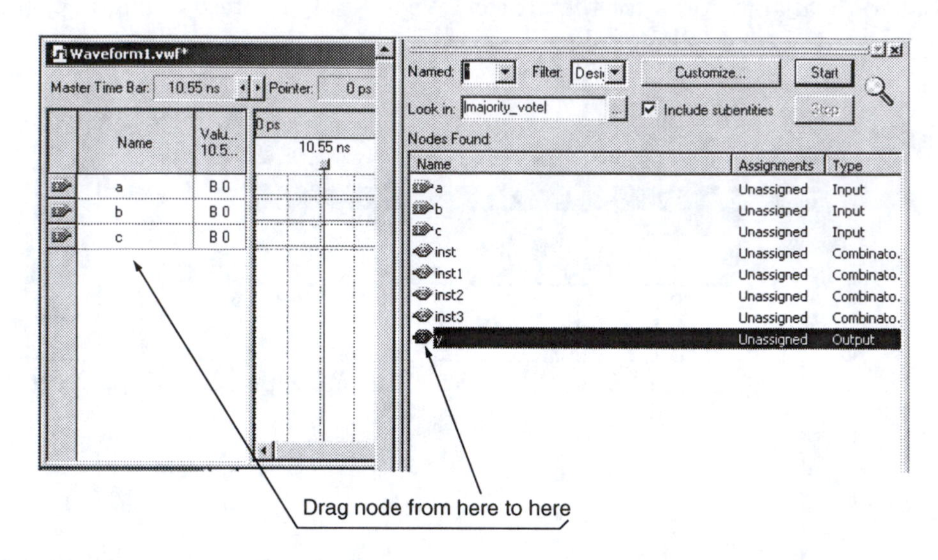

Figure 9.24 Dragging from Node Finder to Vector Waveform File

Once the waveforms have been added to the Waveform Editor, close the Node Finder window and save the Vector Waveform File, using the **Save As** dialog box in Figure 9.25. Make sure that the box labeled **Add file to current project** is checked. (This step could be done earlier, such as immediately after creating a new Vector Waveform File.)

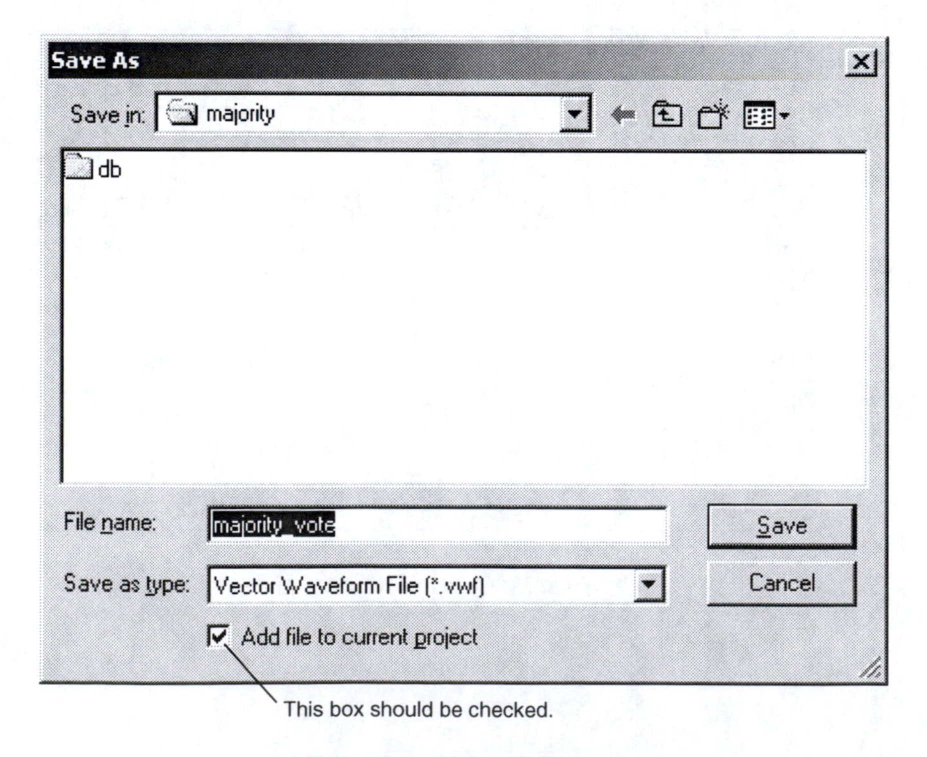

Figure 9.25 Save as Dialog Box (Adding vwf to Project)

Once we have entered the input and output nodes, we must determine the length of time our simulation should run. The default time is 100 ns. The target CPLD has a delay time from input to output of about 7 ns. (This is indicated by the −7 at the end of its part number.) If we changed our inputs every 10 ns, we would have sufficient room in the simulation to display ten input changes, which is enough to fit in the eight changes we need, plus a little more. However, due to the input-to-output delay in the target device, the output would not change until it is nearly time for a new input change. This delayed output would produce an offset waveform that would be confusing to read. In this case, we would be better off with a simulation time that is long compared to the input-to-output delay of the target device.

To change the end time of the simulation, select **End Time** from the **Edit** menu, shown in Figure 9.26. In the **End Time** dialog box, shown in Figure 9.27, change the unit from **ns** to **us** (microseconds). This is an easy change that gives a simulation time that is long compared to device delay time.

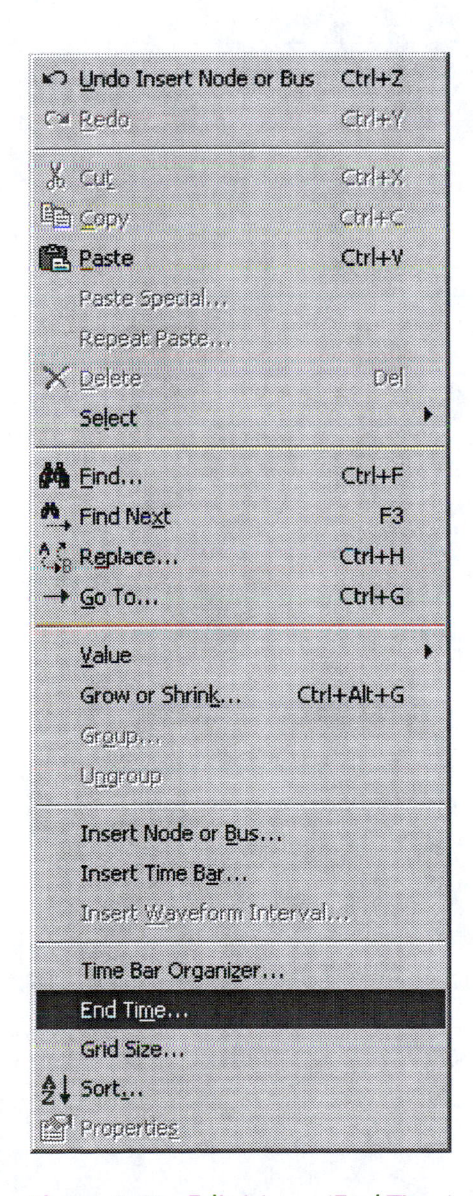

Figure 9.26 Edit Menu (End Time)

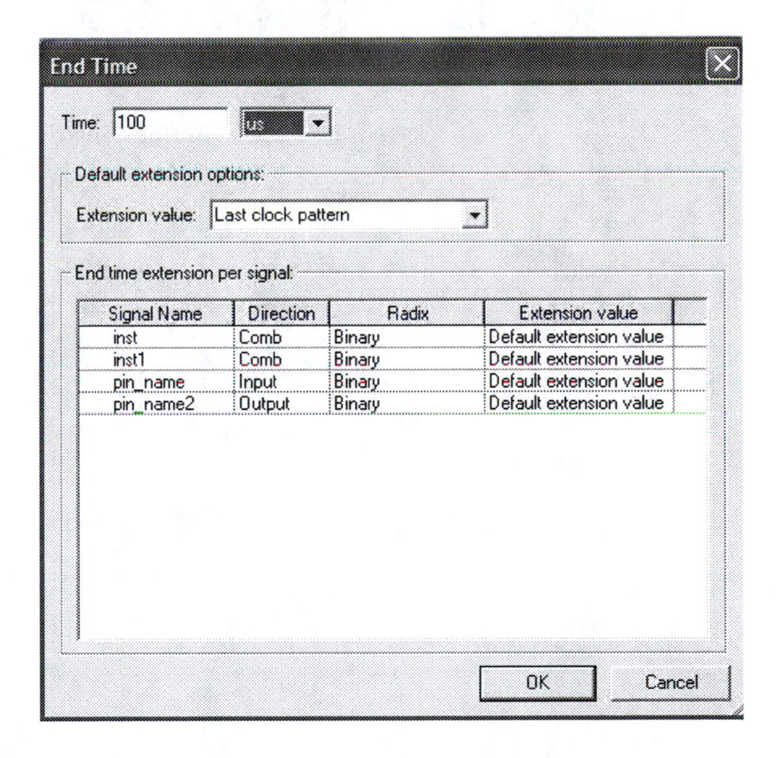

Figure 9.27 End Time Dialog Box

To see the entire waveform file, select **Fit in Window** from the **View** menu.

We could enter waveforms on **a, b,** and **c** individually to create an increasing 3-bit binary sequence, but it is easier and more accurate to group these waveforms together and apply the sequence to the whole group. Figure 9.28 shows a highlighted group of waveforms in the Quartus II Waveform Editor. To highlight a group, click on the top waveform (a), then drag the cursor to the last waveform of the group (c). Right-click on the highlighted group of waveforms and select **Group . . .** from the pop-up menu of Figure 9.29. Type the group name **(Inputs)** in the dialog box of Figure 9.30. Select the radix (i.e., the number system base) as **Binary.** Click **OK.** The grouped waveforms appear as shown in Figure 9.31.

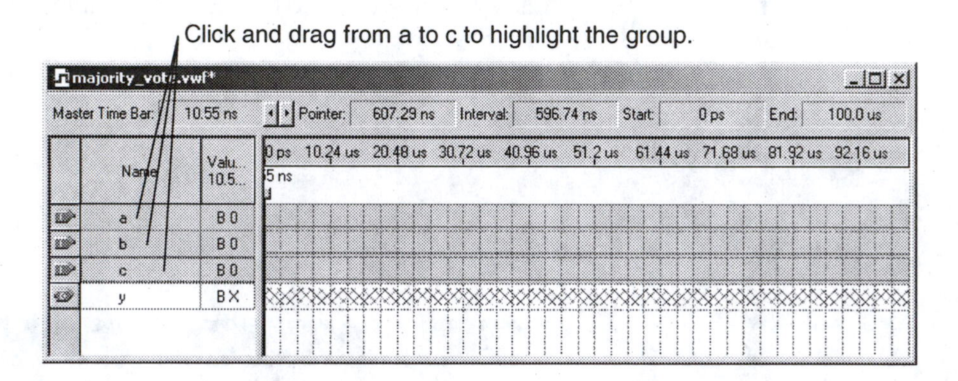

Figure 9.28 Highlighting a Group of Waveforms

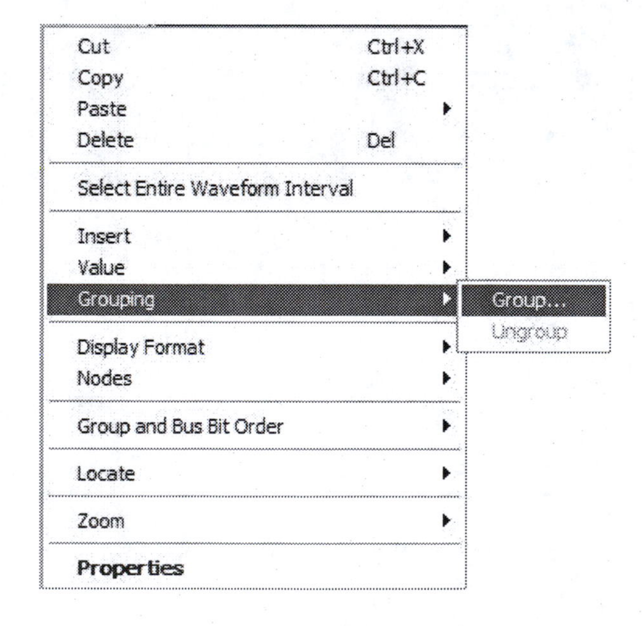

Figure 9.29 Pop-up Menu (Group)

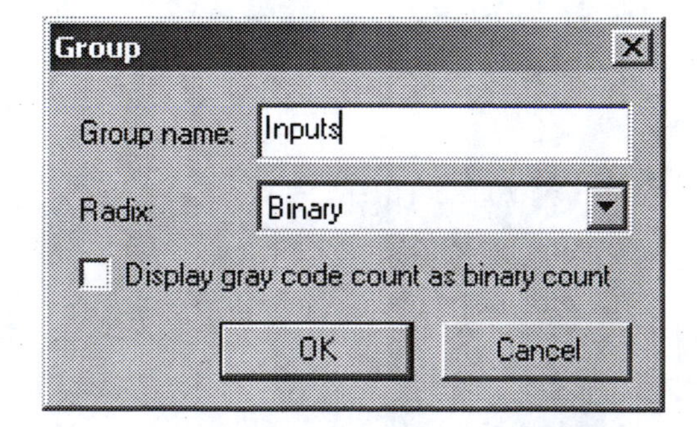

Figure 9.30 Group Name and Radix

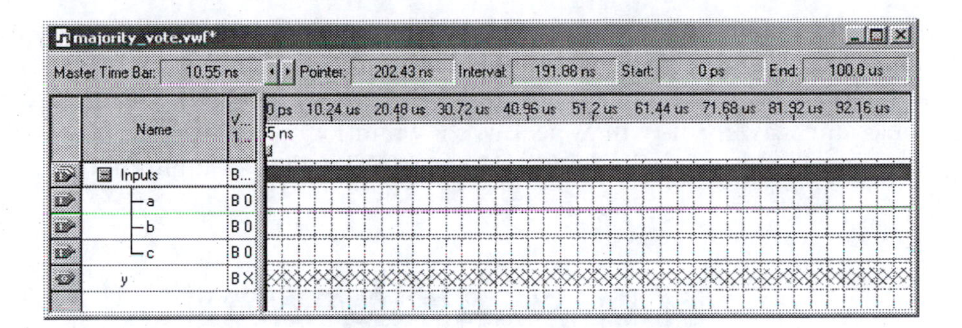

Figure 9.31 Grouped Waveforms

Click on the waveform group (**Inputs**) to highlight it and apply an increasing binary count to the group by clicking the **Count Value** toolbar button, shown in Figure 9.32, or by selecting **Value, Count Value** from the **Edit** menu. In the **Counting** tab of the **Count Value** dialog box, shown in Figure 9.33, select **Radix: Binary; Start value: 000; Increment by: 1;** and **Count type: Binary. End value** is calculated from the other parameters of the simulation.

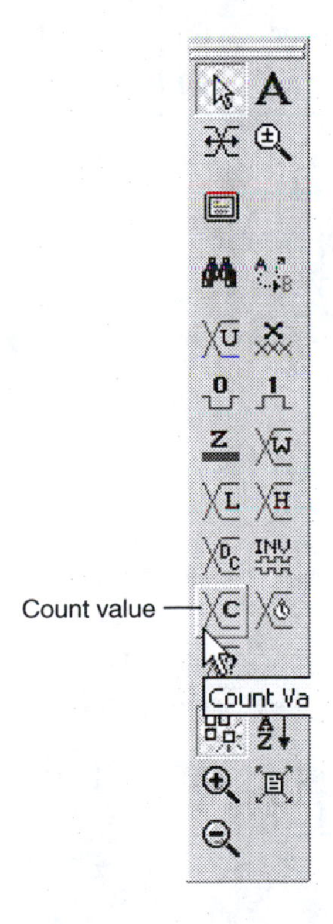

Figure 9.32 Waveform Editor Toolbar (Count Value)

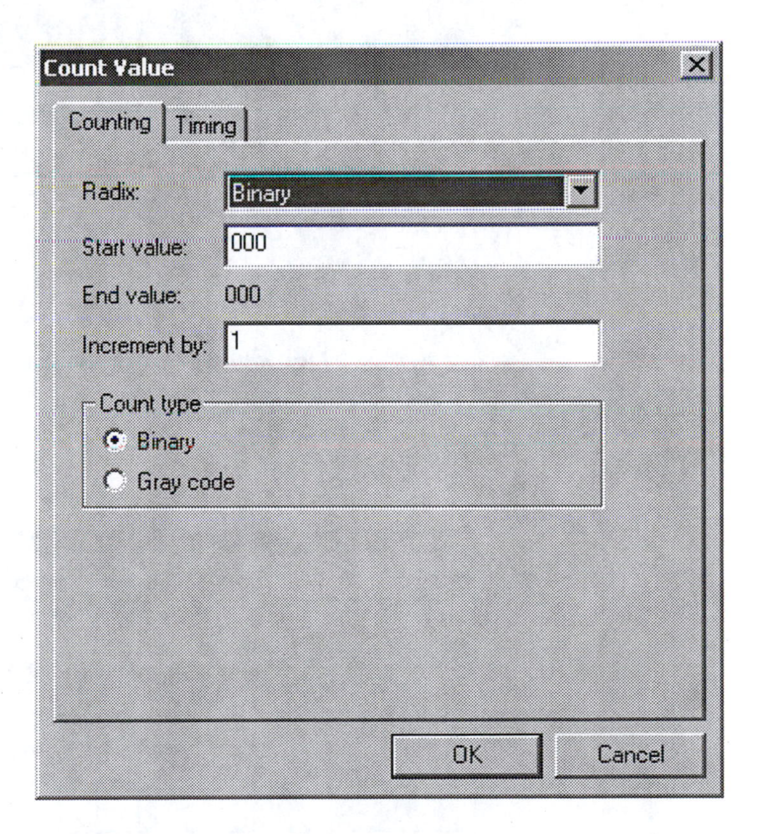

Figure 9.33 Count Waveform Dialog (Counting)

In the **Timing** tab, shown in Figure 9.34, select **Start time: 0 ps; End time: 100 us; Count every: 10.24 us;** and **Multiplied by: 1.** The count interval of 10.24 μs is selected to match the spacing of the Waveform Editor timing grid. Click **OK** to accept the values. The count waveforms will appear as shown in Figure 9.35. The binary value on **Inputs** corresponds to the combined HIGH and LOW values on inputs **a, b,** and **c.**

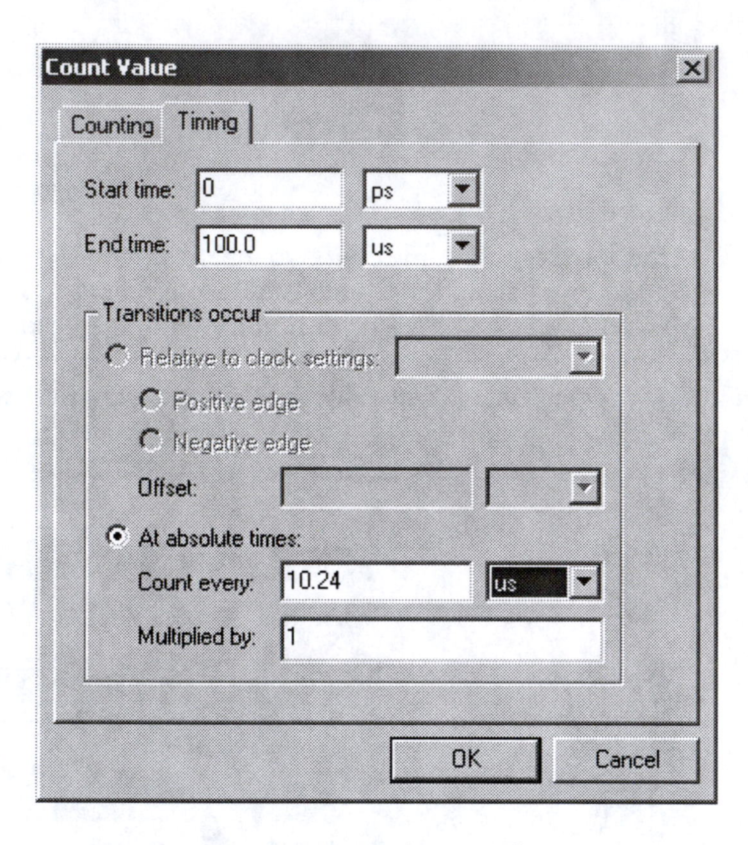

Figure 9.34 Count Waveform Dialog (Timing)

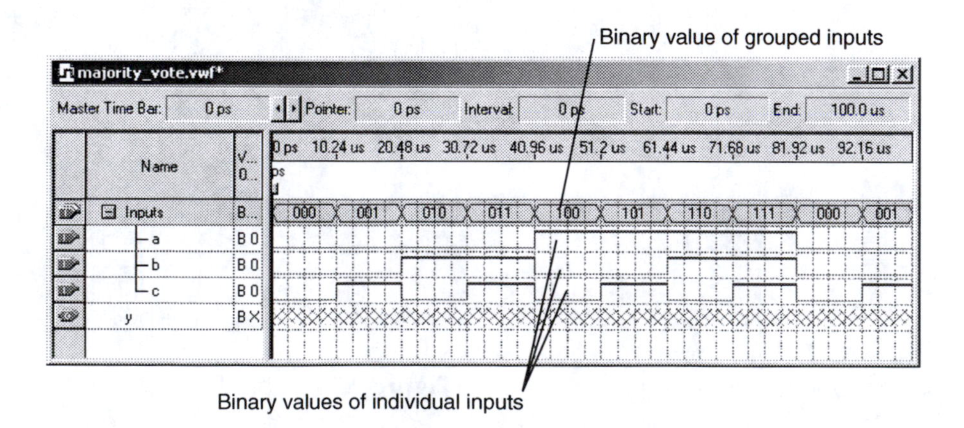

Figure 9.35 Count Waveform on Inputs

Start the simulation by clicking the toolbar button shown in Figure 9.36. The simulation result is shown in Figure 9.37.

Figure 9.36 Start Simulation (Toolbar Button)

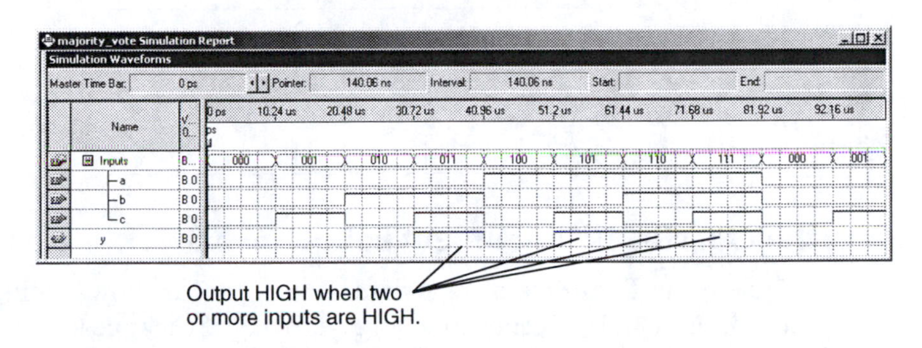

Figure 9.37 Simulation Result

Instructor's Initials ____________________

Transferring a Design to a Target CPLD

After we have completed the compilation and simulation steps and are satisfied that our design is free of conceptual errors, we can convert the project to a physical design within our target CPLD. This involves three steps.

1. We must assign pin numbers to each of the pin names previously assigned.
2. We must recompile the file so as to get the programming information to correspond to the new pin assignments.
3. We must use the Quartus II programming tool to transfer the design from our PC to the target device.

Note In order to assign pins and program the CPLD, your project must have the correct device assignment. Check this by opening the **Device** dialog box from the Quartus II **Assignments** menu. If using the Altera UP-1 or UP-2 board, the correct MAX 7000S device is EPM7128SLC84-7. The RSR PLDT-2 board and the DeVry eSOC board use the EPM7128SLC84-10 or -15. The FLEX 10K device is EPF10K20RC240-4 for the UP-1 board and EPF10K70RC240-4 for the UP-2. For other boards, check the part number on the CPLD, but it is probably EPM7128SLC84-7, -10, or -15. The last numbers (-7, -10, -15) don't matter for programming. They only affect the delay time in the Quartus II simulation.

Assigning Pins

We can assign pins either from a dialog box dedicated to this task or from the Quartus II Assignment Editor. The procedure for using the Assignment Editor is outlined below. For information on using the **Assign Pins** dialog box instead, see the relevant part of Chapter 4 in *Introduction to Digital Electronics.*

We will use the following pin assignments for the majority vote example:

Pin Name	UP-2	PLDT-2	eSOC
a	34	34	50
b	33	33	51
c	36	36	52
y	44	44	4

Using the Quartus II Assignment Editor

Open the **Assignment Editor** from the **Assignments** menu, as shown in Figure 9.38. The **Assignment Editor** default screen, shown in Figure 9.39, will appear. Click the **Pin** button to see the pin assignment screen, shown in Figure 9.40.

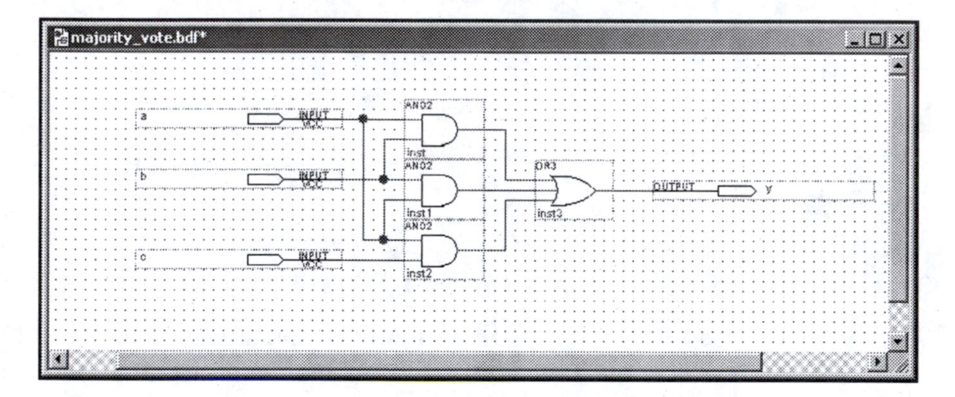

Figure 9.38 Assignments Menu (Assignment Editor)

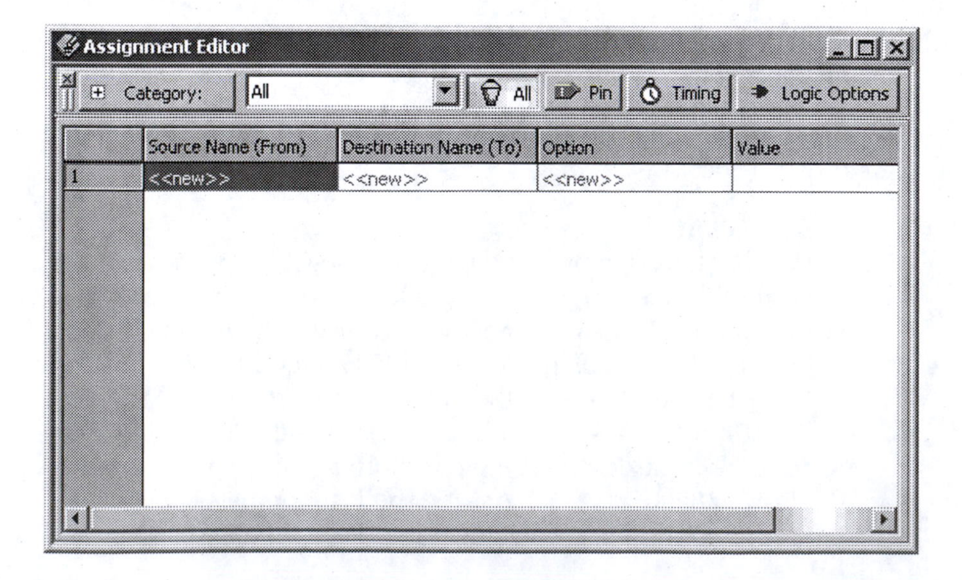

Figure 9.39 Assignment Editor (Default Screen)

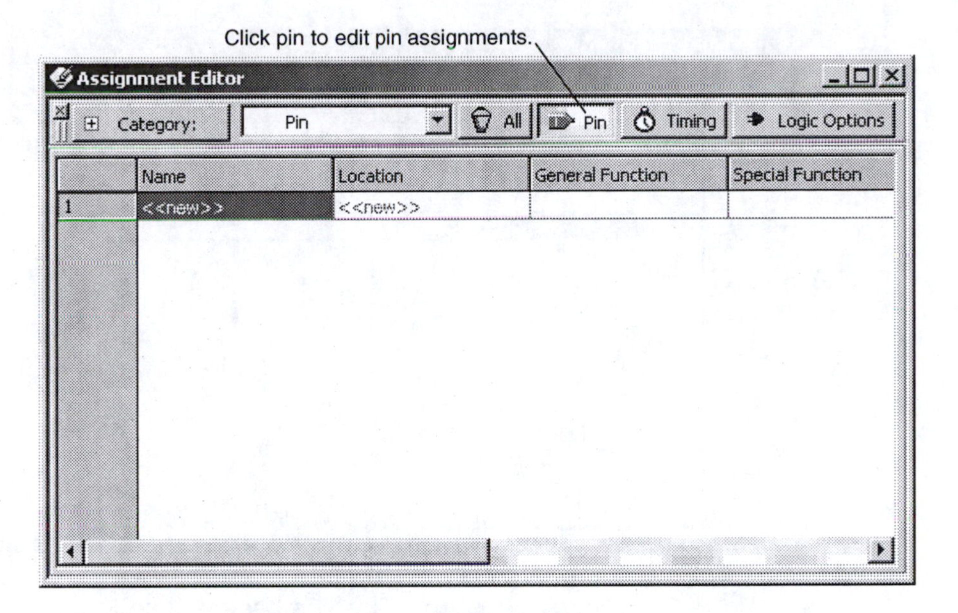

Figure 9.40 Assignment Editor (Pin Assignments)

Under the **Name** column, type a pin name. Select the name from the drop-down box, as shown in Figure 9.41. Figure 9.42 shows the **Assignment Editor** with pins **a** and **b** assigned. When you have made all assignments, close the assignment editor and confirm the assignments by clicking **Yes** in the dialog box shown in Figure 9.43. Recompile the project.

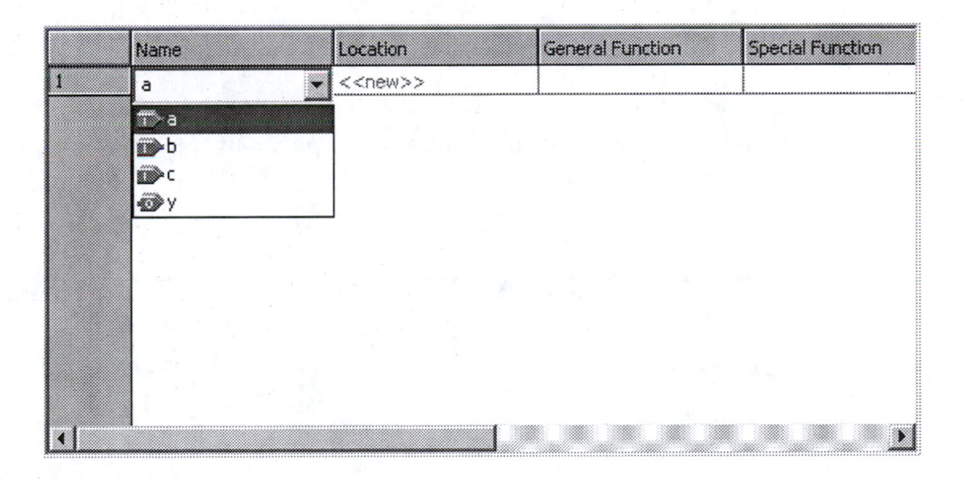

Figure 9.41 Adding Pin Names from a Drop-Down Menu

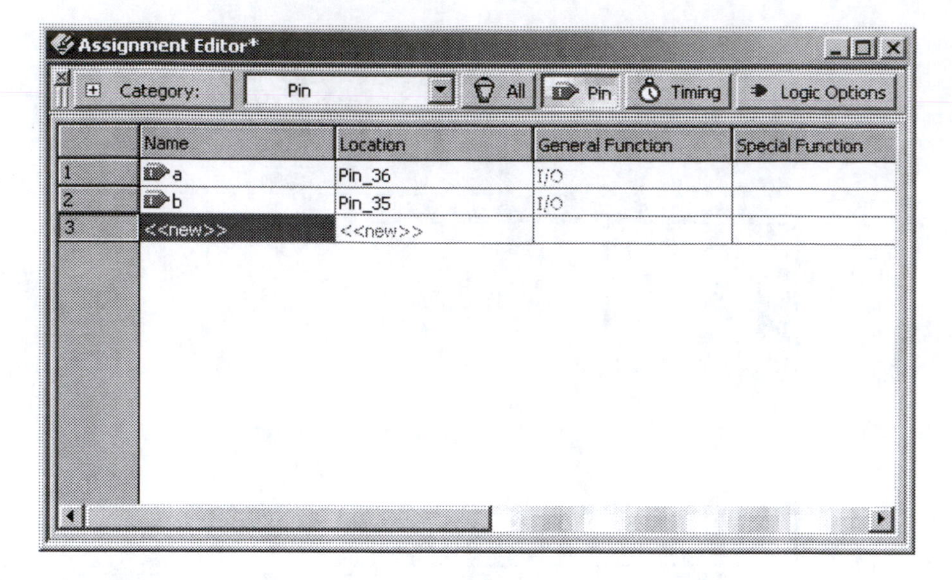

Figure 9.42 Two Pin Numbers Assigned

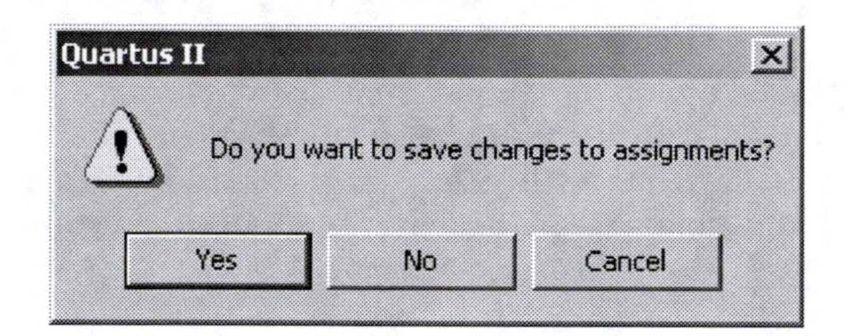

Figure 9.43 Pin Assignment Confirmation Dialog

Once the pins have been assigned, they will appear in the Block Diagram File, as shown in Figure 9.44.

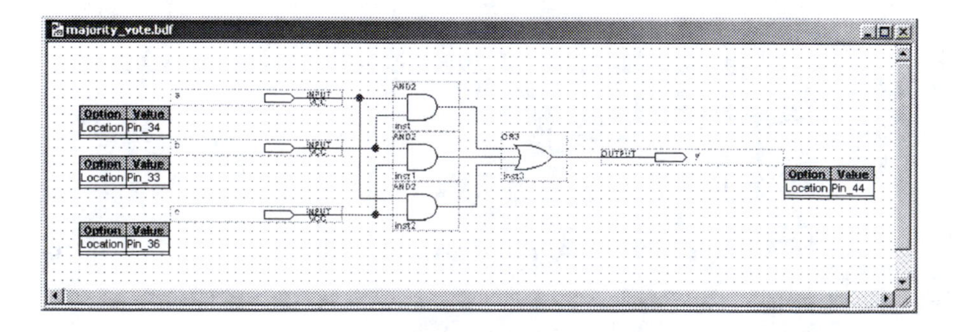

Figure 9.44 Majority Vote Circuit, Showing Pin Numbers

Programming CPLDs on the Altera UP-1 or UP-2 Circuit Board

The CPLDs on the Altera UP-1 and UP-2 circuit boards are programmed via the programming software in Quartus II and a ribbon cable called the **ByteBlaster** or **ByteBlaster MV.** The ByteBlaster, shown in Figure 9.45, connects to the parallel port of the PC running Quartus II to a 10-pin male socket that complies with the **JTAG** standard. The CPLD on the RSR PLDT-2 or DeVry eSOC board is programmed via a special cable that connects the board to the PC parallel port directly. The JTAG interface is included on each of these latter two boards.

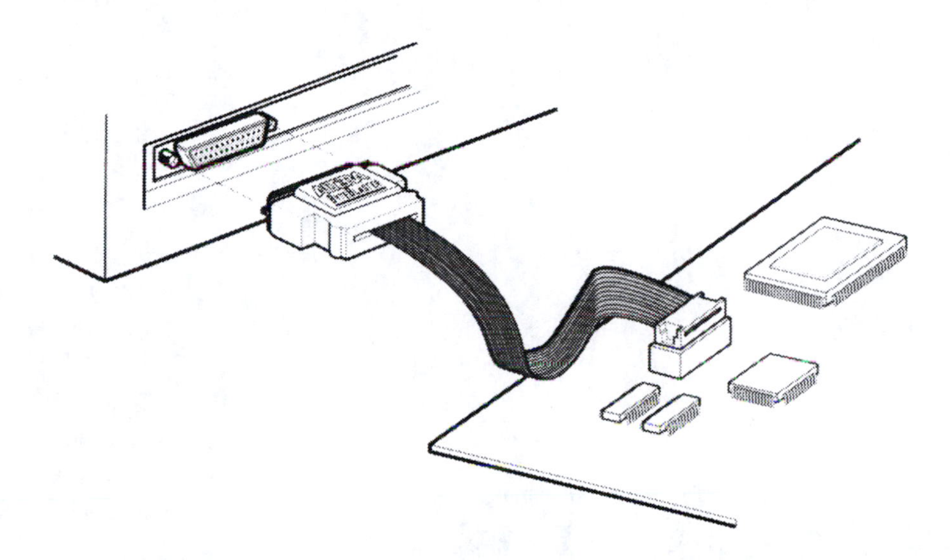

Figure 9.45 ByteBlaster Parallel Port Download Cable

The choice of programming one or more CPLDs, or the CPLDs on one or more UP-1 or UP-2 boards, is determined by the placement of four on-board jumpers. These jumper positions are explained in the *Altera University Program Design Laboratory Package User Guide* that comes with the UP-1 or UP-2 board. A copy of the *User Guide* is available on Altera's web site at http://www.altera.com/literature/univ/upds.pdf. No jumpering is required on the RSR PLDT-2 or DeVry eSOC boards.

Quartus II Programmer

To program a device on the Altera UP-1 or UP-2 board, set the jumpers to program the EPM7128S or configure the EPF10K20 or EPF10K70, as shown in the *Altera University Program Design Laboratory Package User Guide.* Connect the ByteBlaster cable from the parallel port of the PC running Quartus II to the 10-pin JTAG header. (You may have to run a 25-wire straight-through parallel extension cable, with male D-connector to female D-connector, to make it reach.) Plug an AC adapter (7.5- to 9-volt dc output) into the power jack of the UP-1 or UP-2 board.

Note Connection procedures may differ for boards other than the Altera UP-1 or UP-2. For example, some boards incorporate the ByteBlaster circuitry directly on the CPLD board and plug directly into the PC parallel port without a ByteBlaster cable. Consult the manual for your CPLD board for details. Other than connection to the PC, programming procedures are identical for all boards.

Start the Quartus II Programmer, either with the toolbar button shown in Figure 9.46 or by selecting **Programmer** from the **Tools** menu (Figure 9.47). The programmer dialog box (Figure 9.48) will open, showing the programming file for the top-level file of the open project.

Figure 9.46 Programmer Toolbar Button

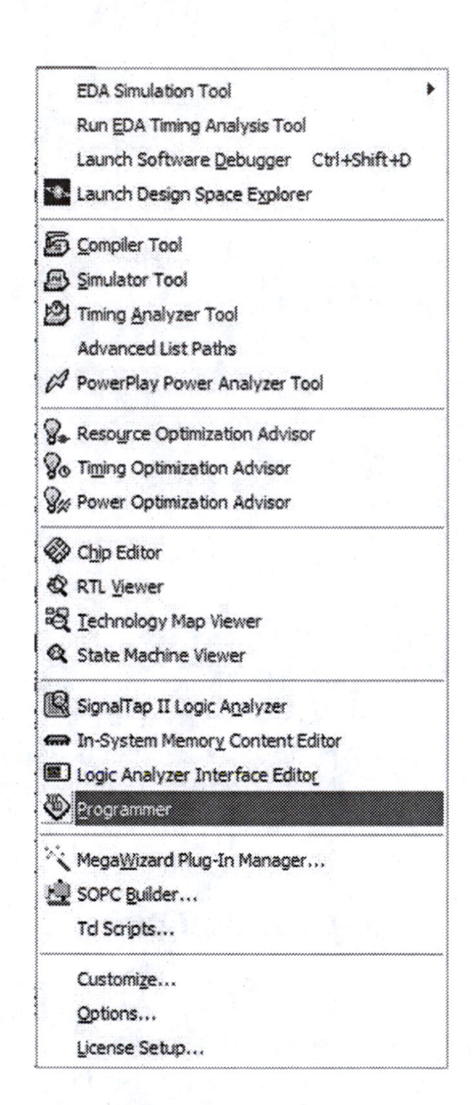

Figure 9.47 Tools Menu (Programmer)

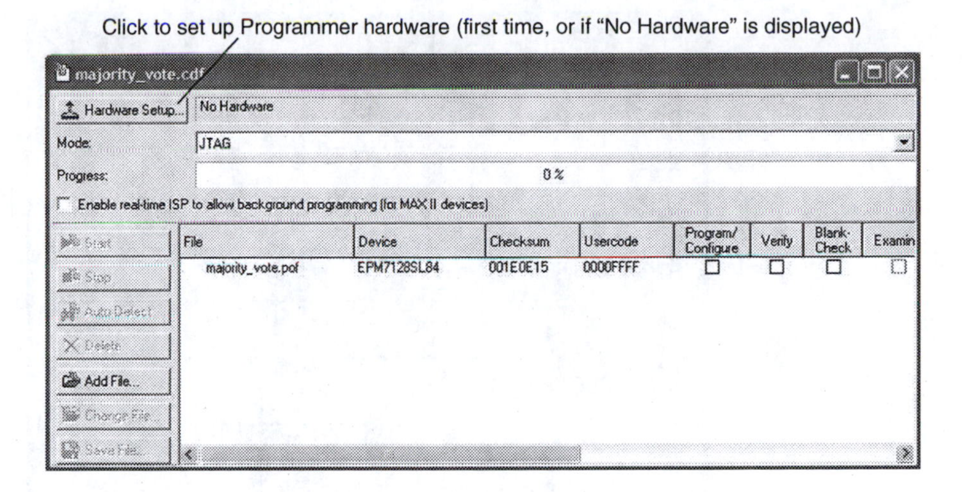

Figure 9.48 Programmer Dialog Box

Note If you have never programmed a CPLD with your particular version of Quartus II, you will need to set up the programming hardware before proceeding. If you have at any time programmed a CPLD with your PC running this version of Quartus II, you do not need to perform the following hardware setup steps.

Click the **Hardware** button at the top left corner of the Programmer dialog box. The **Hardware Setup** dialog box, shown in Figure 9.49, will open. Click the **Add Hardware . . .** button. In the **Add Hardware** dialog in Figure 9.50, select **Hardware type: ByteBlasterMV or ByteBlaster II** and **Port: LPT1** (or another LPT port, if appropriate). Click **OK** to accept the choices and close the box. In the **Hardware Setup** dialog in Figure 9.51, highlight **ByteBlasterMV** in the **Available hardware items** box by clicking the item, then click **Select Hardware.** Click **Close** to return to the Programmer dialog.

Note The selected LPT port must be configured as ECP mode to program a CPLD. This probably will not need to be set up, but if it does, make the changes in the BIOS settings of your PC.

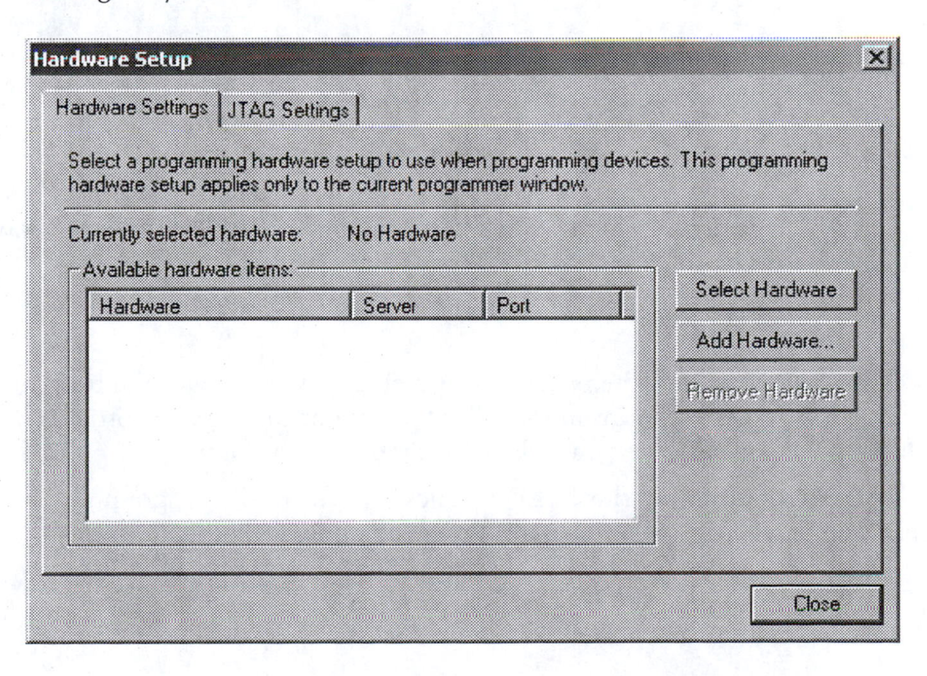

Figure 9.49 Hardware Setup Dialog Box

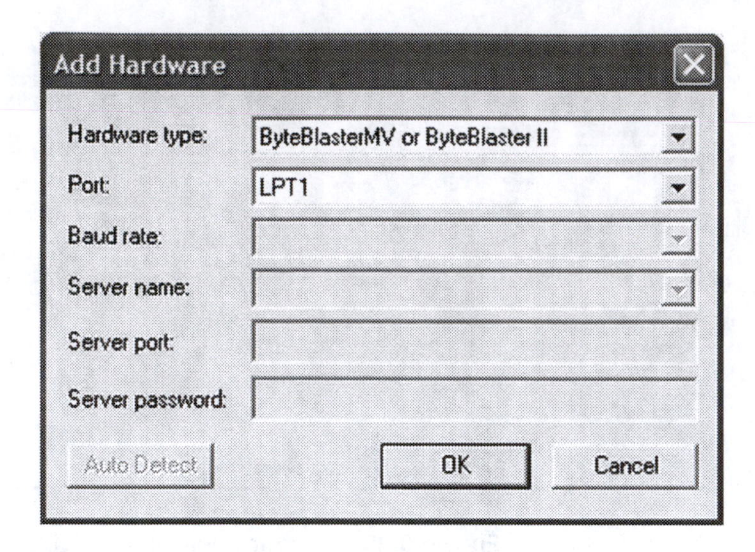

Figure 9.50 Add Hardware Dialog Box

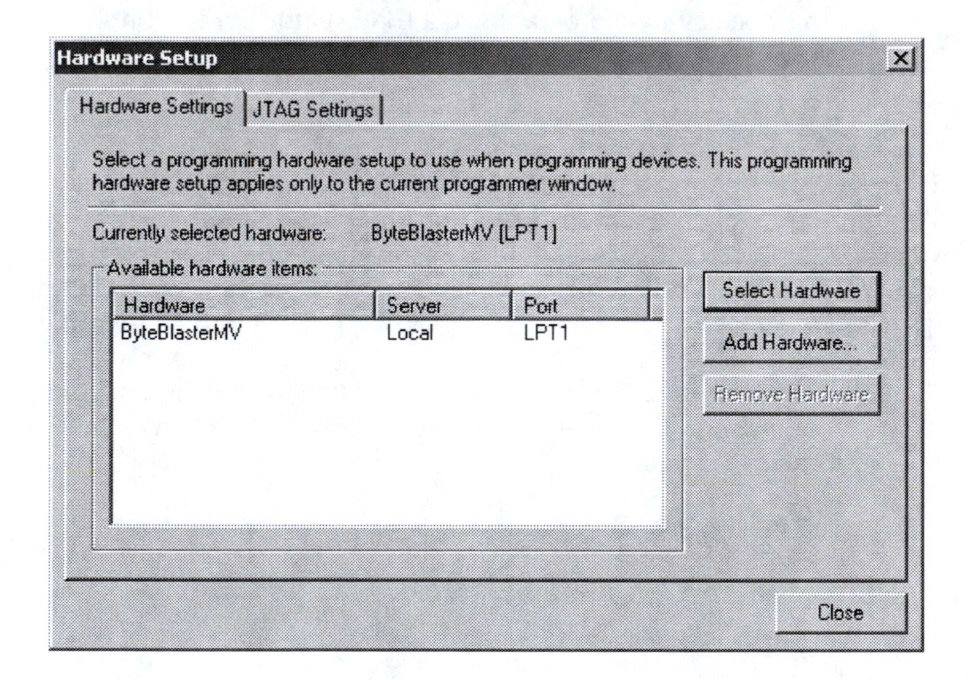

Figure 9.51 Hardware Setup Dialog Box (New Hardware Showing)

Figure 9.52 shows the programmer dialog box, now with the programming hardware selected. In order to program the CPLD, highlight the required programming file by clicking it, then select the checkbox for **Program/Configure**.

Start programming the CPLD by clicking **start** in the Programmer dialog box, or by selecting **Start** from the Quartus II **Processing** menu (Figure 9.53).

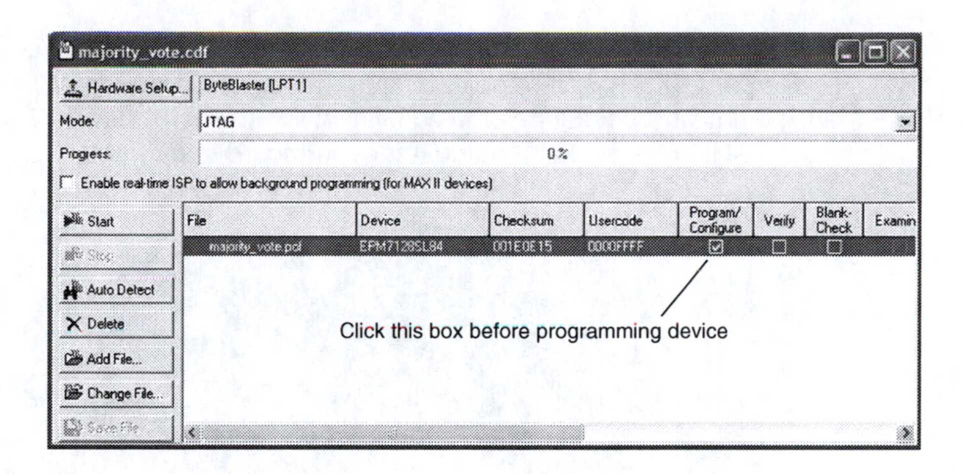

Figure 9.52 Programmer Dialog (Hardware Selected)

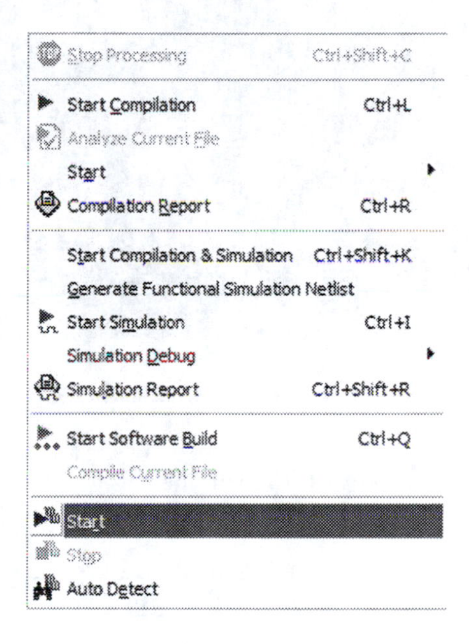

Figure 9.53 Start Programming (Processing Menu)

The majority vote circuit can be tested on a CPLD board, such as the Altera UP-2 board or equivalent. Connect short lengths of #22 wire from three of the input switches on the board to pins 34, 33, and 36 of the CPLD prototyping headers (a series of four dual in-line female connectors surrounding the CPLD). Connect another length of wire from pin 44 of the CPLD header to an LED indicator. These connections are shown in Figure 9.54. When two of the three switches are HIGH, there should be a HIGH at the CPLD output.

Note The LEDs on the Altera UP-1 and UP-2 boards are configured as active-LOW, so when an LED is on, it indicates a LOW at the CPLD output. When it is off, the LED indicates a HIGH. Some equivalent CPLD boards have LEDs that are active-HIGH. Check the manual for your board if in doubt.

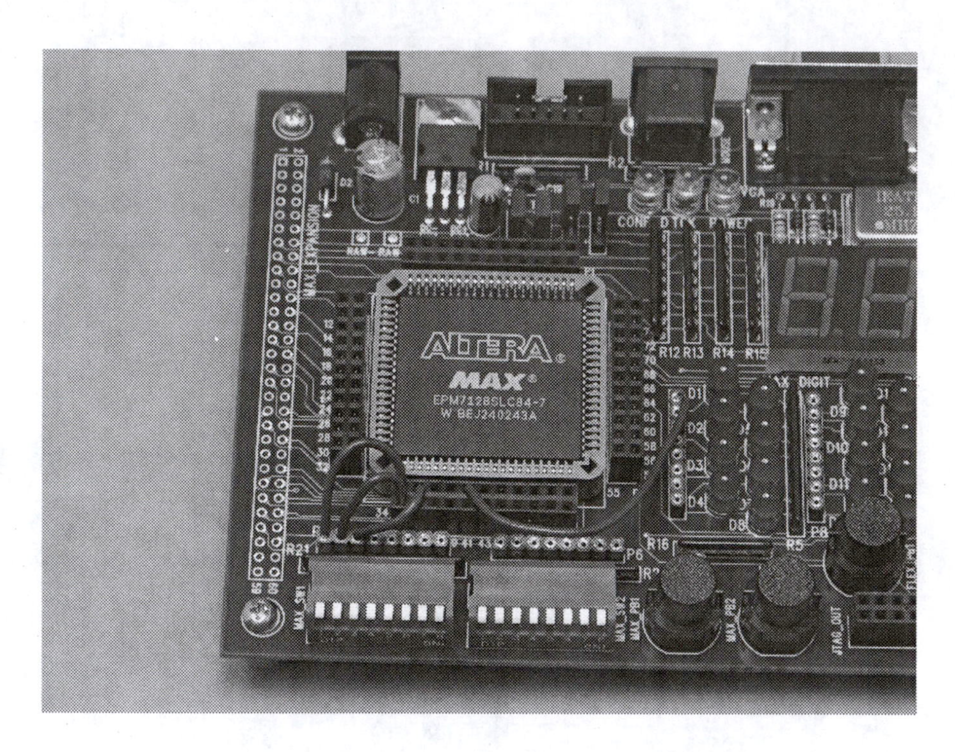

Figure 9.54 Wiring the Altera UP-2 Board

Fill in the following truth table for the design example:

A	*B*	*C*	*Y*

Instructor's Initials ____________________

Lab 10

Standard Wiring Configuration for Altera UP-1/UP-2 and RSR PLDT-2 Boards

Name ______________________ Class ____________ Date ____________

Objectives Upon completion of this laboratory exercise, you should be able to:

- Wire a CPLD trainer board in a standard configuration.
- Test the board to verify the correctness of the configuration.

Reference Ken Reid and Robert Dueck, *Introduction to Digital Electronics*

Chapter 4: Introduction to PLDs and Quartus II

Equipment Required CPLD Trainer:

Altera UP-1 or UP-2 circuit board with ByteBlaster Download Cable, or RSR PLDT-2 circuit board with Straight-Through Parallel Port Cable, or equivalent CPLD trainer board with Altera EPM7128S CPLD

Quartus II Web Edition Software
AC Adapter, minimum output: 7 VDC, 250 mA DC
Anti-static wrist strap
#22 solid-core wire
Wire strippers

Experimental Notes

The CPLD labs in this manual have been designed for use with either an Altera UP-2 board, a DeVry eSOC board, or an RSR PLDT-2 board. Other CPLD boards with an Altera EPM7128SLC84 CPLD can also be used, but may or may not conform to the standard wiring configuration designed for the labs. In this case, you will need to assign pins to your CPLD designs according to the layout of your board.

Labs 11 through 24 in this manual are designed with a standard wiring configuration, so that the board can be wired once (ever) and reprogrammed as necessary for the practical exercises in the various labs. If you are using an Altera UP-2 board or an RSR PLDT-2 board, my recommendation is to spend half an hour or so making the connections as directed in this exercise, then leaving the board alone until after Lab 30, where some minor changes need to be made. Lab boards with USB programming ports, such as the DeVry eSOC board, do not require any external wiring.

The Altera UP-1 board is shown in Figure 10.1 with a standard wiring configuration. The Altera UP-2 is wired identically. The wiring configuration of the RSR PLDT-2 board is shown in Figure 10.2.

Note Documentation with the RSR PLDT-2 board indicates that input jumpers should be removed before programming the board to avoid damaging the CPLD. The main issue here is that you should never assign an output to a pin connected to an input switch. The logic level forced onto the pin by the input switch may conflict with the logic level driven to the pin by the assigned output and cause permanent damage.

This also applies to other CPLD boards: Never assign an output to a pin connected to an input switch. If you observe these precautions, you will probably be fine with a permanent wiring configuration. The author wired an Altera UP-1 board in 1998 and an RSR PLDT-2 board in 1999 and has used them ever since without changing the wiring and have never had any trouble with them.

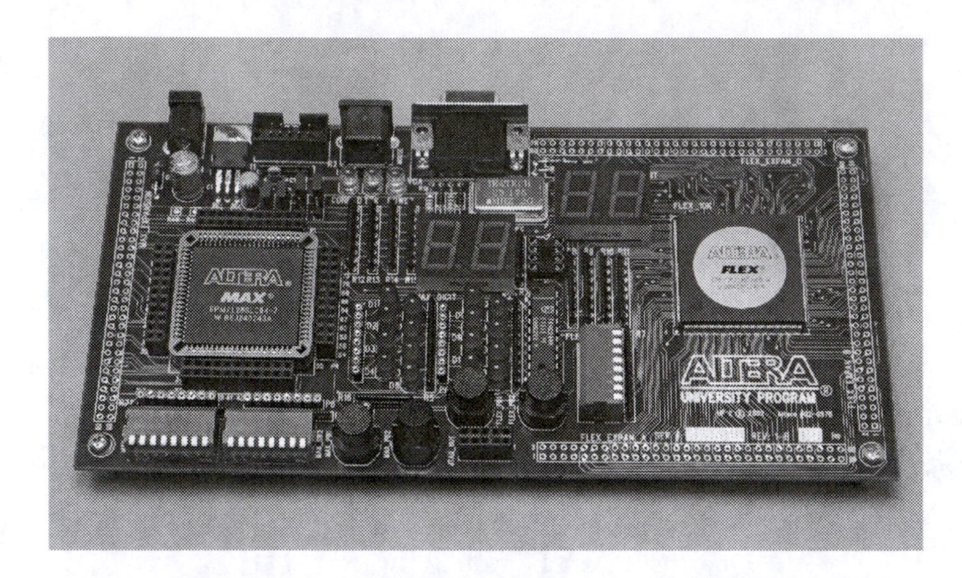

Figure 10.1 Wiring for Altera UP-1/UP-2 Board

Figure 10.2 Wiring for RSR PLDT-2 Board

Procedure

1. For the Altera UP-2 board, the connections should be as indicated in Table 10.1. For the RSR PLDT-2 board, connections to SW1-1 through SW1-8 (labeled S1-1 through S1-8 on the PLDT-2 board) and to LED1 through LED8 are normally made by

removable shorting jumpers. These connections are shown in shaded cells in Table 10.1. If the jumpers are left in place, further wiring is not required for these inputs and outputs. Connection to the seven segment displays and to the clock in input (pin 83) are hardwired on both the Altera and RSR boards.

Pin assignments for the DeVry eSOC board are shown for reference in Table 10.2. All connections on this board are available with switchable jumpers and thus do not need wiring. More information on the DeVry board may be found at http://www.devry.edu/eSOC/.

2. Open Quartus II, then the project for either **board_test_2** (UP-1/UP-2) or **board_test_3** (PLDT-2). Compile the file. When the compile process is complete, connect the CPLD trainer board to your PC. Open the Quartus II programmer window, select **Program/Configure,** and select **Start Programming.**

3. The test files should cause an increasing hexadecimal count to appear on the board's numerical displays. The files also test the DIP switch and LED connections by turning on an LED when a switch is made HIGH. The switches and LEDs operate in the same order as they are laid out on the board. If they operate in a different order, check your wiring.

Note You may wish to print out Table 10.1 or Table 10.2 for future reference.

Table 10.1 EPM7128LC84-7 Pin Assignments Altera UP-2 Board and RSR PLDT-2 Board

Seven-Segment Digits			
Function	**Pin**	**Function**	**Pin**
a1	58	a2	69
b1	60	b2	70
c1	61	c2	73
d1	63	d2	74
e1	64	e2	76
f1	65	f2	75
g1	67	g2	77
dp1	68	dp2	79

Pushbuttons			
Function	**Pin**	**Function**	**Pin**
PB1	11	PB2	1

DIP Switches			
Function	**Pin**	**Function**	**Pin**
SW1-1	34	SW2-1	28
SW1-2	33	SW2-2	29
SW1-3	36	SW2-3	30
SW1-4	35	SW2-4	31
SW1-5	37	SW2-5	57
SW1-6	40	SW2-6	55
SW1-7	39	SW2-7	56
SW1-8	41	SW2-8	54

LED Outputs			
Function	**Pin**	**Function**	**Pin**
LED1	44	LED9	80
LED2	45	LED10	81
LED3	46	LED11	4
LED4	48	LED12	5
LED5	49	LED13	6
LED6	50	LED14	8
LED7	51	LED15	9
LED8	52	LED16	10

Unassigned: Pins 12, 15, 16, 17, 18, 20, 21, 22, 24, 25, 27

Special Function*: Pin 1 (GCLRn); Pin 2 (Input/OE2/GCLK2); Pin 83 (GCLK1; hardwired); Pin 84 (OE1)

Table 10.2 EP2C8Q208C8N Pin Assignments DeVry eSOC Board

Seven-Segment Digits			
Function	**Pin**	**Function**	**Pin**
a1	169	a0	137
b1	170	b0	138
c1	171	c0	139
d1	173	d0	128
e1	175	e0	80
f1	133	f0	81
g1	135	g0	84

Pushbuttons (debounced)			
Function	**Pin**	**Function**	**Pin**
PB0	131	PB1	130

Pushbuttons (non-debounced)			
Function	**Pin**	**Function**	**Pin**
PB0	145	PB2	147
PB1	146	PB3	149

DIP Switches			
Function	**Pin**	**Function**	**Pin**
SW0	160	SW4	168
SW1	161	SW5	141
SW2	162	SW6	142
SW3	163	SW7	144

LED Outputs			
Function	**Pin**	**Function**	**Pin**
RED0	115	GRN0	101
RED1	114	GRN1	99
RED2	113	GRN2	97
RED3	112	GRN3	96
RED4	110	GRN4	95
RED5	102	GRN5	92
RED6	151	GRN6	118
RED7	150	GRN7	117

Special Function: Pin 24 (On board 24 MHz clock), Pin 27 (External Clock)

Lab 11

Binary Decoders (Block Diagram File)

Name ______________________ Class ____________ Date ____________

Objectives Upon completion of this laboratory exercise, you should be able to:

- Enter the design for a binary decoder in Quartus II as a Block Diagram File.
- Create a Quartus II simulation of a binary decoder.
- Test the binary decoders on a CPLD board.

Reference Ken Reid and Robert Dueck, *Introduction to Digital Electronics*

Chapter 5: Combinational Logic Functions

Equipment Required CPLD Trainer:

DeVry eSOC board and USB cable, or
Altera UP-2 board with ByteBlaster download cable, or
RSR PLDT-2 board with straight-through parallel port cable, or
equivalent CPLD trainer board with Altera EPM7128S CPLD

Quartus II Web Edition software
AC adapter, minimum output: 7 VDC, 250 mA DC
Anti-static wrist strap
#22 solid-core wire
Wire strippers

Experimental Notes

Binary Decoder

A decoder is a combinational circuit with one or more outputs, each of which activates in response to a unique binary input value. For example, a 2-line-to-4-line decoder, shown in Figure 11.1, has two inputs, D_1 and D_0, and four outputs, Y_0, Y_1, Y_2, and Y_3. Y_0 is active when $D_1D_0 = 00$, Y_1 activates when $D_1D_0 = 01$, Y_2 is active for $D_1D_0 = 10$, and Y_3 activates for $D_1D_0 = 11$. Only one output is active at any time.

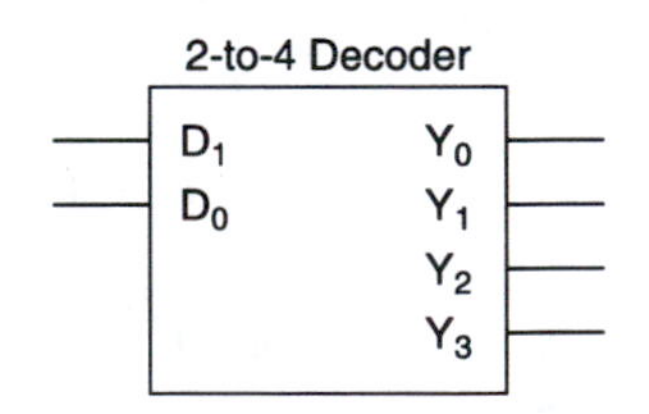

Figure 11.1 2-Line-to-4-Line Decoder

The circuit for the decoder is shown in Figure 11.2. Each AND gate is configured so that its output goes HIGH with a particular value of D_1D_0. In general, the active output is the one whose subscript is equivalent to the binary value of the input. For example, if $D_1D_0 = 10$, only the AND gate for output Y_2 has two HIGH inputs and therefore a HIGH output.

A standard part of the CPLD design cycle is to create a simulation of a design before programming it into a CPLD. A simulation is a timing diagram which is generated by specifying a set of input waveforms to the design under test. The simulation software examines the design equations and input logic waveforms and calculates the response of the digital circuit, which is displayed as a set of output waveforms. The simulation allows us to determine if the design is working as planned by observing the response of the

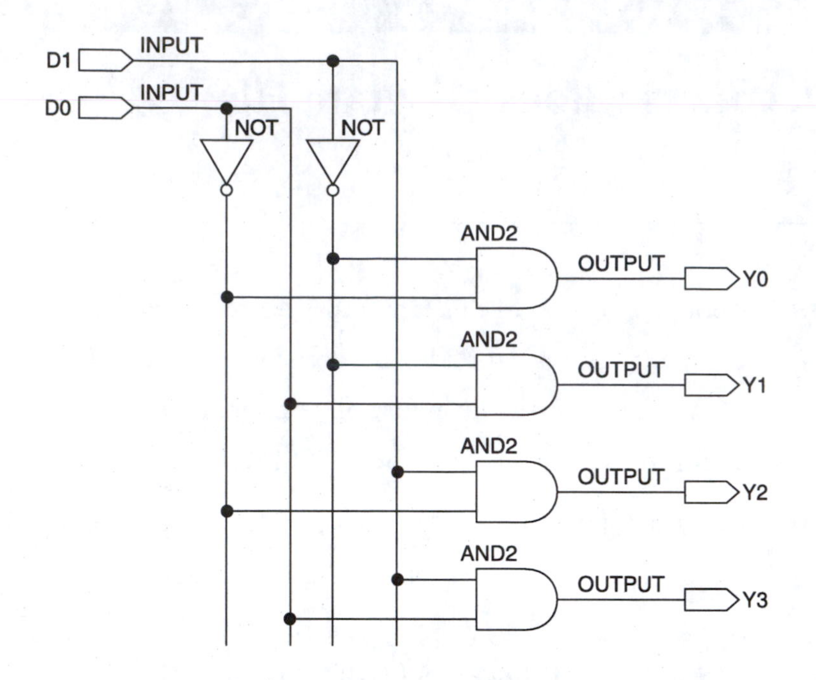

Figure 11.2 2-Line-to-4-Line Decoder Circuit

design to a defined input. A good simulation will test the design under all possible input conditions. We can specify a set of **simulation criteria** that will help us construct a proper simulation.

What are simulation criteria? They are just a series of input and output conditions that can be used to fully test a design to see if our design logic is correct. They answer the questions:

- What input waveforms should I apply to the circuit in a Vector Waveform File to see if it is correctly designed?
- If the design is correct, what output waveforms should I expect to see in the Simulation Report window?

We can use the following set of simulation criteria to verify the correctness of the 2-to-4 decoder design.

Simulation Criteria

- Each output must respond to its appropriate binary input by going HIGH when selected.
- Only one output must be HIGH at any time.
- If an ascending 2-bit binary count is applied to inputs *D*1 and *D*0, the outputs will go HIGH in the sequence *Y*0, *Y*1, *Y*2, and *Y*3, then repeat.

Figure 11.3 shows a simulation of a 2-line-to-4-line decoder, generated in Quartus II. The inputs *D*1 and *D*0 are grouped together as a single 2-bit value that can range from 00 to 11 in binary. The *D* inputs have an increasing 2-bit binary count applied to them: the inputs start at 00, increase up to 11, then go back to 00 and repeat. In response, the *Y* outputs activate one at a time by going HIGH in a sequence that corresponds to the change on the *D* inputs. A procedure for creating a simulation is shown in Section 4.8 in Chapter 4 of *Introduction to Digital Electronics* and are reviewed in this laboratory exercise.

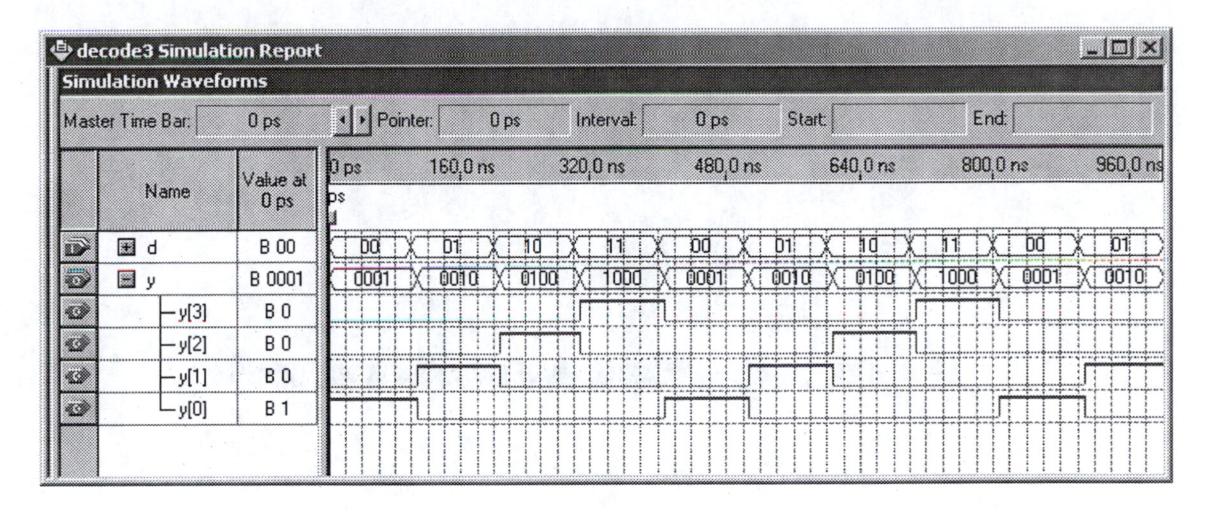

Figure 11.3 Simulation Waveforms for a 2-Line-to-4-Line Decoder

Procedure

Binary Decoder

1. Use the Block Editor in Quartus II to create a 3-line-to-8-line decoder with active-HIGH outputs. Designate the inputs *D*2, *D*1, and *D*0 and the outputs *Y*0 through *Y*7. Save the file as ***drive:*\qdesigns\labs\lab11\decode3to8\decode3to8.bdf.** Use the file to create a new project.

2. Compile the project and follow the procedure outlined next to create a simulation to show that the design is correct.

Creating a Quartus II Simulation

Before we create the simulation for the 3-line-to-8-line decoder, we must write an appropriate set of simulation criteria. In the following space, write the criteria for the 3-line-to-8-line decoder. These will be very similar to the criteria for the 2-line-to-4-line decoder described in the Experimental Notes for this laboratory exercise.

Simulation Criteria

Create a Vector Waveform File and Set the Simulation End Time

1. Select **New** from the Quartus II **File** menu or click the **New File** icon. From the resultant dialog box, shown in Figure 11.4, select **Vector Waveform File,** from the **Other Files** tab.

   The waveform editor window will appear, as shown in Figure 11.5

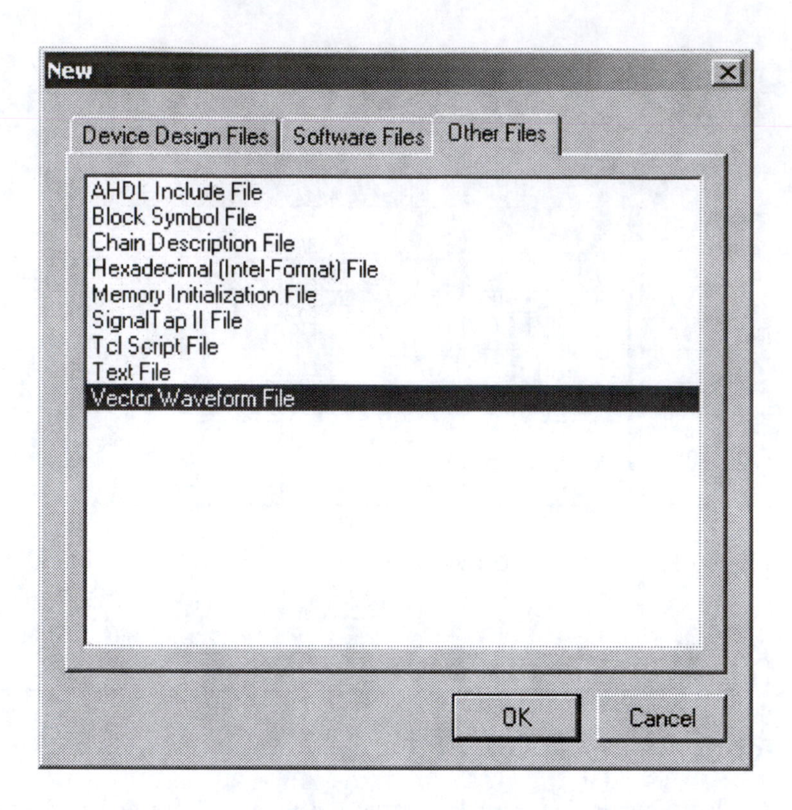

Figure 11.4 New File Dialog (Vector Waveform File)

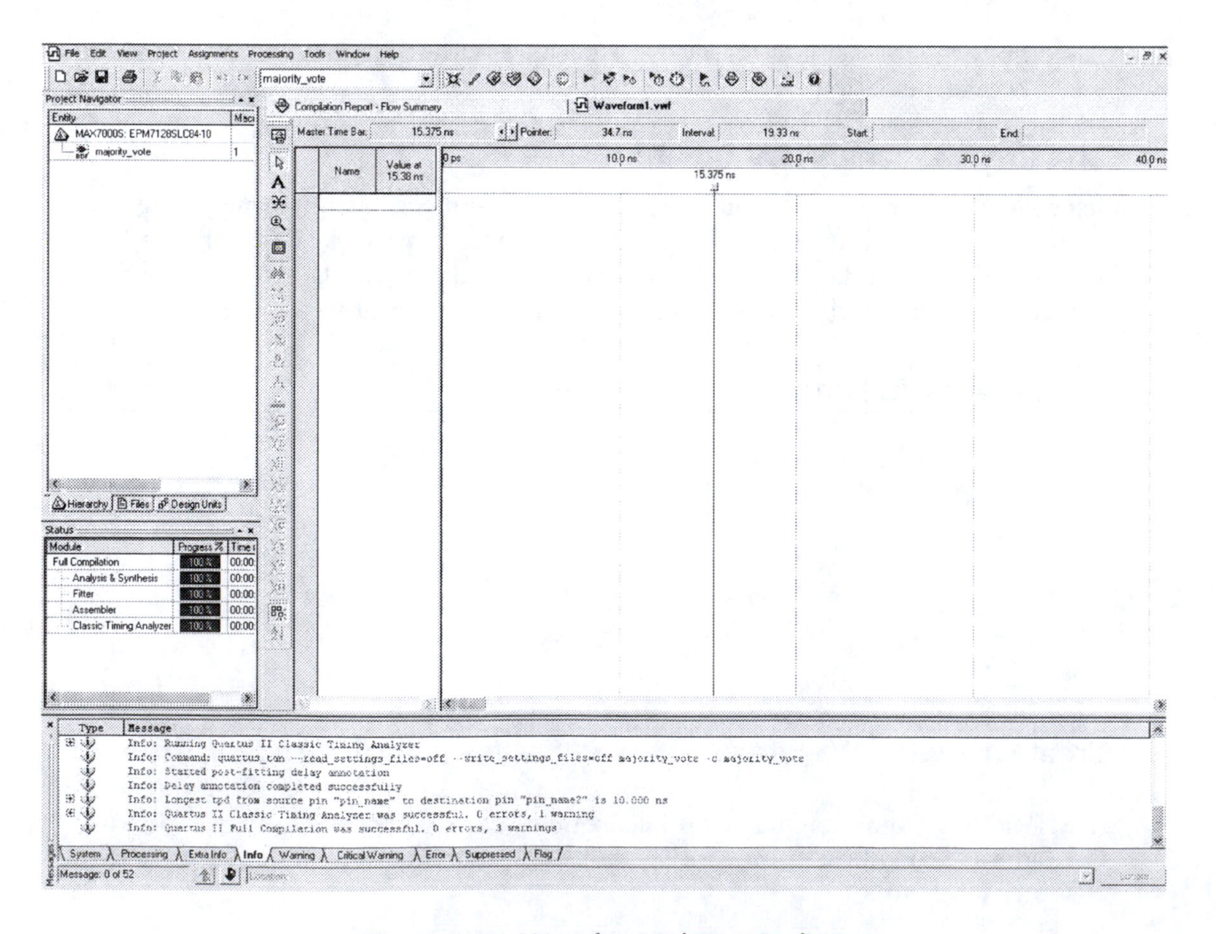

Figure 11.5 Waveform Editor Window

2. The default view only shows a small part of the Waveform Editor. Adjust the view to see the entire range of the window by selecting **Fit in Window** from the **View** menu, as shown in Figure 11.6. This action can also be performed by the keyboard combination **Ctrl+W**.

   The adjusted waveform editor window, showing the default end time of 1 μs, is shown in Figure 11.7.

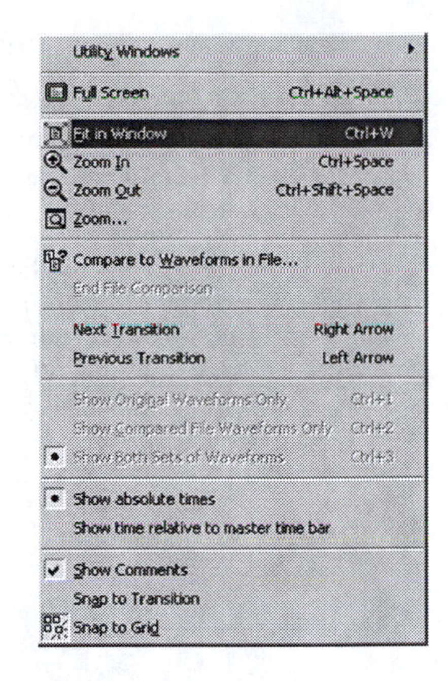

Figure 11.6 Fit in Window (View Menu)

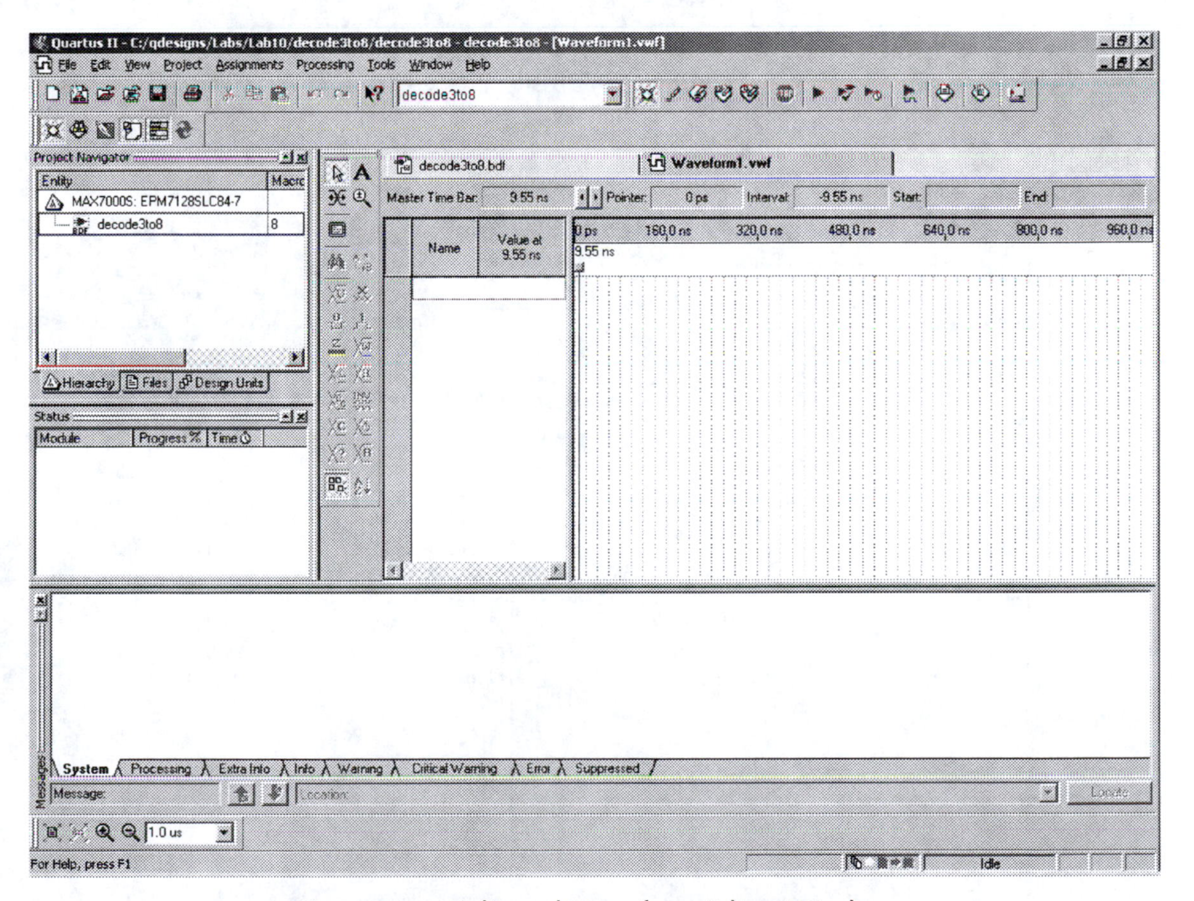

Figure 11.7 Adjusted Waveform Editor Window

Add Nodes to the Simulation

A node in a simulation is a point for a which a waveform is created by the user (i.e., an input) or determined by the simulator software (i.e., an output or an internal point in the design). We can use the **Node Finder** utility to enter input and output nodes into our simulation.

1. To start the Node Finder, place the mouse cursor in the **Name** column of the Waveform Editor and double-click. A dialog box labeled **Insert Node or Bus** will appear, as shown in Figure 11.8. Click **Node Finder** to start the Node Finder utility, shown in Figure 11.9.

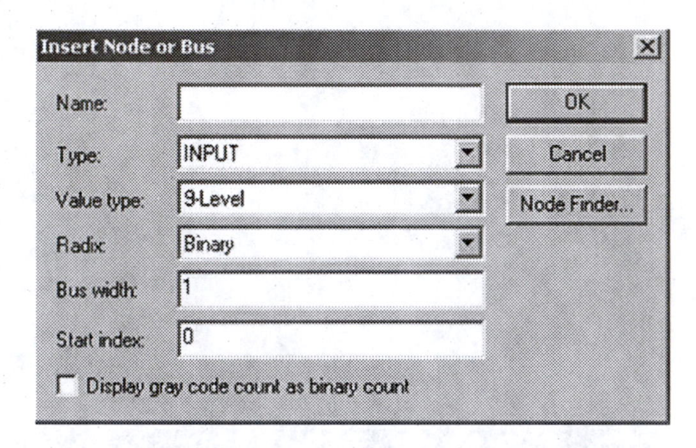

Figure 11.8 Insert Node or Bus Dialog

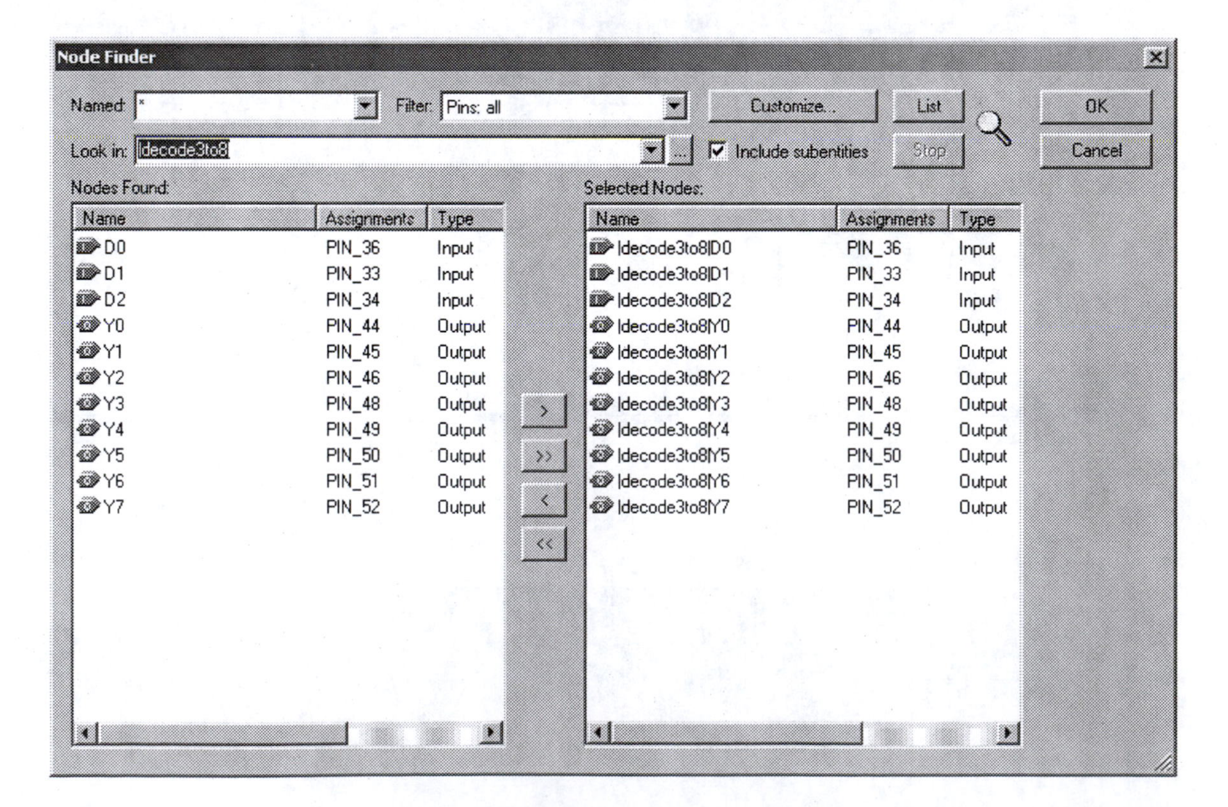

Figure 11.9 Node Finder Window Showing All Pin Names

2. In the Node Finder dialog box, select **Pins: all** from the drop-down menu labeled **Filter.** Click **List** to see a list of all input and output nodes in the design. Click the right-going double-arrow button (shown at the left) to transfer all the pin names from the left side of the Node Finder window (shown in Figure 11.9 as **Nodes Found**) to the right side (**Selected Nodes**). Click **OK** to close the Node Finder. Click **OK** to close the Insert Node or Bus dialog. The selected nodes will appear in the Waveform Editor, as shown in Figure 11.10. (To select only one node or a group of specifically chosen nodes, use the single-arrow button.)

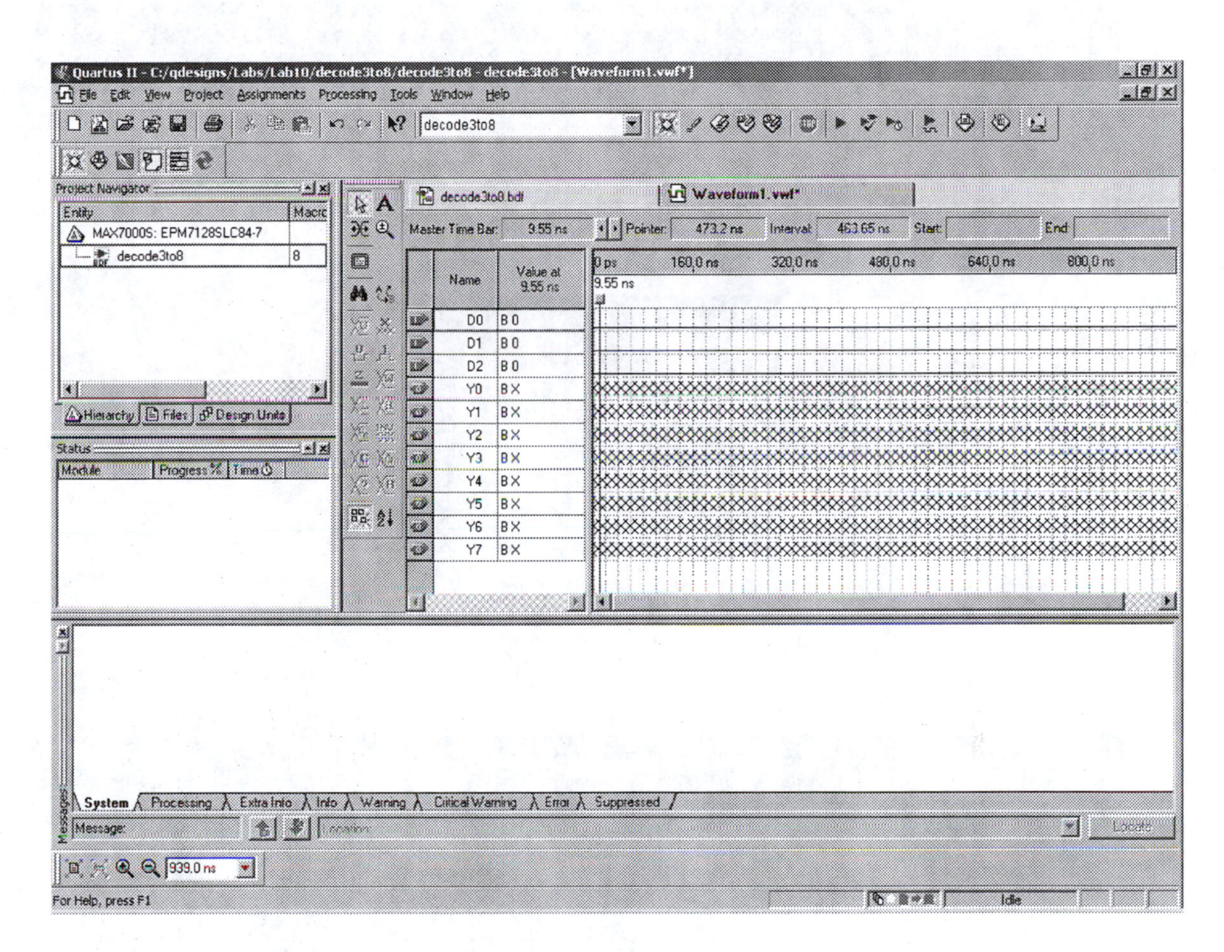

Figure 11.10 Waveform Editor Showing Input and Output Nodes at Default Levels

Create an Input Group

1. Highlight *D*0, *D*1, and *D*2, as shown in Figure 11.11, by dragging across these waveforms.

2. Sort the *D* waveforms from highest to lowest (i.e., from Most to Least Significant Bit) by clicking the **Sort** toolbar button, shown at the left. In the **Sort** dialog box that appears, as shown in Figure 11.12, select **Descending** and click **OK.**

Figure 11.13 shows the *D* input waveforms sorted in the proper order. This order is necessary for the next step, where we group the waveforms and apply an ascending sequence of binary numbers to the inputs of the decoder.

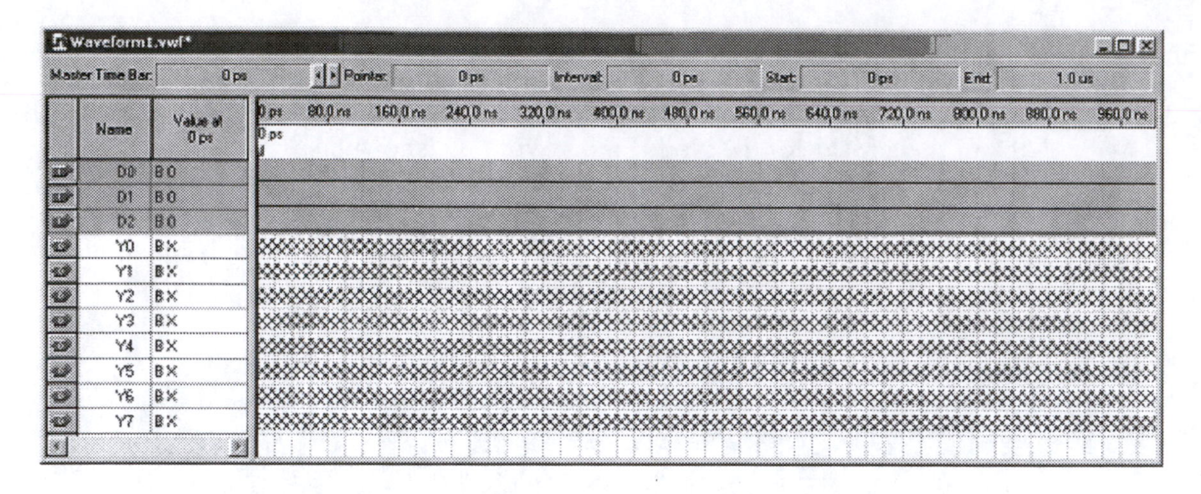

Figure 11.11 Highlighting a Group of Waveforms

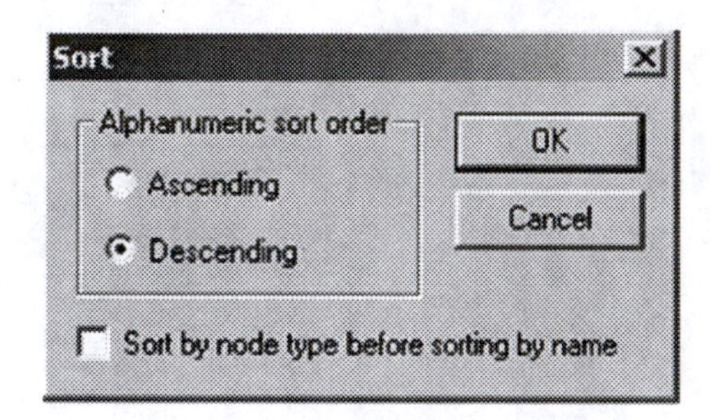

Figure 11.12 Sort Dialog Box

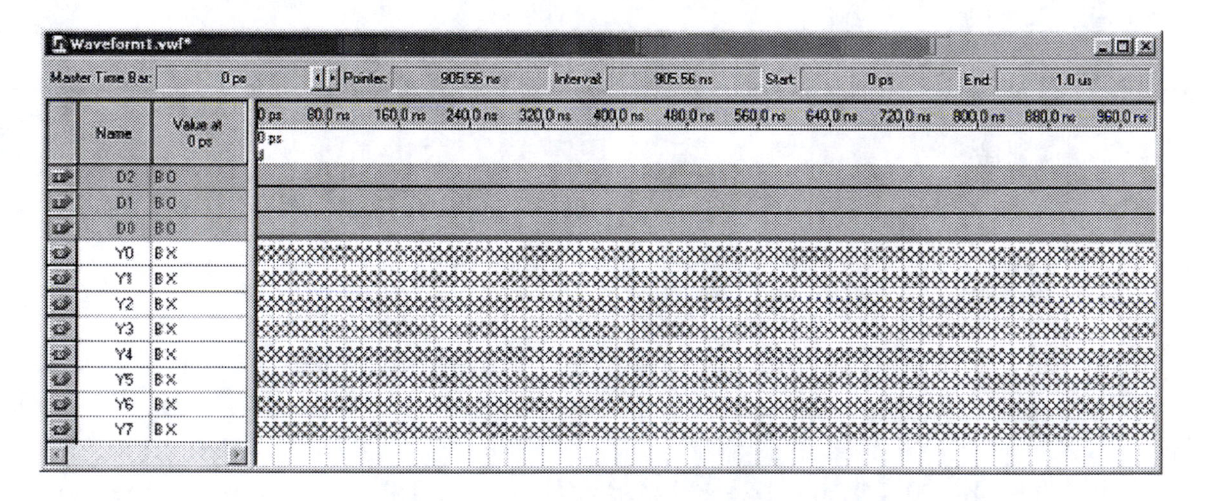

Figure 11.13 Waveform Editor Showing Sorted D Input Waveforms

3. Right-click on the highlighted group to get the pop-up menu shown in Figure 11.14. Select **Group** from the menu.

4. In the **Group** dialog box (Figure 11.15), type **D** in the **Group name** box and select **Binary** for the **Radix** box. Click **OK** to accept the choices and close the box. The waveforms will appear as shown in Figure 11.16.

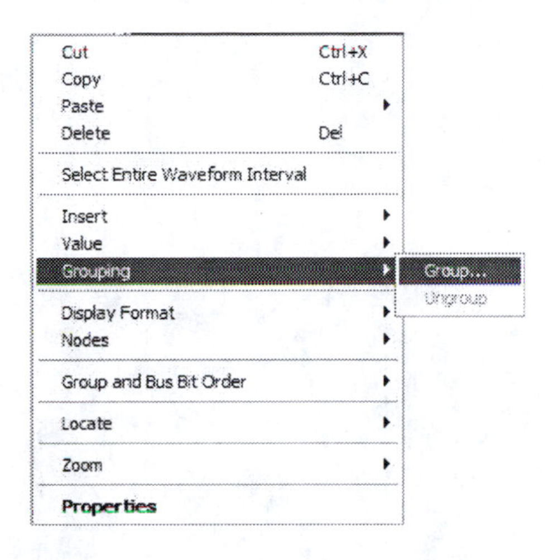

Figure 11.14 Pop-up Menu Showing Group Function

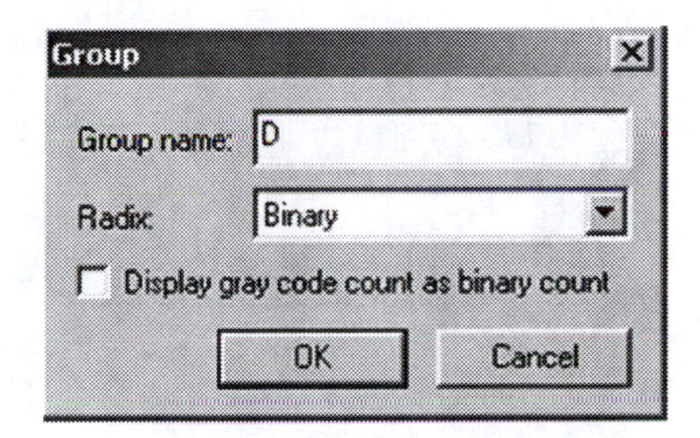

Figure 11.15 Group Dialog Box

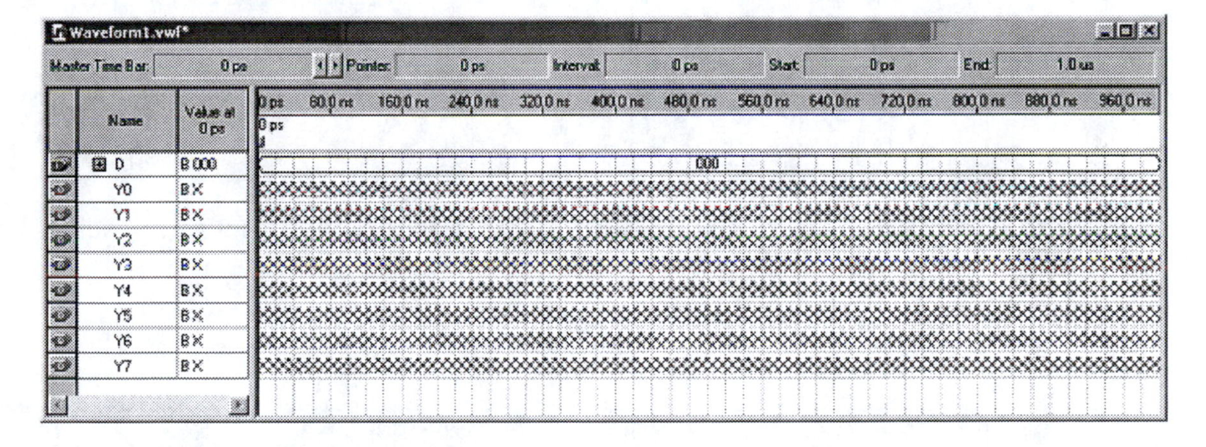

Figure 11.16 Waveform Editor Showing Input Group

Add a Binary Count to the Input Group

1. Click on the D waveform to highlight it, then click the **Count Value** button from the Waveform Editor toolbar, as shown at the left, or select **Value, Count Value** from either the **Edit** menu or from the pop-up menu that appears when you highlight **D** and right-click.

2. Fill in the **Counting** tab of the **Count Value** dialog box, shown in Figure 11.17, as follows:

   Radix: Binary
   Start Value: 000
   Increment by: 1
   Count type: Binary

3. Fill in the **Timing** tab of the **Count Value** dialog box, shown in Figure 11.18, as follows:

   Start time: 0 ps
   End time: 1.0 us
   Transitions occur: At absolute times
      Count every: 100.0 ns
      Multiplied by: 1

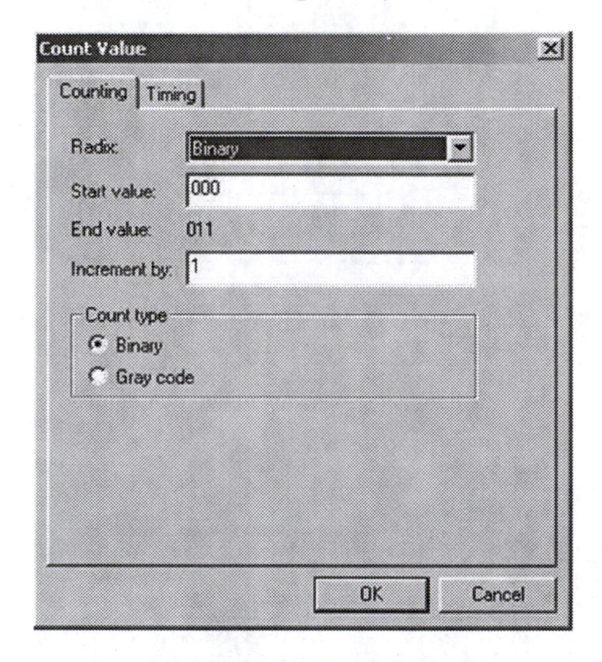

Figure 11.17 Count Value Dialog (Counting)

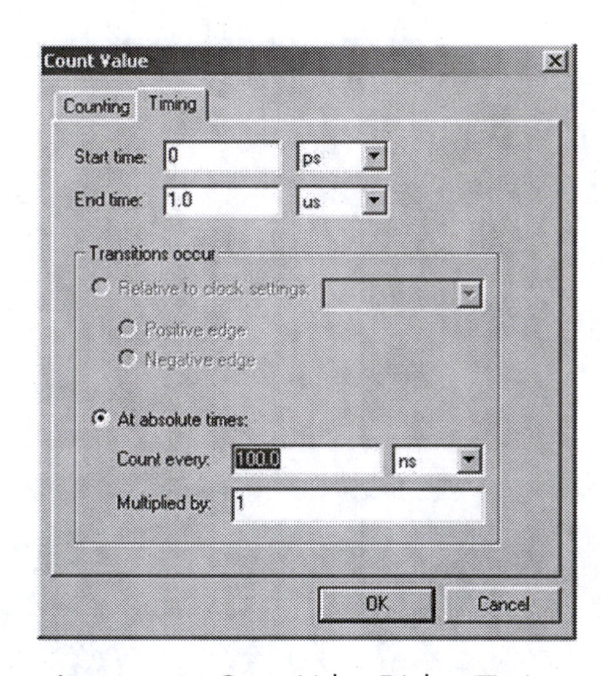

Figure 11.18 Count Value Dialog (Timing)

Click **OK** to close the box. The binary count appears on input waveform **D,** as shown in Figure 11.19.

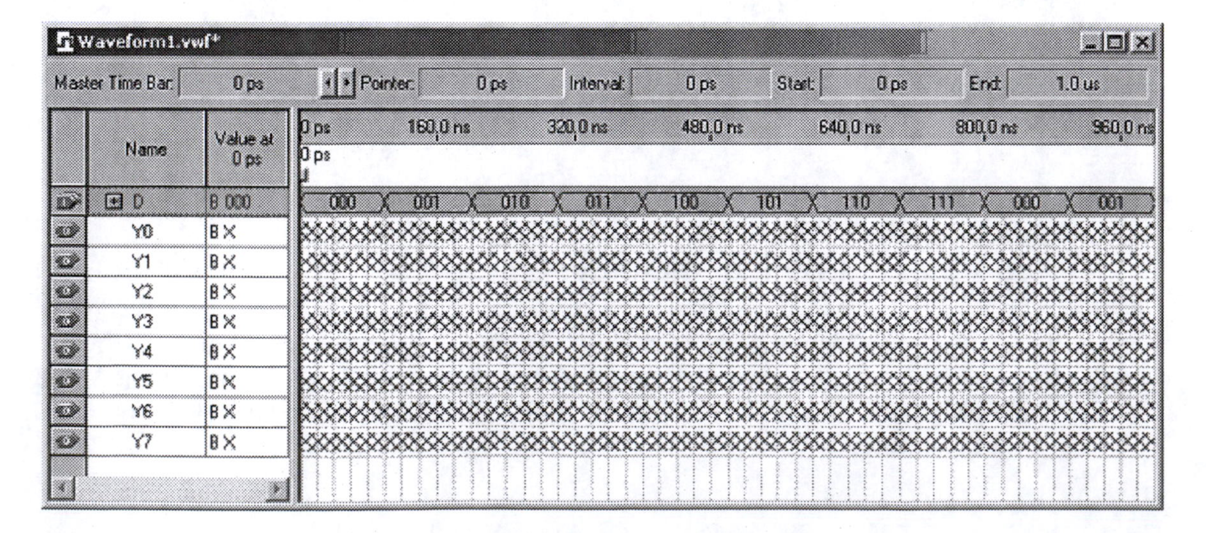

Figure 11.19 Waveform Editor with Binary Input Count

Run Simulation and Save Vector Waveform File

1. Save the Vector Waveform File as **decode3to8.vwf** in your working folder. Start the simulation by clicking the toolbar button shown at the left or by selecting **Start Simulation** from the **Processing** menu.

   Note If the simulator cannot locate your Vector Waveform File, you may need to do one additional step. Open the **Assignments** menu and select **Settings** from the list of options. In the box that opens, select **Simulator Settings** from the list of **Categories.** Make sure that the box labeled **Simulation input** shows the name of your Vector Waveform File. Add the file name if necessary and click **OK** to close the dialog box.

2. After the simulation runs, the simulation report waveforms should appear, as shown in Figure 11.20. To see the whole simulation, click the **Fit In Window** toolbar button. To see the simulation in a larger scale, select **Full Screen** from the **View** menu, or click the **Full Screen** toolbar button, shown at the left. To restore the screen to the original view, select **Full Screen** or click the toolbar button again.

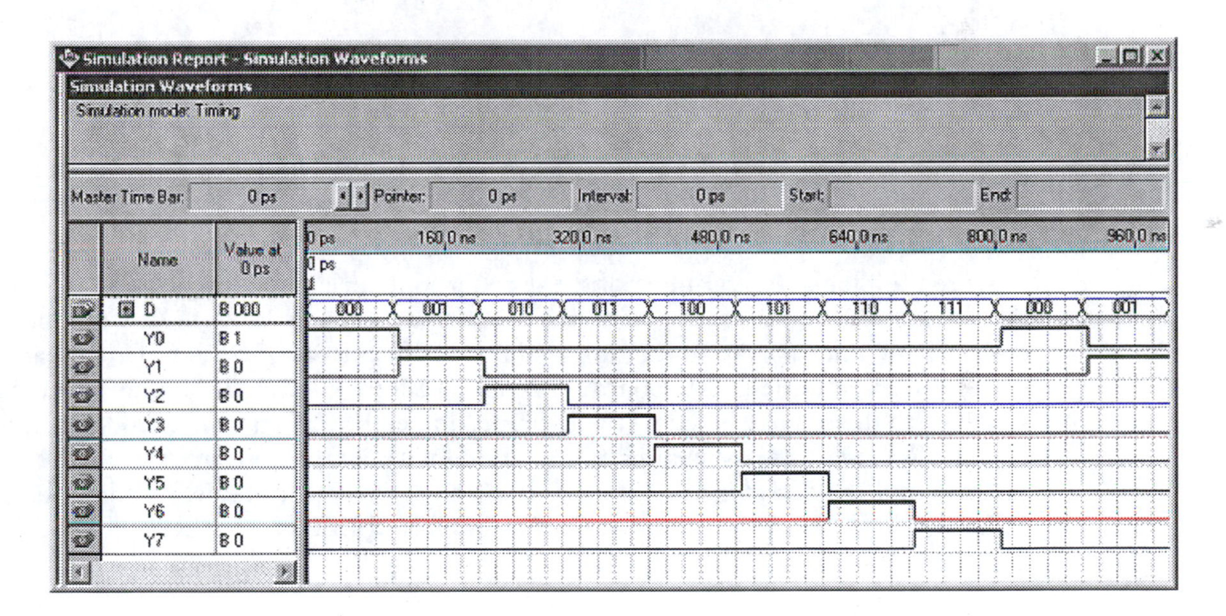

Figure 11.20 Simulation Report Waveforms

Show the simulation result to your instructor.

Instructor's Initials ____________

Simulation as a Diagnostic Tool

Simulations are very useful for detecting design errors. For example, the simulation in Figure 11.21 shows that one of the logic gates in the decoder has been incorrectly connected. Which one? ________

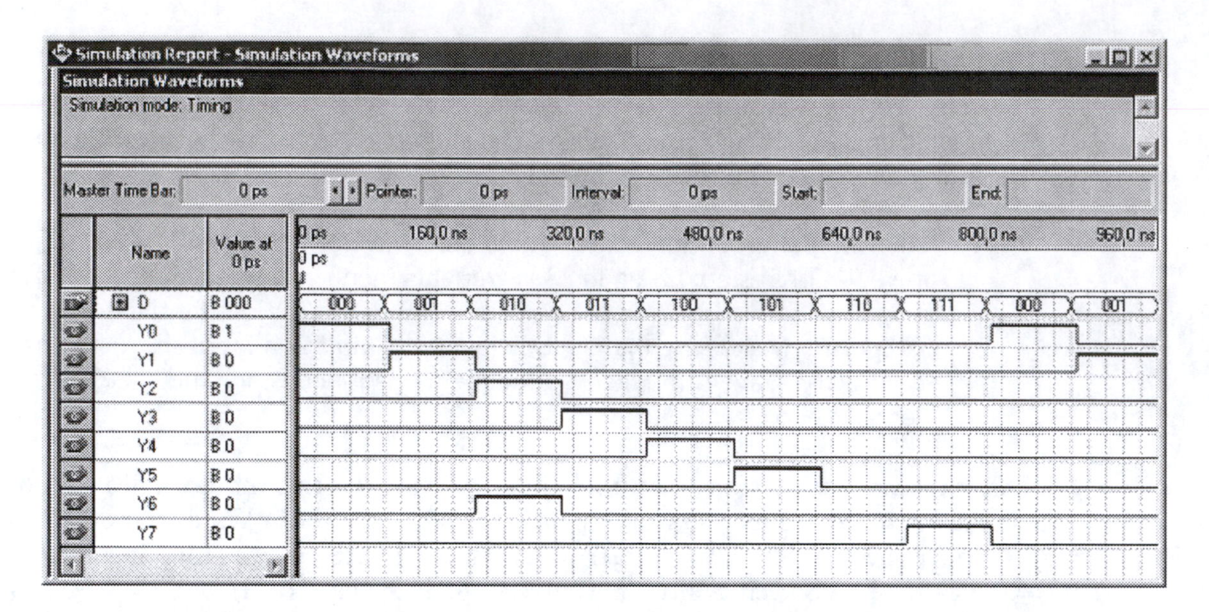

Figure 11.21 Simulation Report Showing Design Error

Modify your Block Diagram File so that it replicates the error shown in the simulation of Figure 11.21. Compile the file and run the simulation again. Show the revised simulation to your instructor.

Instructor's Initials ____________________

Programming and Testing the Binary Decoder

1. Restore the Block Diagram File of the decoder to its original configuration (without the design error) and assign pins to the decoder, as listed in Table 11.1. Compile the design again and run the simulation again to verify correct operation. Download the decoder design to the CPLD board. If using the Altera UP-2 board, connect three DIP switches to the pins assigned to the D inputs and eight LEDs to the decoder outputs. (The RSR PLDT-2 board is already configured with jumpers.) Demonstrate the operation of the decoder to your instructor and take its truth table, recalling that the LEDs on the Altera UP-2 board are active-LOW and LEDs on the RSR PLDT-2 board are active-HIGH.

Table 11.1 Pin Assignments for a 3-Line-to-8-Line Decoder

Pin Name	Pin Number (UP-2 or PLDT-2 board)	Pin Number (DeVry eSOC board)
D2	34	160
D1	33	161
D0	36	162
Y0	44	115
Y1	45	114
Y2	46	113
Y3	48	112
Y4	49	110
Y5	50	102
Y6	51	151
Y7	52	150

Truth Table:

D2	*D1*	*D0*	*Y0*	*Y1*	*Y2*	*Y3*	*Y4*	*Y5*	*Y6*	*Y7*
0	0	0								
0	0	1								
0	1	0								
0	1	1								
1	0	0								
1	0	1								
1	1	0								
1	1	1								

Instructor's Initials ____________

Lab 12

Seven-Segment Decoders

Name ______________________________ Class ____________ Date ____________

Objectives Upon completion of this laboratory exercise, you should be able to:

- Create a seven-segment decoder as a Block Diagram file.
- Test the seven-segment decoder on a CPLD test board.

Reference Ken Reid and Robert Dueck, *Introduction to Digital Electronics*
Chapter 5: Combinational Logic Functions

Equipment Required CPLD Trainer:
DeVry eSOC board and USB cable, or
Altera UP-2 circuit board with ByteBlaster download cable, or
RSR PLDT-2 circuit board with straight-through parallel port cable,
or equivalent CPLD trainer board with Altera EPM7128S CPLD
Quartus II Web Edition software
AC adapter, minimum output: 7 VDC, 250 mA DC
Anti-static wrist strap
#22 solid-core wire
Wire strippers

Experimental Notes

Seven-Segment Decoder

A seven-segment display, shown in Figure 12.1, consists of seven luminous segments, such as LEDs, arranged in a figure-8 pattern. The segments are conventionally designated *a* through *g*, beginning at the top and moving clockwise around the display.

When used to display decimal digits, the various segments are illuminated as shown in Figure 12.2. For example, to display digit 0, all segments are on except *g*. To display digit 1, only segments *b* and *c* are illuminated.

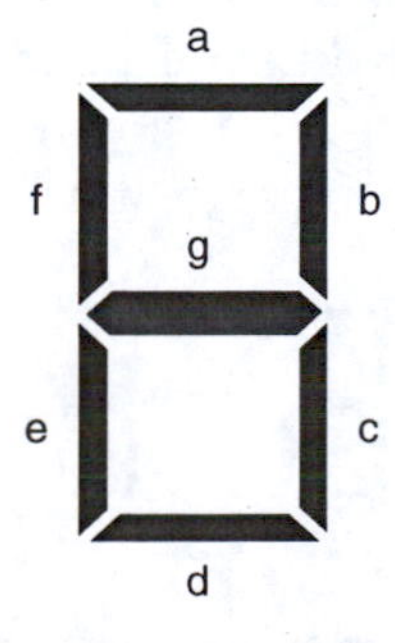

Figure 12.1 Seven-Segment Numerical Display

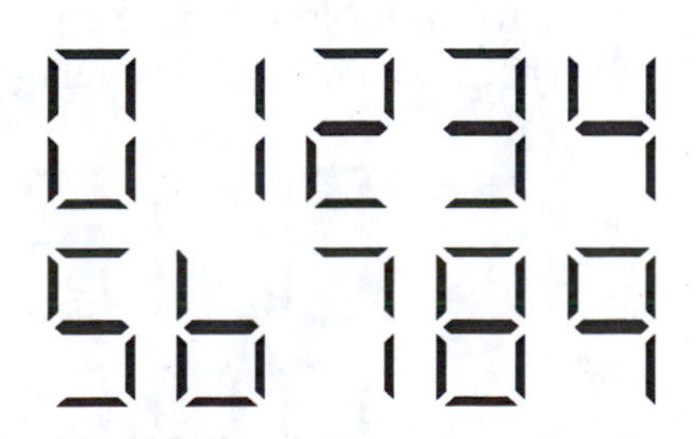

Figure 12.2 Convention for Displaying Decimal Digits

The seven-segment displays on the Altera UP-2 and DeVry eSOC boards are configured as common anode, meaning that the anodes of all LEDs are tied together and connected to the board power supply, V_{CC}. (Refer to Chapter 5, *Introduction to Digital Electronics*.) To turn on a segment, the cathode end of the LED is set to logic 0 through a current-limiting series resistor. This is illustrated for digits 0 and 1 in the partial truth table shown in Table 12.1.

Table 12.1 Partial Truth Table for a Common Anode BCD-to-7-Segment Decoder

D_3	D_2	D_1	D_0	*a*	*b*	*c*	*d*	*e*	*f*	*g*
0	0	0	0	0	0	0	0	0	0	1
0	0	0	1	1	0	0	1	1	1	1

The seven-segment displays on the RSR PLDT-2 circuit board are common cathode, or active HIGH. The cathodes of all segment LEDs are tied to ground and each individual segment is illuminated by a HIGH applied to the segment anode through a series resistor. Table 12.2 illustrates a partial truth table for a common cathode decoder, showing the output values for digits 0 and 1.

Table 12.2 Partial Truth Table for a Common Cathode BCD-to-7-Segment Decoder

D_3	D_2	D_1	D_0	*a*	*b*	*c*	*d*	*e*	*f*	*g*
0	0	0	0	1	1	1	1	1	1	0
0	0	0	1	0	1	1	0	0	0	0

The seven-segment displays and series resistors on the Altera UP-2 board, the DeVry eSOC board, and the RSR PLDT-2 board are hardwired to the board's CPLD. Thus, all that is required to turn on the illuminated segments for each digit is to make the appropriate CPLD pins LOW (Altera UP-2 or DeVry eSOC) or HIGH (RSR PLDT-2).

Procedure

Hexadecimal-to-Seven-Segment Decoder

1. Complete truth tables for both the common-anode and common-cathode hexadecimal-to-seven-segment decoder.

2. Create a hexadecimal-to-seven-segment decoder in a Block Diagram file, using the segment patterns in Figure 12.3 as a model. The decoder for the Altera UP-2 or DeVry eSOC board will be common anode. For the RSR PLDT-2 board, it will be common cathode. Note that the patterns for digits 6 and 9 are shown differently than in Figure 12.2. Save the Block Diagram file. Use the file to create a new project.

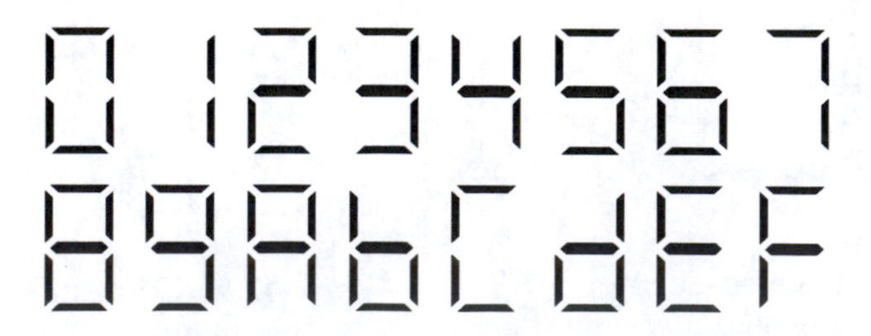

Figure 12.3 Hexadecimal Digit Display Format

3. Assign pin numbers as shown in Table 12.3. Compile the file and download it to the CPLD board. Demonstrate its operation to your instructor.

Table 12.3 Pin Assignments for a Hexadecimal-to-Seven-Segment Decoder

	Pin Number		
Pin Name	**UP-2**	**PLDT-2**	**eSOC**
d3	34	34	163
d2	33	33	162
d1	36	36	161
d0	35	35	160
a	69	69	169
b	70	70	170
c	73	73	171
d	74	74	173
e	76	76	175
f	75	75	133
g	77	77	135

Instructor's Initials ____________

Lab 13

Priority Encoders

Name ______________________ Class __________ Date __________

Objectives Upon completion of this laboratory exercise, you should be able to:

- Enter a Block Diagram design for a BCD priority encoder.
- Write simulation criteria for the BCD priority encoder and create a simulation in Quartus II.
- Create a test circuit for the BCD priority encoder in the Quartus II Block Editor.
- Test the priority encoder on a CPLD test board.

Reference Ken Reid and Robert Dueck, *Introduction to Digital Electronics*
Chapter 5: Combinational Logic Functions

Equipment Required CPLD Trainer:
Altera UP-2 board with ByteBlaster download cable, or
DeVry eSOC board with USB cable, or
RSR PLDT-2 board with straight-through parallel port cable, or
Equivalent CPLD trainer board with Altera EPM7128S CPLD
Quartus II Web Edition software
AC adapter, minimum output: 7 VDC, 250 mA DC
Anti-static wrist strap
#22 solid-core wire
Wire strippers

Experimental Notes

In this lab, we will examine the behavior of a **priority encoder.** A priority encoder becomes much more complex with each additional input bit, so it is often easiest to use an existing *device* in a *Block Diagram* file, rather than its specific Boolean equations.

A priority encoder is a circuit that generates a binary or BCD code that corresponds to the active input with the highest priority, which generally means the active input with the highest decimal subscript. Figure 13.1 shows an example of the operation of a BCD priority encoder.

In Figure 13.1a, only input D_2 is active, so the output code generated is 0010, the binary equivalent of 2. In Figure 13.1b, input D_2 remains active, but now D_7 is also active. Since D_7 has the higher priority, the output code is now 0111, the binary equivalent of 7. In Figure 13.1c, input D_9 is HIGH, in addition to the other two inputs. Since D_9 has the highest priority, the output is 1001. There is no input for D_0 since the output 0000 is the default value when no inputs are active.

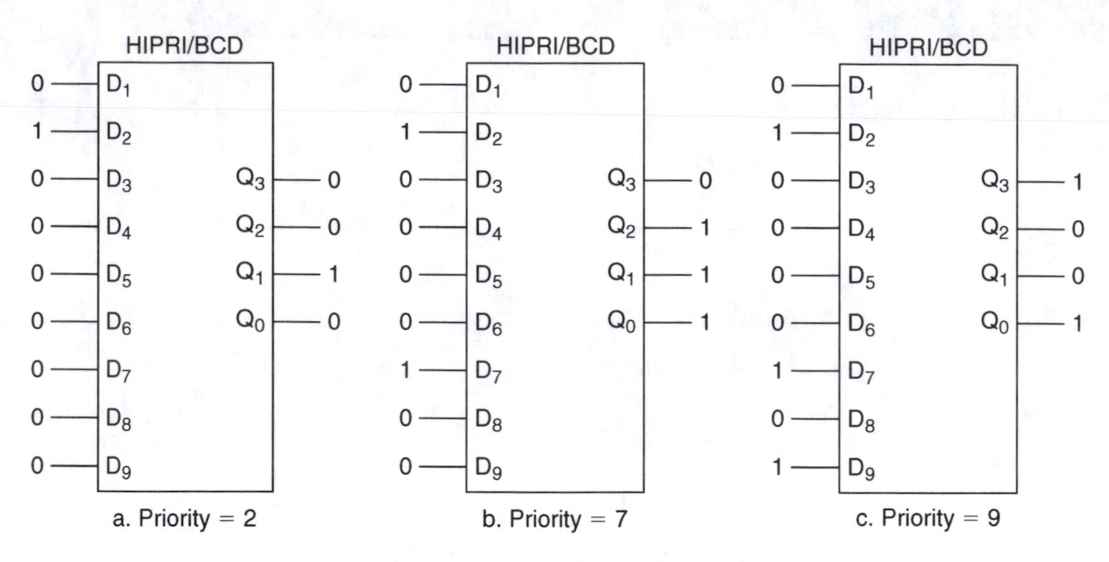

Figure 13.1 BCD Priority Encoder

Procedure

Design Entry and Simulation

1. Start a Block Diagram file and add a 74147 component. Find a datasheet for this device to verify its operation. Build a circuit using a 74147 and input and output pins.

2. Save and compile the project.

3. Write a set of simulation criteria that will test the correctness of the priority encoder design.

Simulation Criteria

Instructor's Initials ______________

4. Use the simulation criteria you wrote in step 2 to create a Quartus II simulation of the BCD priority encoder.

Instructor's Initials ______________

Test Circuit for Priority Encoder

Figure 13.2 shows a test circuit with active-LOW LED outputs for the BCD priority encoder that can be used with the Altera UP-2 or DeVry eSOC boards. Figure 13.3 shows a test circuit with active-HIGH LED outputs for the RSR PLDT-2 board. The test circuit displays the output of the priority encoder both as a binary value and as a decimal digit on the board's seven-segment numerical display. Unused LEDs are disabled by tying them

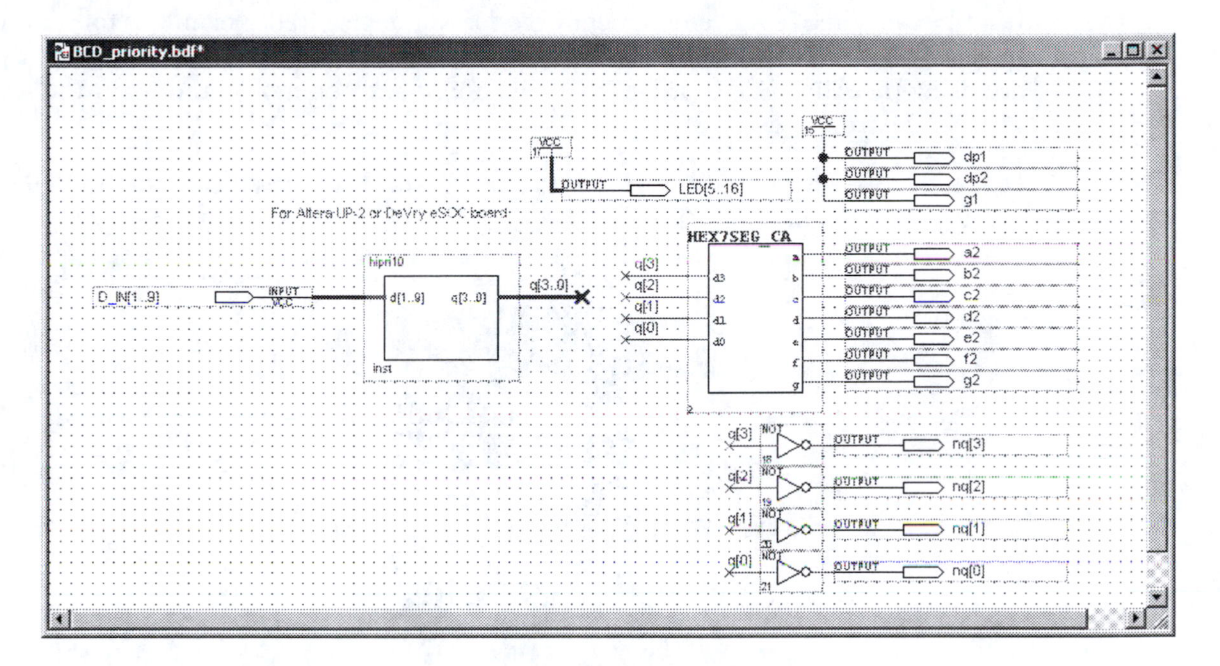

Figure 13.2 Priority Encoder Test Circuit (Altera UP-2 and DeVry eSOC)

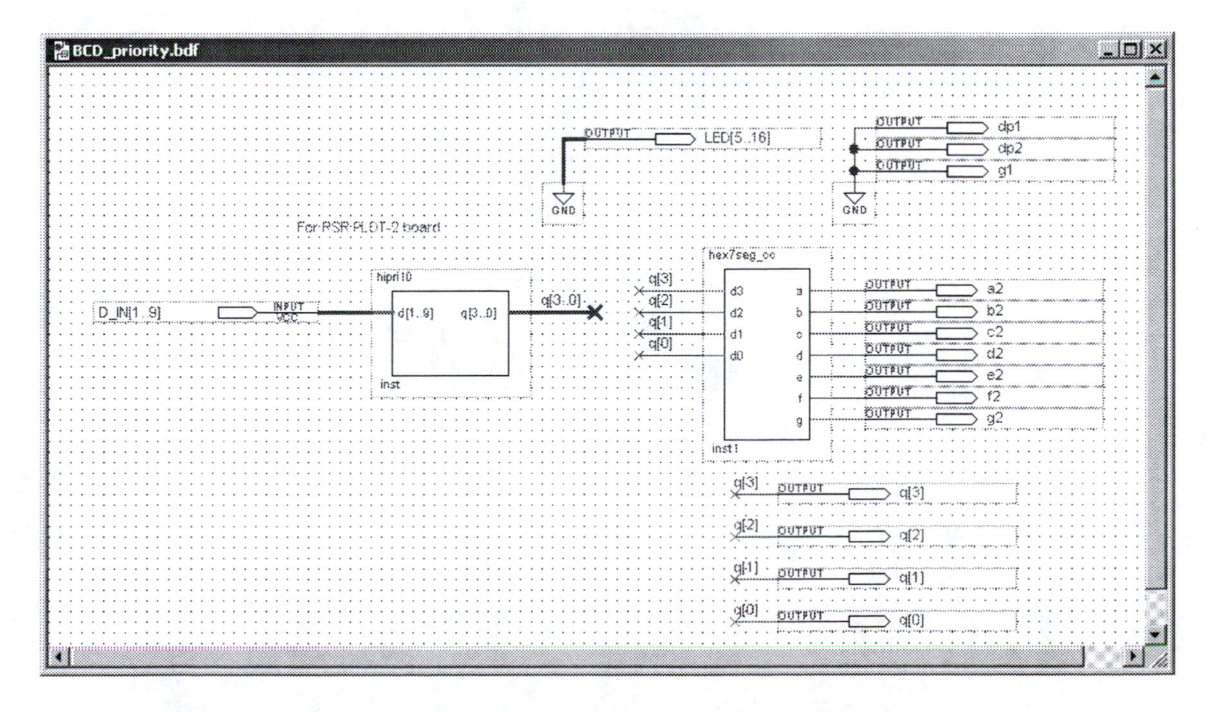

Figure 13.3 Priority Encoder Test Circuit (RSR PLDT-2)

either to VCC or ground, as appropriate for the selected board. The symbols for the priority encoder and the seven-segment decoder are created from VHDL files for those components.

1. Create a new folder for the test circuit and name it:

   ***drive:*\qdesigns\labs\lab13\BCD_priority**

2. Copy the .bdf file for the priority encoder to the new folder. Also copy the .bdf file for a hexadecimal-to-seven-segment decoder to the new folder from the folder used for Lab 12 (Seven-Segment Decoders). The decoder should be common anode if you are using the Altera UP-2 or DeVry eSOC board and common cathode for the RSR PLDT-2 board.

3. Make a project for the test circuit by selecting **New Project Wizard** from the **File** menu, as shown in Figure 13.4

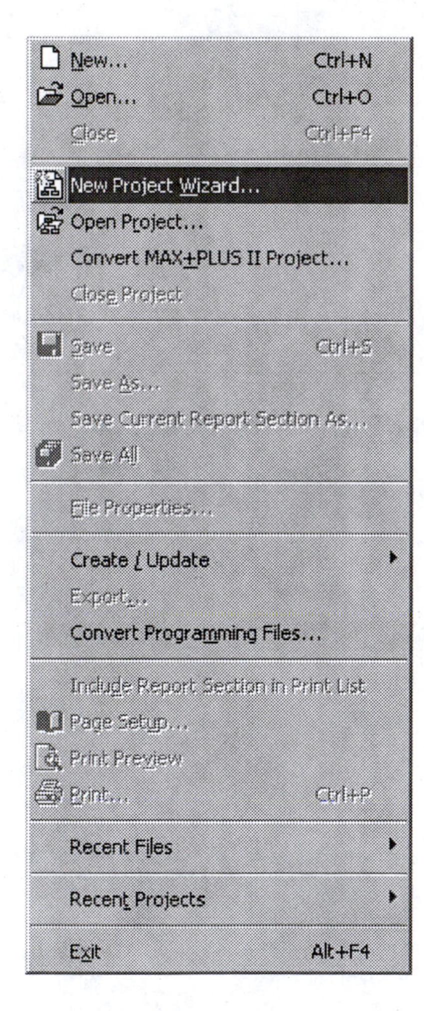

Figure 13.4 New Project Wizard (File Menu)

4. In the first screen of the New Project Wizard, shown in Figure 13.5, select the (. . .) button to browse to the new folder, as shown in Figures 13.5 and 13.6. Click **Open** in the dialog box shown in Figure 13.6 to select the working directory.

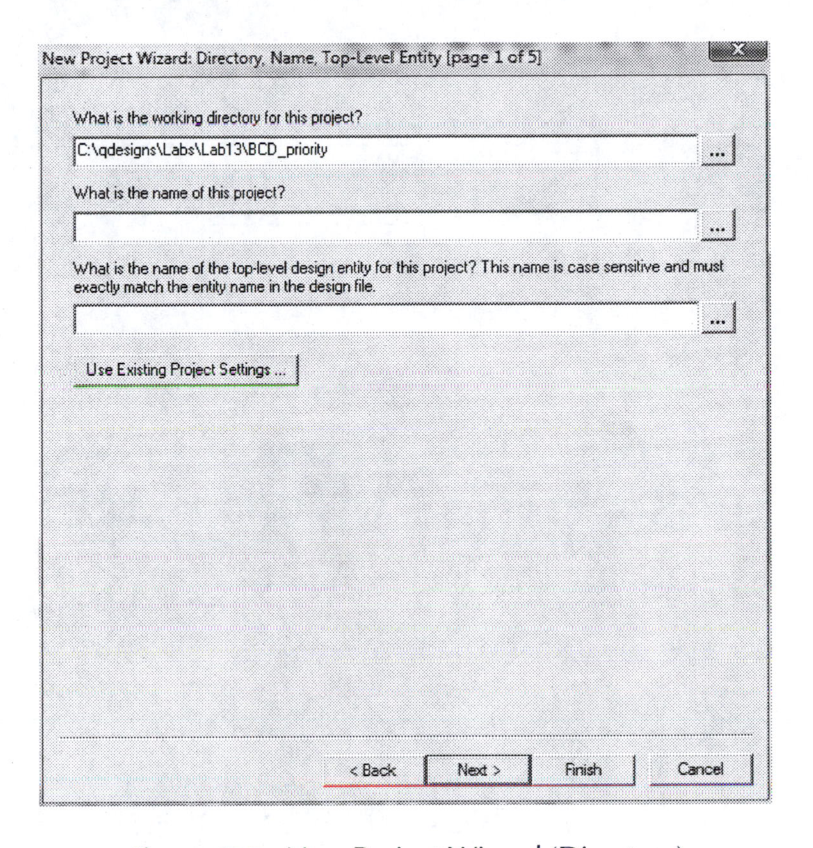

Figure 13.5 New Project Wizard (Directory)

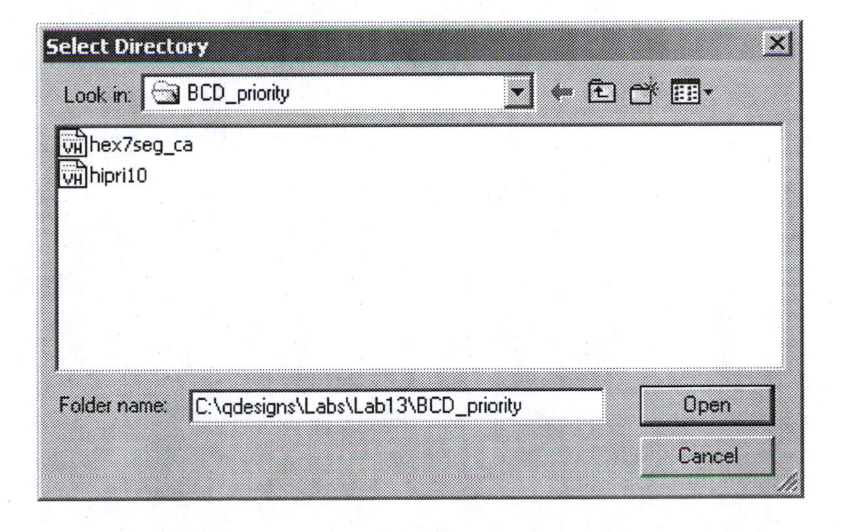

Figure 13.6 Select Directory Dialog Box

Figure 13.7 New Project Wizard (Directory Selected)

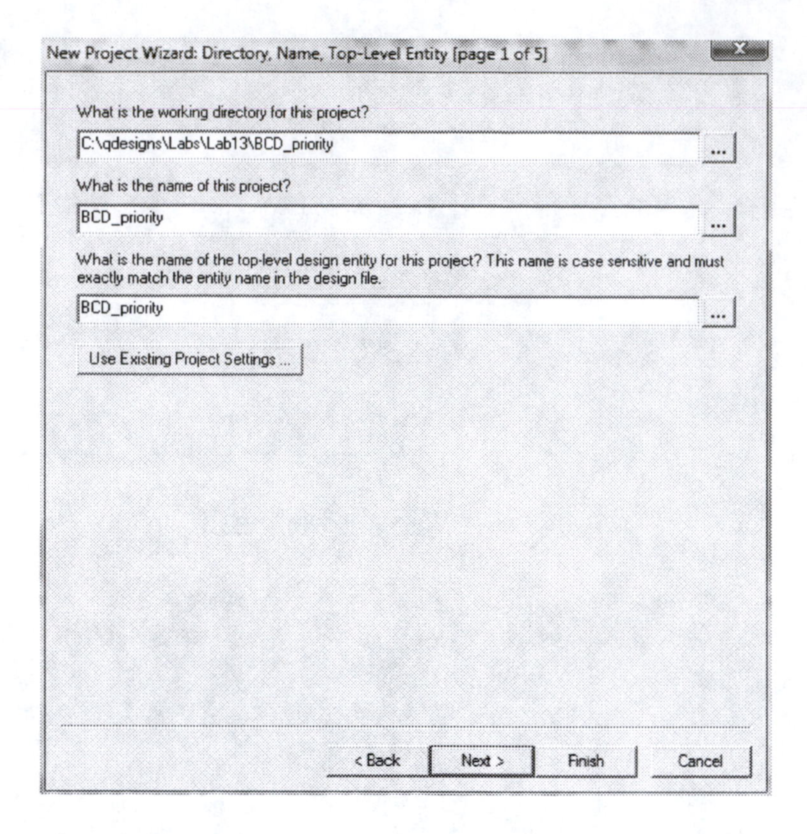

Figure 13.7 shows the New Project Wizard screen with the completed directory and entity information for the project.

5. Figure 13.8 shows the second screen of the New Project Wizard. Click **Add All** to add the VHDL files for the encoder and decoder to the working project.

Figure 13.8 New Project Wizard (Add Files)

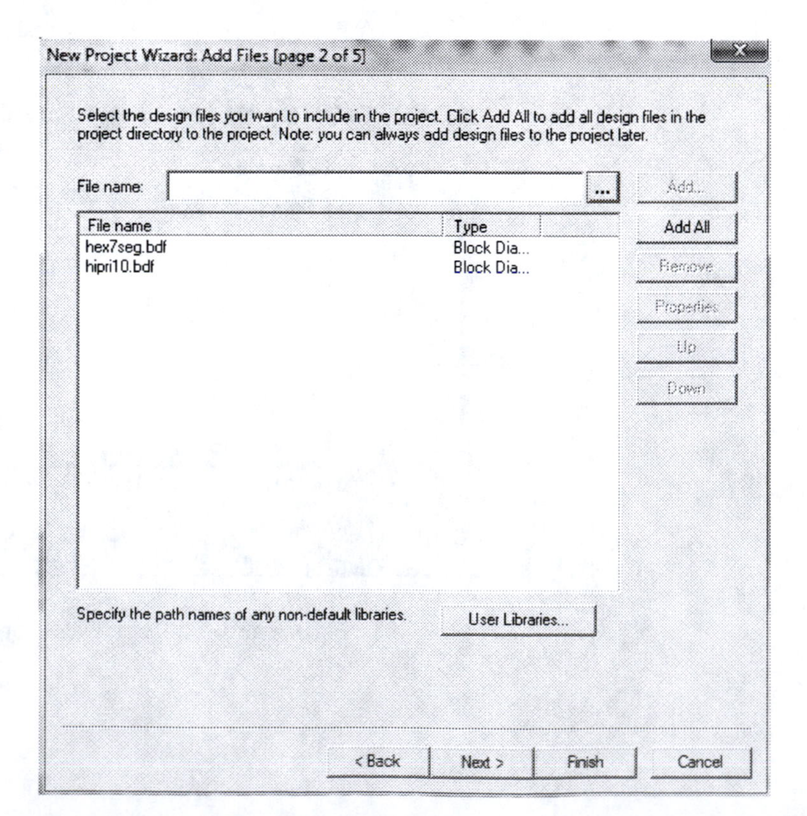

6. Continue through the screens of the New Project Wizard to complete the project initialization. If necessary select the device family (MAX 7000S) and the target device (EPM7128SLC84-7, -10, or -15, depending on your board).

7. Open the file for the priority encoder in Quartus II. Create a symbol for the encoder by selecting **Create/Update, Create Symbol Files for Current File** from the **File** menu, as shown in Figure 13.9.

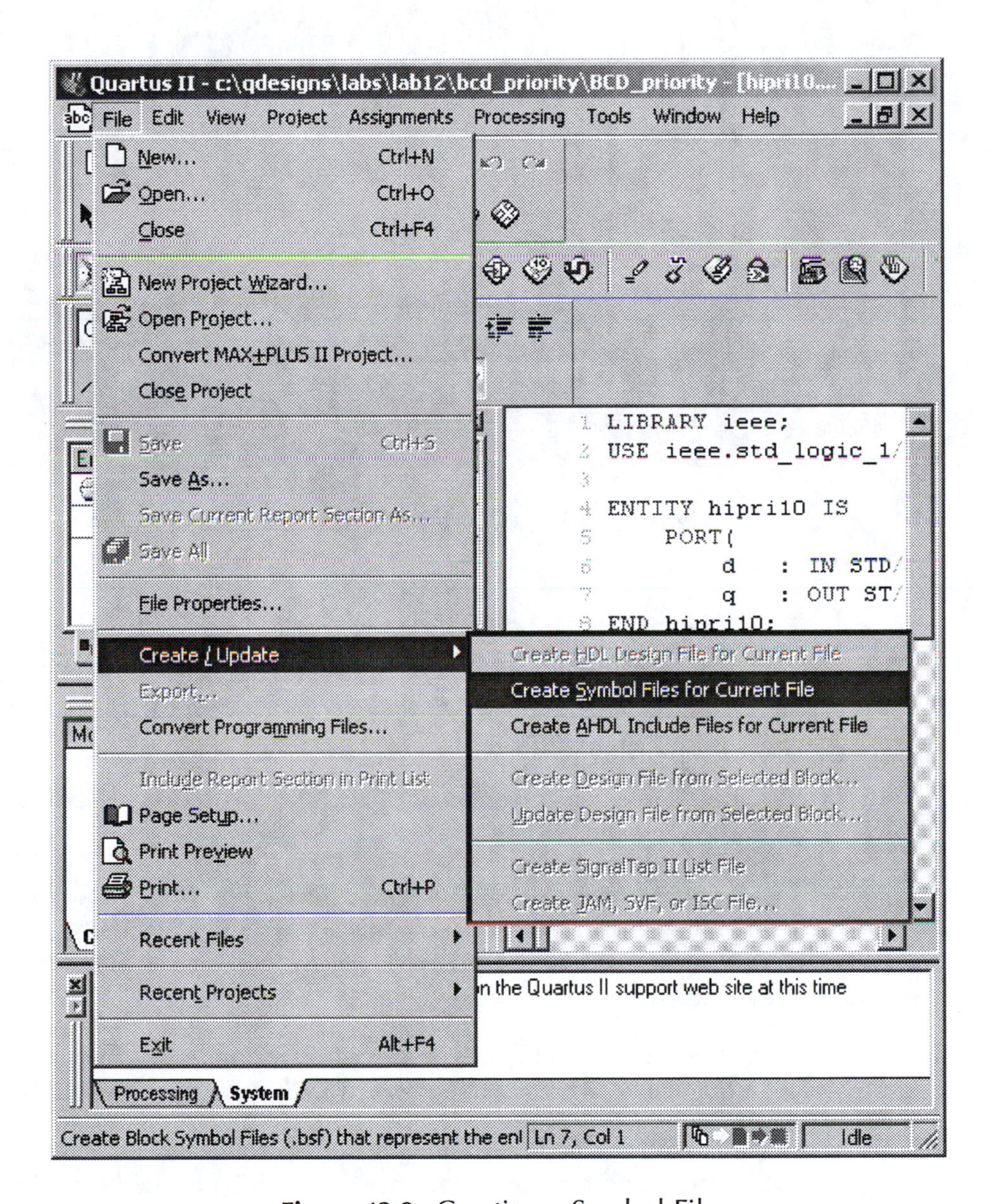

Figure 13.9 Creating a Symbol File

8. Open the Block Diagram file for the seven-segment decoder and create a symbol, as described in step 7.

9. Create a new block diagram file and save the file as:

   ***drive:*\qdesigns\labs\lab13\BCD_priority\BCD_priority.bdf**

Figure 13.10 shows the **Save As** dialog box. Make sure the box labeled **Add file to current project** is checked.

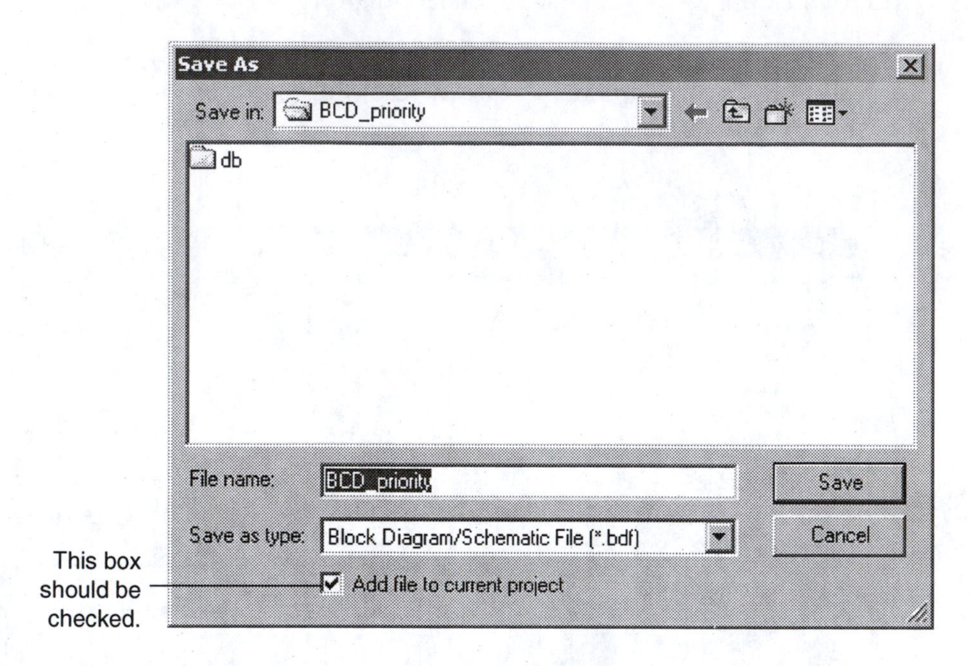

Figure 13.10 Saving the Block Diagram File

10. Add the symbol for the priority encoder to the Block Diagram File. To do so, double-click on the Block Editor desktop and select the symbol from the **Project** folder of the **Symbol** dialog box, shown in Figure 13.11.

11. Add the symbol for the seven-segment decoder, using the procedure described in step 10.

12. Figure 13.12 shows how a bus and a related series of nodes can be connected by name. Any two points in a block diagram are deemed to be connected if they are both labeled with the same name. Thus, the bus labeled **q[313..0]** is connected to the nodes labeled **q[3], q[2], q[1],** and **q[0],** even though there are no lines connecting these points in the diagram.

    To label the bus, click on it to highlight it, then right-click and select **Properties** from the pop-up menu. In the dialog box shown in Figure 13.13, type **q[3..0]** in the **Name** box.

13. To label a node, click on the node to highlight it, then right-click to get a pop-up menu. Select **Properties** and type the node name (e.g., **d[3]**) in the **Name** box, as shown in Figure 13.14.

14. Complete the Block Diagram File, as shown in Figure 13.2 or 13.3, as appropriate for your board. Compile the project.

15. Add pin numbers to the priority encoder test circuit, as listed at the end of this lab exercise in Table 13.3 through Table 13.5. Compile the file again and use it to program your CPLD board.

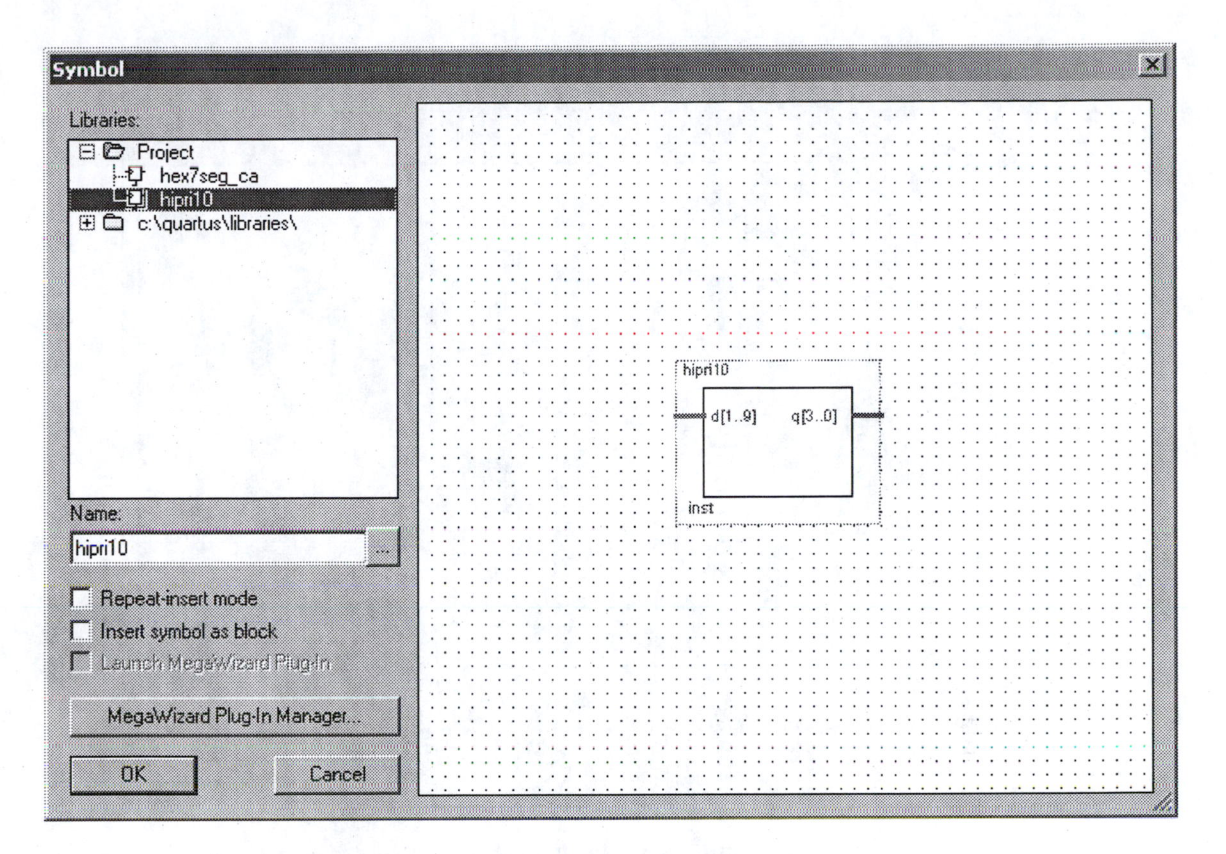

Figure 13.11 Entering the Symbol for the Priority Encoder

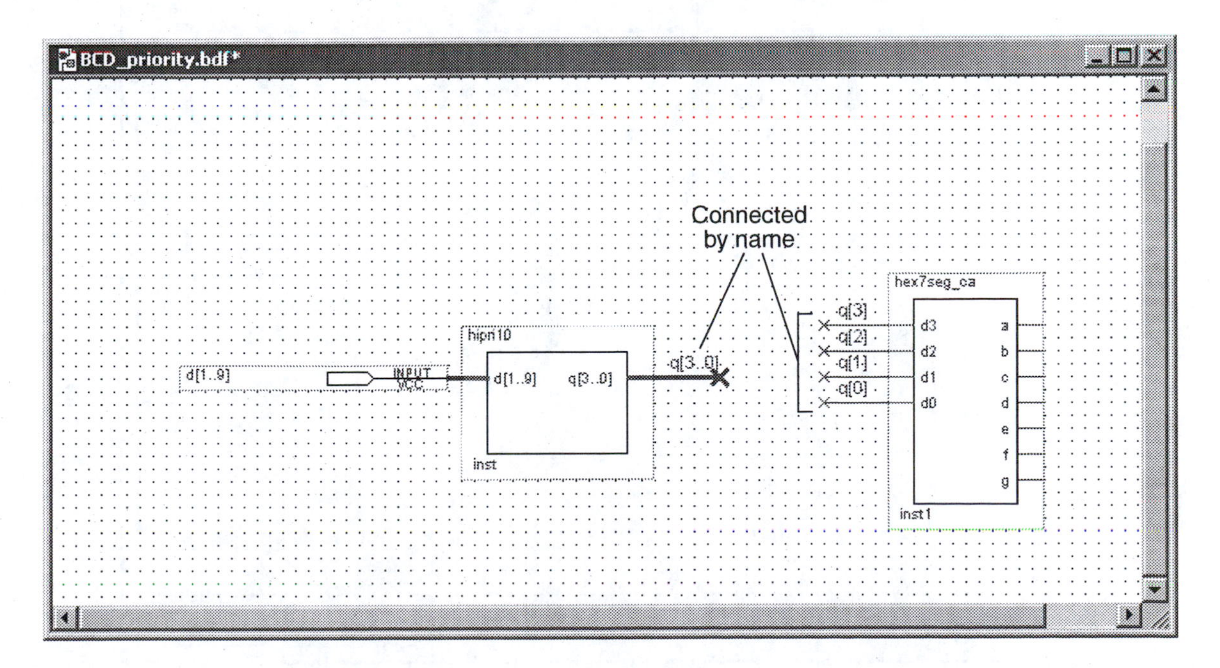

Figure 13.12 Connecting Components by Name

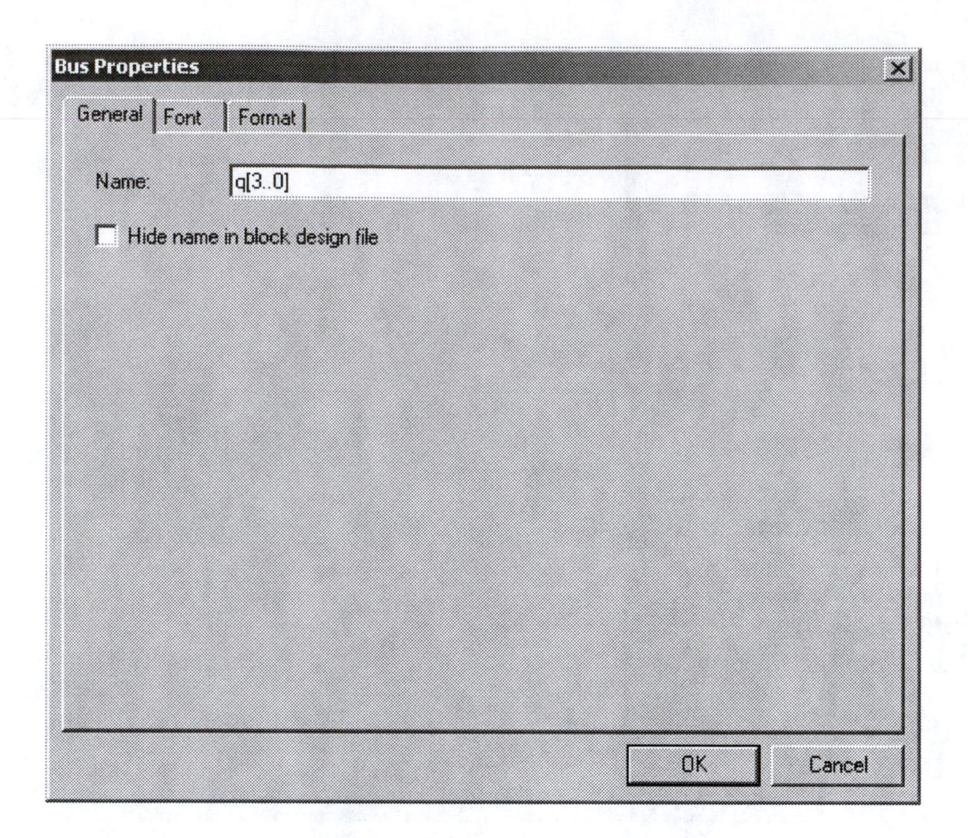

Figure 13.13 Labeling a Bus

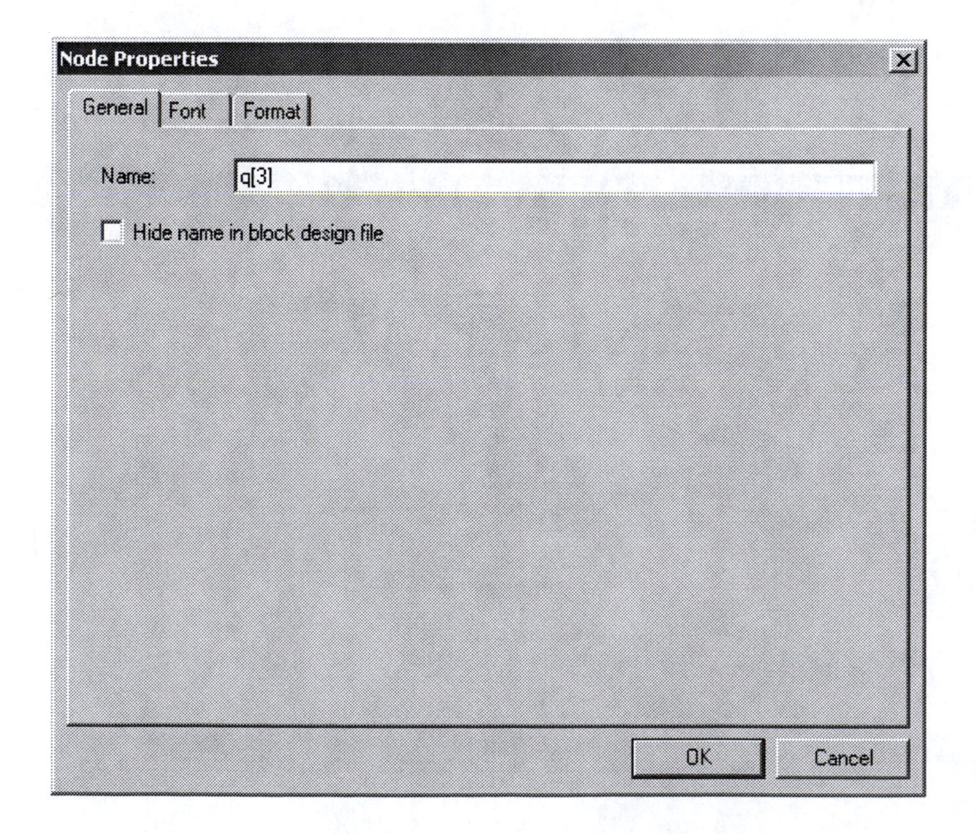

Figure 13.14 Labeling a Node

Testing the Priority Encoder

1. Set all DIP switches on the CPLD board LOW. Set the switches HIGH in the following sequence: D_9, D_8, D_7, D_6, D_5, D_4, D_3, D_2, D_1. Fill in Table 13.1 with the binary values of the priority encoder, as displayed by the LEDs on the CPLD test board. For each line also write the displayed decimal digit on the seven-segment display.

Table 13.1 Inputs LOW, Then HIGH

Last Switch Set HIGH	Highest Priority Input	q[2] q[1] q[0]	Digit
none			
D_9			
D_8			
D_7			
D_6			
D_5			
D_4			
D_3			
D_2			
D_1			

2. Make sure all switches are set HIGH. Set the switches LOW in the following sequence: D_9, D_8, D_7, D_6, D_5, D_4, D_3, D_2, D_1. Fill in Table 13.2 with the binary values of the priority encoder, as displayed by the LEDs on the CPLD test board. For each line also write the displayed decimal digit on the seven-segment display.

Table 13.2 Inputs HIGH, Then LOW

Last Switch Set Low	Highest Priority Input	q[2] q[1] q[0]	Digit
none			
D_9			
D_8			
D_7			
D_6			
D_5			
D_4			
D_3			
D_2			
D_1			

3. Explain the difference between the two tables.

4. Explain the operation of a priority encoder.

Instructor's Initials ________________

Table 13.3 EPM7128LC84-7 Pin Assignments Altera UP-2 Board

Seven-Segment Digits			
Function	**Pin**	**Function**	**Pin**
		a2	69
		b2	70
		c2	73
		d2	74
		e2	76
		f2	75
g1	67	g2	77
dp1	68	dp2	79

DIP Switches			
Function	**Pin**	**Function**	**Pin**
d_in[1]	34	d_in[9]	28
d_in[2]	33		
d_in[3]	36		
d_in[4]	35		
d_in[5]	37		
d_in[6]	40		
d_in[7]	39		
d_in[8]	41		

LED Outputs			
Function	**Pin**	**Function**	**Pin**
nq[3]	44	LED9	80
nq[2]	45	LED10	81
nq[1]	46	LED11	4
nq[0]	48	LED12	5
LED5	49	LED13	6
LED6	50	LED14	8
LED7	51	LED15	9
LED8	52	LED16	10

Table 13.4 EPM7128LC84-7 Pin Assignments RSR PLDT-2 Board

Seven-Segment Digits			
Function	**Pin**	**Function**	**Pin**
		a2	69
		b2	70
		c2	73
		d2	74
		e2	76
		f2	75
g1	67	g2	77
dp1	68	dp2	79

DIP Switches			
Function	**Pin**	**Function**	**Pin**
d_in[1]	34	d_in[9]	28
d_in[2]	33		
d_in[3]	36		
d_in[4]	35		
d_in[5]	37		
d_in[6]	40		
d_in[7]	39		
d_in[8]	41		

LED Outputs			
Function	**Pin**	**Function**	**Pin**
q[3]	44	LED9	80
q[2]	45	LED10	81
q[1]	46	LED11	4
q[0]	48	LED12	5
LED5	49	LED13	6
LED6	50	LED14	8
LED7	51	LED15	9
LED8	52	LED16	10

Table 13.5 EPM7128LC84-7 Pin Assignments DeVry eSOC Board

Seven-Segment Digits			
Function	**Pin**	**Function**	**Pin**
		a2	137
		b2	138
		c2	139
		d2	128
		e2	80
		f2	81
g1	135	g2	84

DIP Switches			
Function	**Pin**	**Function**	**Pin**
d_in[1]	160		
d_in[2]	161		
d_in[3]	162		
d_in[4]	163		
d_in[5]	168		
d_in[6]	141		
d_in[7]	142		
d_in[8]	144		

Pushbutton (Debounced)			
Function	**Pin**	**Function**	**Pin**
d_in[9]	131		

LED Outputs			
Function	**Pin**	**Function**	**Pin**
nq[3]	115	LED9	101
nq[2]	114	LED10	99
nq[1]	113	LED11	97
nq[0]	112	LED12	96
LED5	110	LED13	95
LED6	102	LED14	92
LED7	151	LED15	118
LED8	150	LED16	117

Note d_in[9] will utilize a pushbutton instead of a toggle switch on the eSOC board.

Lab 14

Multiplexers

Name ____________________ Class __________ Date __________

Objectives Upon completion of this laboratory exercise, you should be able to:

- Enter the logic circuit of a 4-to-1 multiplexer (MUX) as a Block Diagram file, using Altera's Quartus II CPLD design software.
- Create a Quartus II simulation file for the 4-to-1 multiplexer described above.
- Create a hierarchical design in the Quartus II Block Editor that contains a multiplexer and other components.
- Download the 4-to-1 MUX to a CPLD test board and test its function.

Reference Ken Reid and Robert Dueck, *Introduction to Digital Electronics*

Chapter 5: Combinational Logic Functions

Equipment Required CPLD Trainer:

Altera UP-2 board with ByteBlaster download cable, or
DeVry eSOC board, with parallel port cable, or
RSR PLDT-2 board with straight-through parallel port cable, or
equivalent CPLD trainer board with Altera EPM7128S CPLD

Quartus II Web Edition software
AC adapter, minimum output: 7 VDC, 250 mA DC
Anti-static wrist strap
#22 solid-core wire
Wire strippers
Oscilloscope

Experimental Notes

A multiplexer (abbreviated MUX) is a device for switching one of several digital signals to an output, under the control of another set of binary inputs. The inputs to be switched are called the **data inputs;** those that determine which signal is directed to the output are called the **select inputs.**

Figure 14.1 shows the logic diagram of a 4-to-1 multiplexer, with data inputs labeled D_3 to D_0 and the select inputs labeled S_1 and S_0. By examining the circuit, we can see that the 4-to-1 MUX is described by the following Boolean equation:

$$Y = D_0\overline{S}_1\overline{S}_0 + D_1\overline{S}_1S_0 + D_2S_1\overline{S}_0 + D_3S_1S_0$$

For any given combination of S_1S_0, only one of the four product terms will be enabled. For example, when $S_1S_0 = 10$, the equation evaluates to:

$$Y = (D_0 \cdot 0) + (D_1 \cdot 0) + (D_2 \cdot 1) + (D_3 \cdot 0) = D_2$$

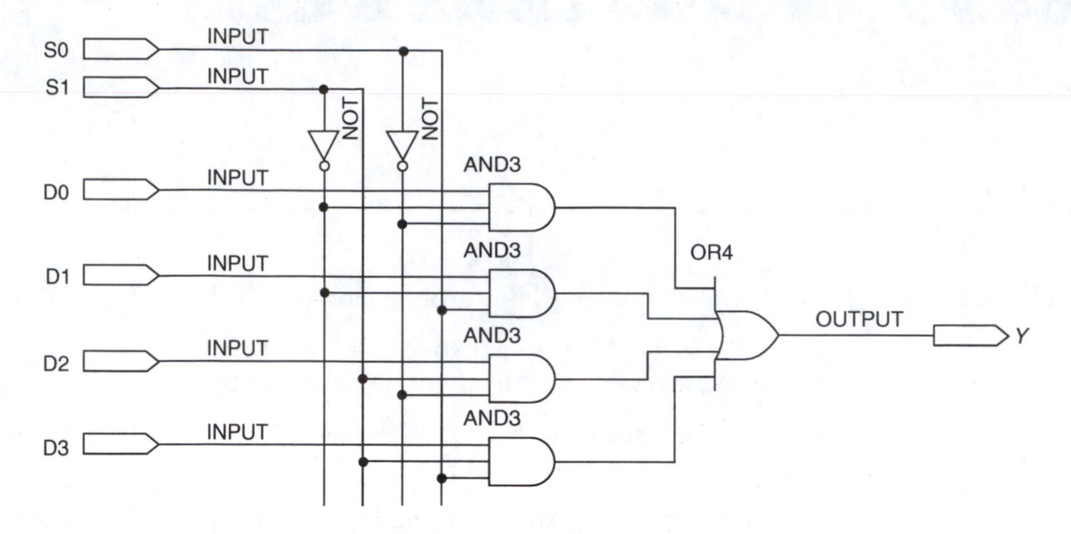

Figure 14.1 4-to-1 Multiplexer

The MUX equation can be described by a truth table, as in Table 14.1. The subscript of the selected data input is the decimal equivalent of the binary combination S_1S_0.

Multiplexers can be implemented in Quartus II as a Block Diagram file similar to Figure 14.1

Table 14.1 Truth Table of a 4-to-1 MUX

S_1	S_0	Y
0	0	D_0
0	1	D_1
1	0	D_2
1	1	D_3

Procedure

Block Diagram File and Simulation for 4-to-1 Multiplexer

1. Create a Block Diagram File for a 4-to-1 multiplexer as shown in Figure 14.1. Save the file as ***drive:*****\qdesigns\labs\lab14\4to1mux\4to1mux.bdf** (**Tip:** You can place the inverters vertically, as shown in Figure 14.1, by entering the NOT symbol, right-clicking on the symbol, and choosing **Rotate, 270°** from the pop-up menu.)

   Compile the design.

2. We can use the following set of simulation criteria to test the 4-to-1 multiplexer.

Simulation Criteria

- Each data input channel of the multiplexer will be selected in an ascending sequence by applying a binary count to the combined select inputs.
- Each data input should be easily recognizable by having a "signature" waveform applied to it. Each channel should be selected for a period no less than about two or three cycles of the signature waveform.
- The output waveform should display a series of unique signature waveforms, indicating the selection of the data channels in the correct sequence.

3. A set of signature waveforms can be created by applying waveforms of different frequencies that are related by increasing binary multipliers. If, for example, we have a count waveform with a base period of 20 ns, we can create other waveforms with periods of 40 ns, 80 ns, and 160 ns, which represent multipliers of 2×, 4×, and 8× the base period. In addition, we should choose a period for the select inputs such that we see about three cycles of the slowest waveform, in this case 960 ns. This set of waveforms is shown in Figure 14.2. The waveforms are staggered in their frequencies so as to make the largest contrast between adjacent waveforms.

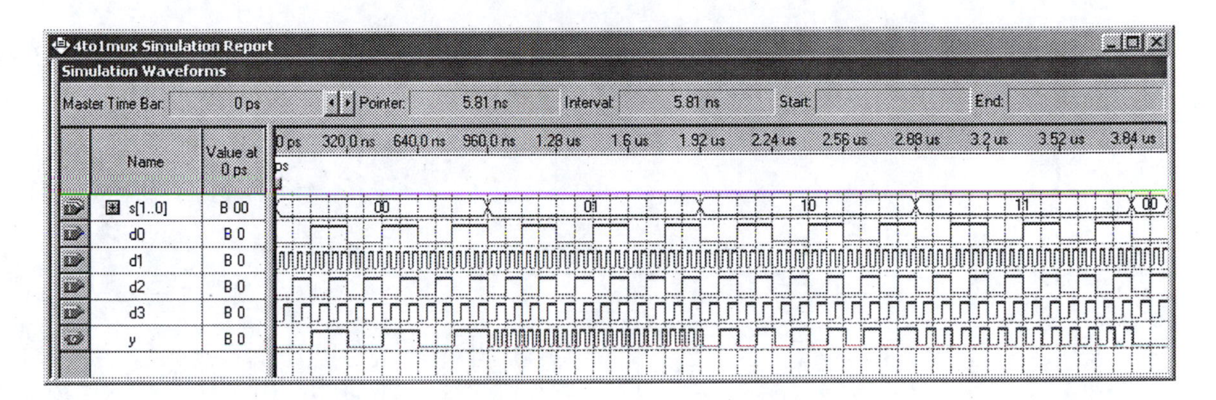

Figure 14.2 Simulation for a 4-to-1 Multiplexer

4. Create the simulation waveforms of Figure 14.2, using an end time of 4 μs. Apply a count value to each of the inputs using the count value specified in Table 14.2.

   To apply a count value to a waveform, highlight the waveform and click the **Count Value** button on the Waveform Editor toolbar, as shown in Figure 14.3. In the **Count Value** dialog box, select the **Timing** tab and fill in the value for **Count every:** as shown in Table 14.2.

Table 14.2 Simulation Specifications

Input	Count every:
s[2..0]	960 ns
d0	160 ns
d1	20 ns
d2	80 ns
d3	40 ns

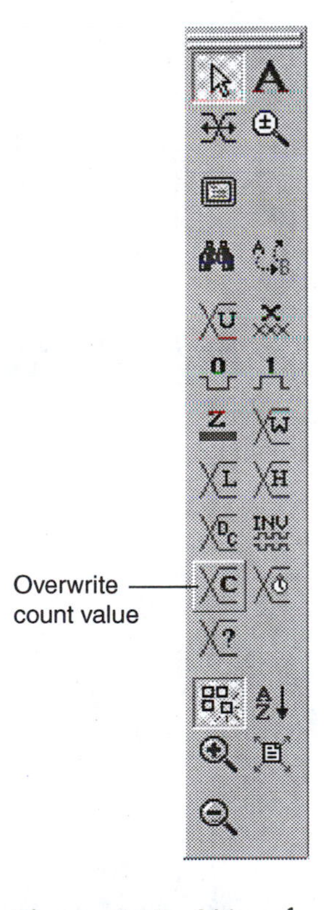

Figure 14.3 Waveform Editor Toolbar

5. When you have completed the input waveforms, run the simulation and show your instructor.

Instructor's Initials ____________

Testing the 4-to-1 MUX

One way to test the MUX function on the CPLD board is to apply a known signal to each data input, as we did in our simulation, and manually change the values of the select inputs with DIP switches. The output signal can be observed on a monitoring device such as an LED or an oscilloscope. The LED or oscilloscope will tell you which MUX input channel has been selected.

Two MUX test circuits are shown in Figures 14.4 and 14.5. The test circuits are examples of **hierarchical design.** All this means is that the test circuit (called the **top level**

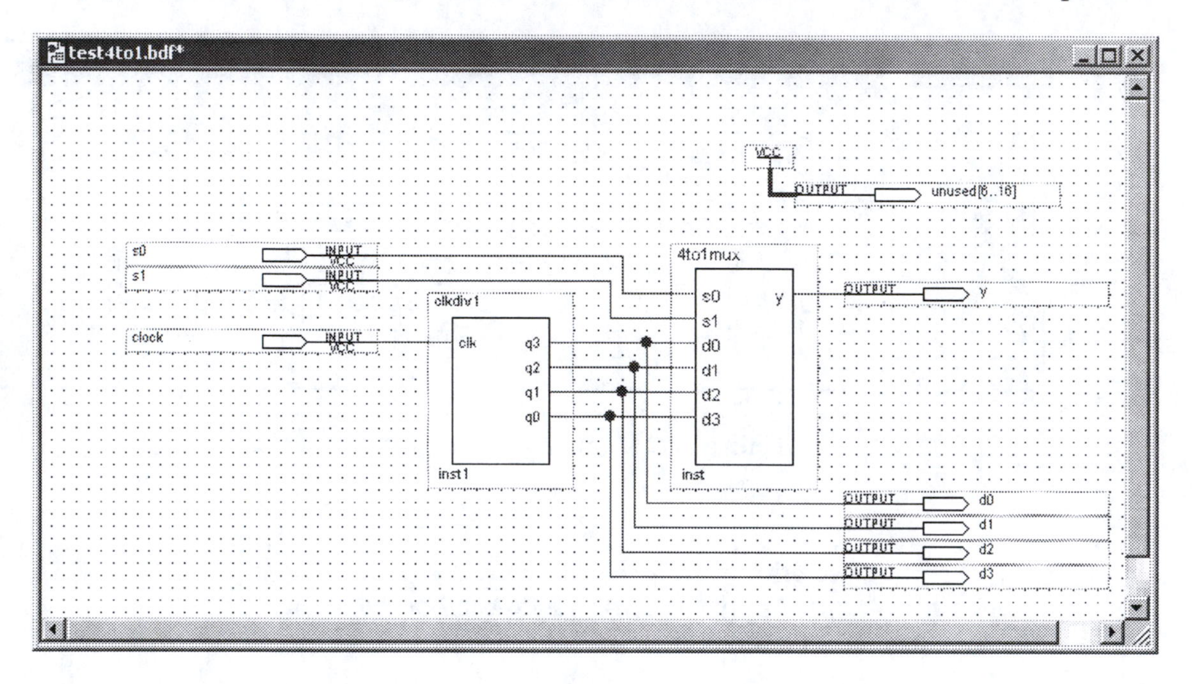

Figure 14.4 Test Circuit for a 4-to-1 MUX (Altera UP-2 or DeVry eSOC)

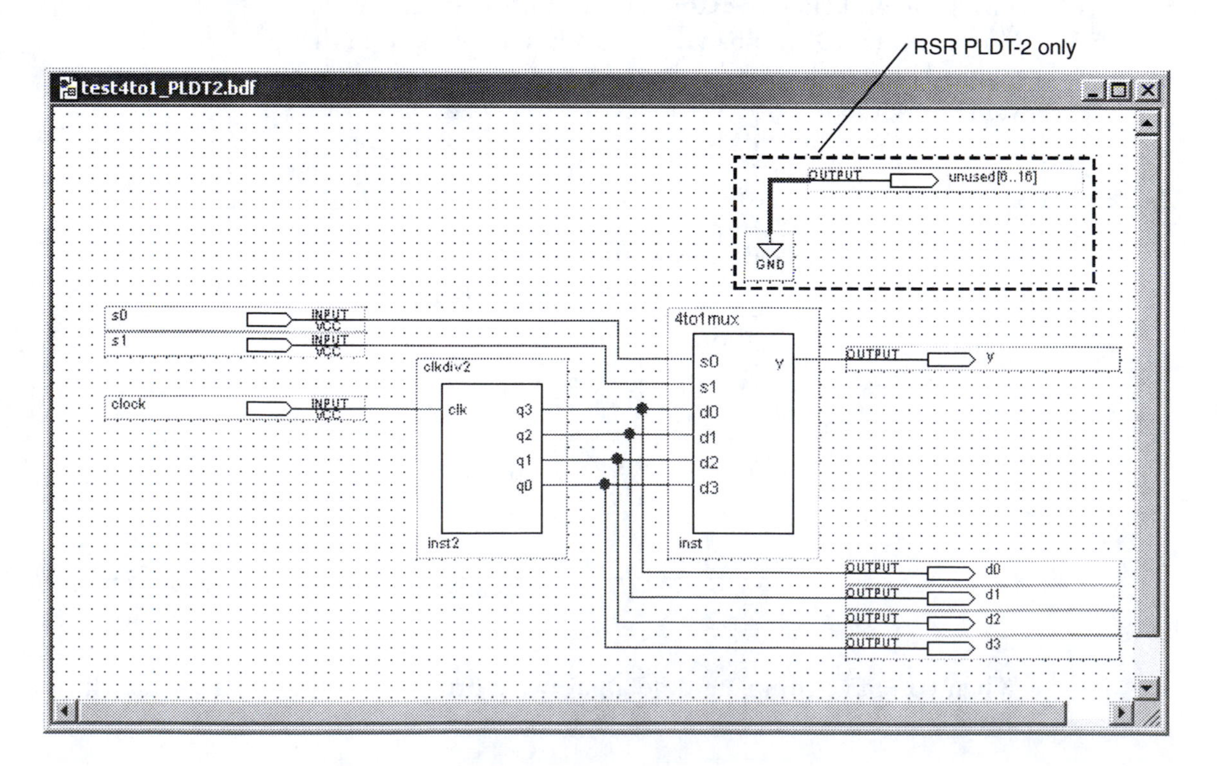

Figure 14.5 Test Circuit for a 4-to-1 MUX (RSR PLDT-2)

of the hierarchy) contains components that are complete designs in and of themselves. The design hierarchy shown contains two components: a 4-to-1 multiplexer, created from the block diagram file in Figure 14.1, and a predesigned clock divider, **clkdiv1** or **clkdiv1a** for the Altera UP-2 or DeVry eSOC board or **clkdiv2** or **clkdiv2a** for the RSR PLDT-2 board.

The clock dividers in Figures 14.4 and 14.5 (**clkdiv1** (Altera UP-2 or DeVry eSOC) or **clkdiv2** (RSR PLDT-2 or DeVry eSOC) provide digital square wave signals of binary-multiple frequencies at four different outputs (Altera UP-2 and DeVry eSOC: 1.5 Hz, 3 Hz, 6 Hz, and 12 Hz; RSR PLDT-2: 1 Hz, 2 Hz, 4 Hz, 8 Hz). These frequencies are slow enough to be observed visually.

An alternate pair of clock dividers, **clkdiv1a** (Altera UP-2 and DeVry eSOC) and **clkdiv2a** (RSR PLDT-2) are used to view the output waveforms on an oscilloscope. **Clkdiv1a** has a base MSB frequency of about 1.5 kHz, derived from the 25.175 MHz oscillator on the Altera UP-2 board or the 24 MHz oscillator on the DeVry eSOC board. **Clkdiv2a** has a base MSB frequency of about 1 kHz, derived from the 4 MHz oscillator on the RSR PLDT-2.

To create the test circuit, we must first create the symbols for the components, as follows.

1. Create a new folder for the project and call it: ***drive:*****\qdesigns\labs\lab14\test4to1**
2. Copy the files **4to1mux.bdf** and **clkdiv1.vhd** to the new folder.
3. Create symbols for the MUX and clock divider as follows. Open the file **4to1mux.bdf** in Quartus II. From the **File** menu, select **Create/Update, Create Symbol Files for Current File.** Repeat this procedure for the file **clkdiv1.vhd**
4. From the **File** menu, shown in Figure 14.6, select **New Project Wizard.**
5. In the first screen of the New Project Wizard, browse to select the new folder as shown in Figure 14.7.
6. In the second screen of the New Project Wizard, click **Add All.**
7. Continue through the screens of the New Project Wizard to select the Device Family (MAX 7000S) and target device (EPM7128SLC84-7, -10, or -15).
8. Click the **New File** icon or select **New** from the **File** menu. From the resultant dialog box, shown in Figure 14.8, select **Block Diagram/Schematic File** and click **OK.**

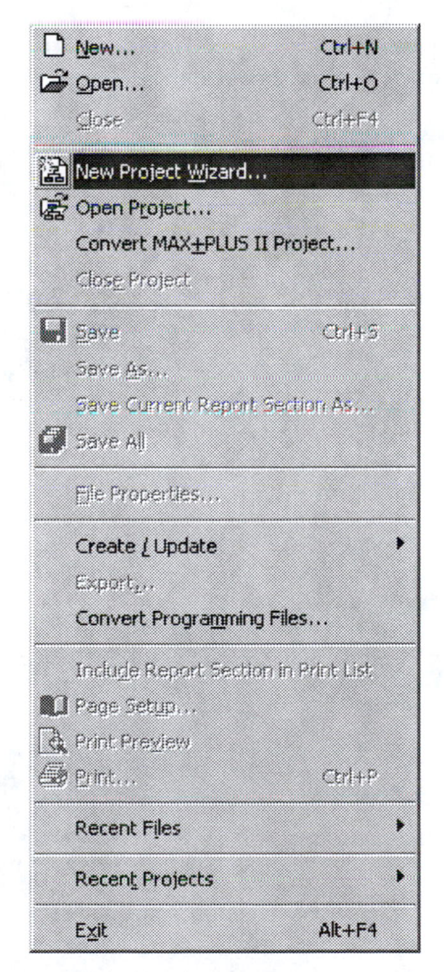

Figure 14.6 New Project Wizard (File Menu)

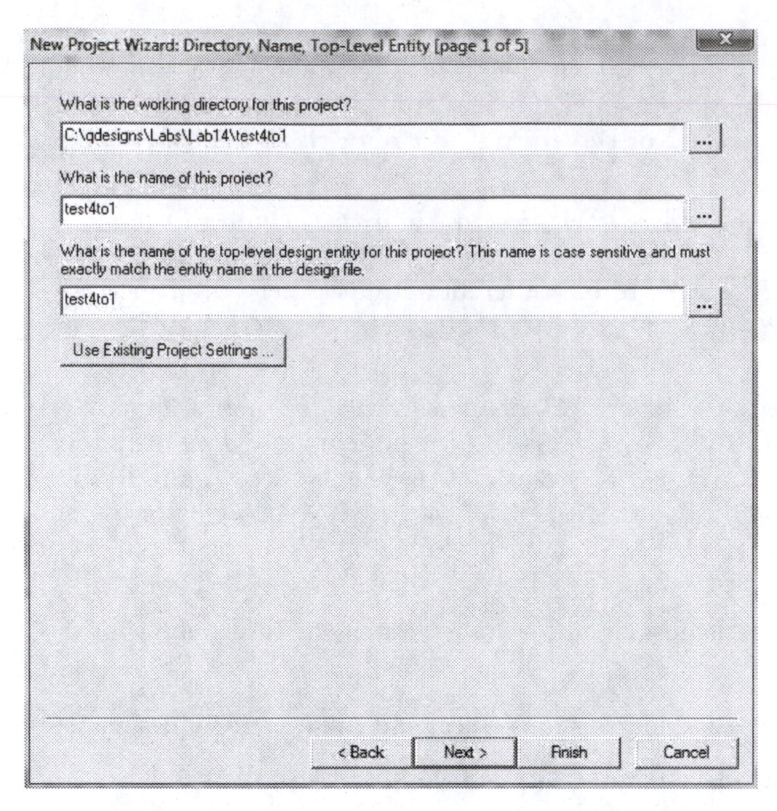

Figure 14.7 New Project Wizard (Directories)

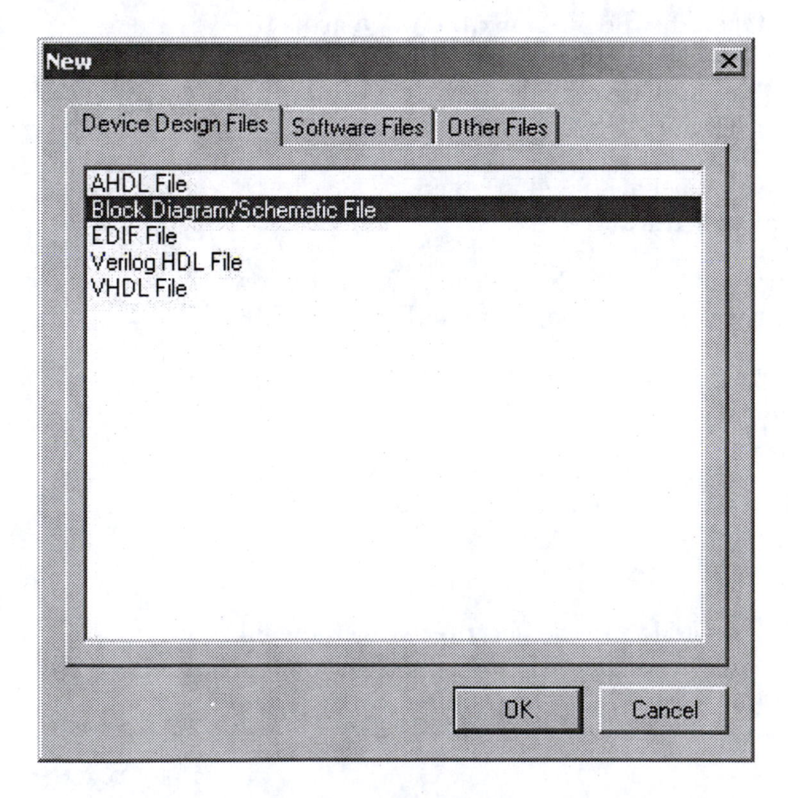

Figure 14.8 New File Dialog Box (Block Diagram File)

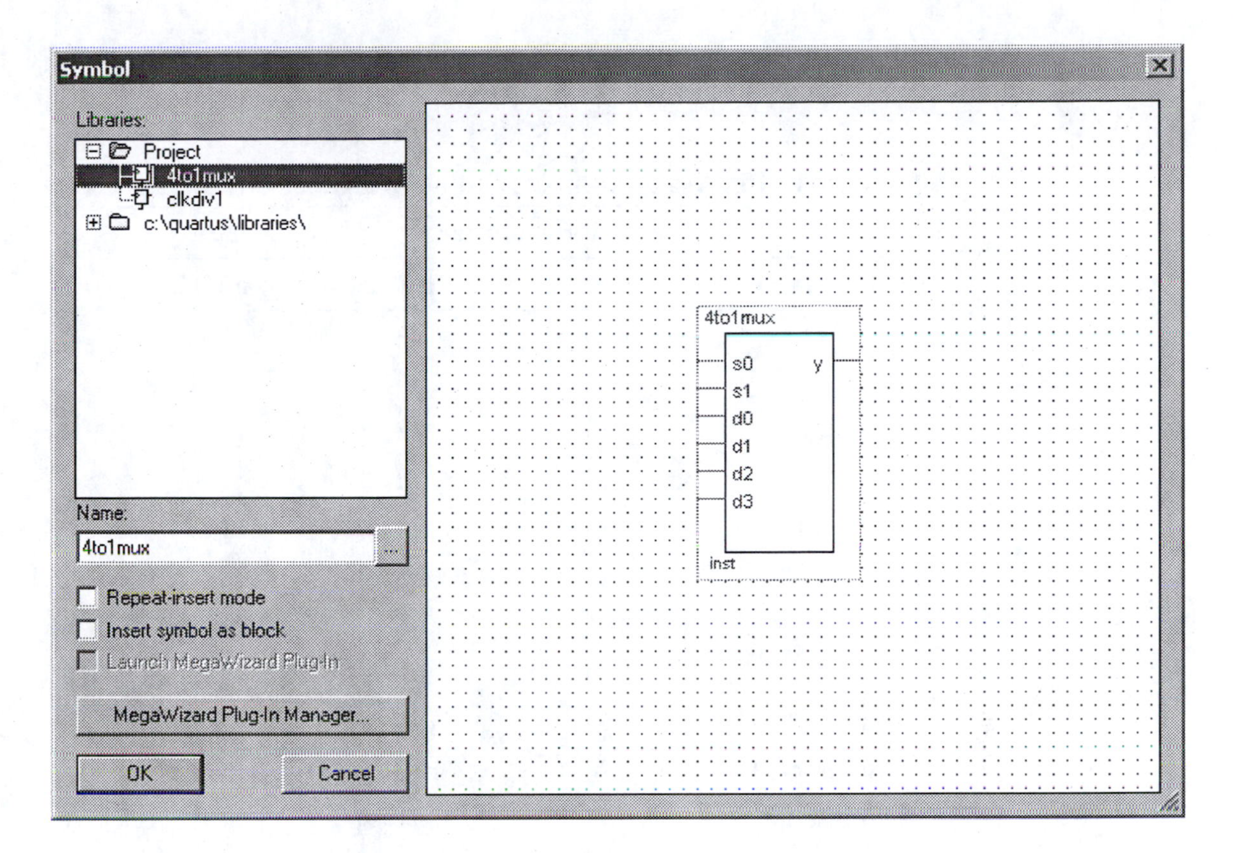

Figure 14.9 Symbol Dialog Box

9. Double-click on the Block Editor desktop to get the **Symbol** dialog box, as shown in Figure 14.9. Enter the symbol for the file **4to1mux,** as found in the **Project** folder of the **Symbol** dialog box. Repeat the procedure for the clock divider component.

   Add input and output pins, as shown in Figure 14.4 or 14.5.

   Create an output for disabling unused LEDs: **unused [6..16].** For the Altera UP-2 board, connect these LED outputs to V_{CC}. For the RSR PLDT-2 board, connect the LED outputs to ground. Draw the connecting line from the output pin symbol to the VCC or GND symbol, not the other way around. Make sure that the connecting line is a thick line, indicating a connection to multiple output lines for one pin symbol. The line thickness can also be changed by highlighting the line, then right-clicking it and choosing Bus Line from the pop-up menu. Save and compile the file.

10. Once the design hierarchy has been created, the component files can be accessed for editing by double-clicking. Try double-clicking on the component called **4to1mux.**

    What do you see? ____

    ____

11. Assign pin numbers as shown in Table 14.3. When you have assigned the pin numbers, save and compile the file again.

Table 14.3 Pin Assignments for 4-to-1 MUX Test Circuit

	Pin Number		
Pin Name	**UP-2**	**PLDT-2**	**eSOC**
s1	34	34	160
s0	33	33	161
clock	83	83	24
y	52	52	150
d0	44	44	115
d1	45	45	114
d2	46	46	113
d3	48	48	112
unused[6]	49	49	
unused[7]	50	50	
unused[8]	51	51	
unused[9]	80	80	
unused[10]	81	81	
unused[11]	4	4	
unused[12]	5	5	
unused[13]	6	6	
unused[14]	8	8	
unused[15]	9	9	
unused[16]	10	10	

12. Connect short lengths of #22 solid-core wire from the prototyping headers around the EPM7128S chip to two DIP switches (for S_1 and S_0) and the LEDs, as required.

    On the RSR board, if no wire connections are in place, it is not necessary to do any wiring. On the Altera UP-2 board, if the wire connections are not in place, it is only necessary to connect S_1, S_0, d_0, d_1, d_2, d_3, and Y. The CLOCK connection is hardwired on both boards.

    The pin assignments for the unused LEDs are there only so that these LEDs are not lit if they have been wired for another project, so don't connect them if they are not already connected.

13. Download the test circuit to the CPLD test board. Set the S_1, S_0 switches to 00 and observe the output of the MUX on the output LED. Repeat for values of 01, 10, and 11. Explain your observations to your instructor. Compare the **y** output to the MUX inputs displayed on the **d** LEDS.

Instructor's Initials ____________

Monitoring the MUX Output with an Oscilloscope

1. Modify the test circuit in Figure 14.4 or 14.5 so that the clock divider is replaced by either **clkdiv1a** or **clkdiv2a.** (You will have to create the component first.) Compile the project again.

2. Program your CPLD test board with the new circuit. With an oscilloscope, monitor the waveforms at the MUX inputs, one at a time, and measure the frequency of each one. Fill in Table 14.4 with the results of your measurements.

Table 14.4 Frequencies of Waveforms Applied to MUX Inputs

Pin Name	Frequency
d0	
d1	
d2	
d3	

3. Monitor the output **y** with the oscilloscope. Change the values of S_1 and S_0 to get all combinations of these inputs. For each combination, fill in Table 14.5 with the frequency of the waveform at output **y** and which selected input waveform this corresponds to.

Table 14.5 Frequency of MUX Output with Different Selected Inputs

S_1	S_0	Frequency at output Y	Selected input
0	0		
0	1		
1	0		
1	1		

Instructor's Initials ____________

Lab 15

Demultiplexers

Name ______________________ Class ____________ Date ____________

Objectives Upon completion of this laboratory exercise, you should be able to:

- Create a demultiplexer.
- Write a set of simulation criteria to test the demultiplexer and use the criteria to create a simulation in Quartus II.
- Test the demultiplexer on a CPLD test board.

Reference Ken Reid and Robert Dueck, *Introduction to Digital Electronics*

Chapter 5: Combination Logic Functions

Equipment Required CPLD Trainer:

Altera UP-2 circuit board with ByteBlaster download cable, or
DeVry eSOC board with USB cable, or
RSR PLDT-2 circuit board with straight-through parallel port cable, or
equivalent CPLD trainer
board with Altera EPM7128S CPLD
Quartus II Web Edition software
AC adapter, minimum output: 7 VDC, 250 mA DC
Anti-static wrist strap
#22 solid-core wire
Wire strippers
Oscilloscope

Experimental Notes

Recall from *Introduction to Digital Electronics* that a decoder and a demultiplexer can be created using the same circuitry. In a decoder, the decoding inputs determine which of several outputs is active and an enable input is used to determine whether any outputs are active. With a little manipulation, we can use the same device as a demultiplexer.

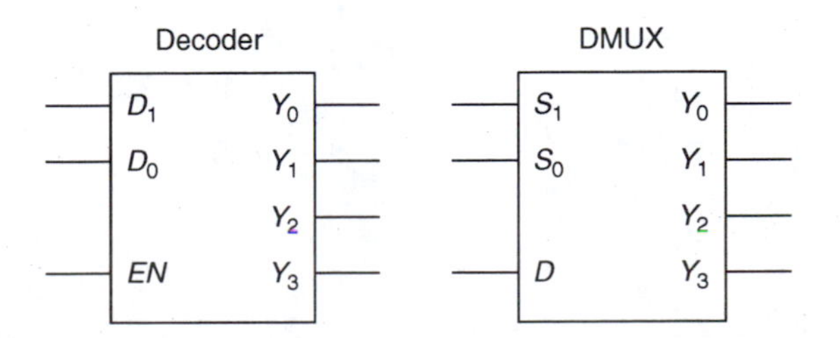

Figure 15.1 Decoder and Demultiplexer Symbols

Figure 15.1 shows the symbols for a 2-to-4 decoder with active-HIGH enable and for a 1-to-4 demultiplexer. Data to be demultiplexed is applied to the same input that is used as the enable input of the decoder circuit. In the demultiplexer, this input is labeled at D for "data." The selected channel is chosen by a binary combination of select inputs S_1 S_0. When a channel is selected, it follows the data on the D input. When it is not selected, it is held LOW.

One way to describe the operation of the DMUX is by a truth table, shown in Table 15.1. We can use this table to describe the operation of the DMUX in VHDL.

Table 15.1 Truth Table for a 1-to-4 Demultiplexer

D	S_1	S_0	Y_0	Y_1	Y_2	Y_3
0	X	X	0	0	0	0
1	0	0	1	0	0	0
1	0	1	0	1	0	0
1	1	0	0	0	1	0
1	1	1	0	0	0	1

Procedure

Design Entry and Simulation

1. Create a new folder for the demultiplexer design and call it ***drive:*\qdesigns\labs\lab15\dmux4**

2. Use the Quartus II Editor to create a Block Diagram file for a 1-to-4 demultiplexer and save it as ***drive:*\qdesigns\labs\lab15\dmux4\dmux4.bdf.** Create a project based on the design and compile.

3. Assign input and output pins, and then download the design to a protoboard and test it.

4. Write a set of simulation criteria that tests the correctness of the demultiplexer design. Use the Quartus II Waveform Editor to create a simulation of the design based on your criteria. Show the simulation criteria and the simulation to your instructor.

Simulation Criteria

Instructor's Initials ______________

Lab 16

MUX and Decoder Lab

Name ______________________ Class ____________ Date ____________

Objectives Upon completion of this laboratory exercise, you should be able to:

- Enter a circuit consisting of several TTL-equivalent components in Quartus II as a Block Diagram File.
- Test the circuit on a CPLD board.

Reference Ken Reid and Robert Dueck, *Introduction to Digital Electronics*
Chapter 5 Combinational Logic Functions

Equipment Required
Altera UP-1/UP-2 circuit board with ByteBlaster download cable or
RSR PLDT-2 Circuit Board with straight-through parallel port and cable
DeVry eSOC board with USB cable
Quartus II Web Edition Software
AC Adapter, minimum output: 7 VDC, 250 mA DC
Anti-static wrist strap
#22 solid-core wire
Wire strippers

Experimental Notes

Quartus II has a number of components that perform functions equivalent to standard TTL 74-series components. In this lab exercise, we will examine the operation of several decoder and multiplexer components to see how they can be used in a small system that selects one of two 4-bit data channels and directs it to one of two seven-segments displays.

Procedure

1. Enter one of the following diagrams using the Quartus II Block Editor.

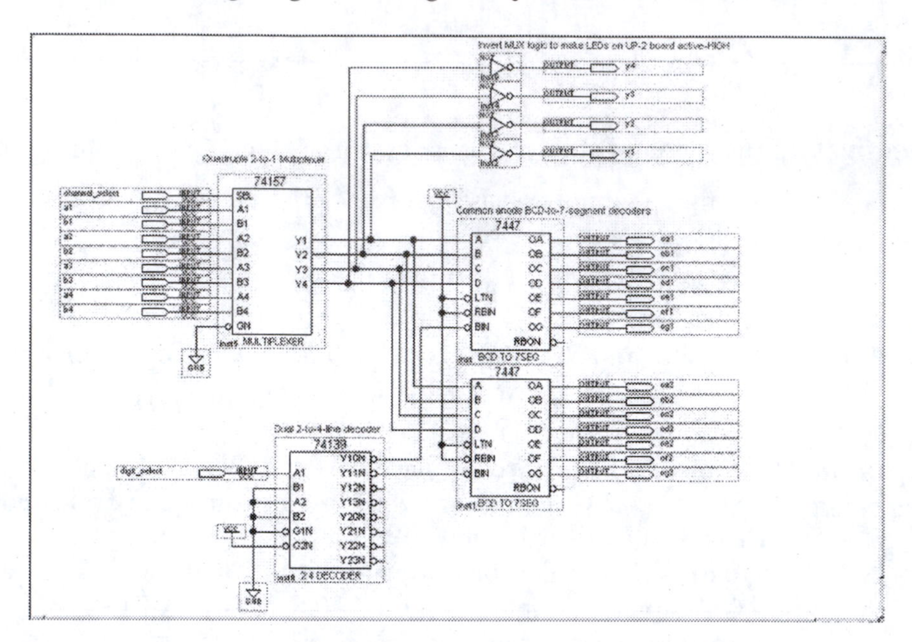

Figure 16.1a Multiplexed Display Digits (for UP-2 or DeVry eSOC Board)

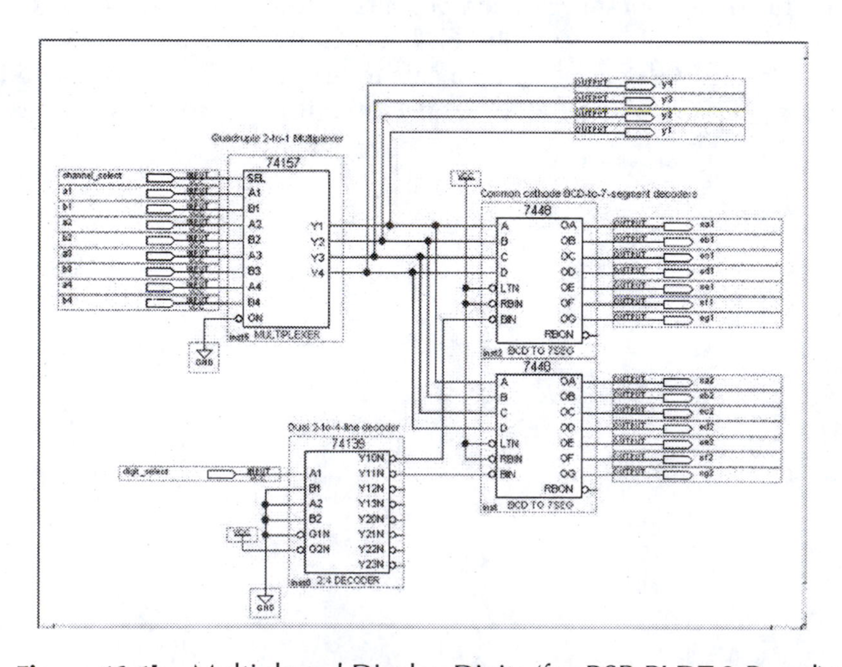

Figure 16.1b Multiplexed Display Digits (for RSR PLDT-2 Board)

The circuit performs the following functions:

- Applies two 4-bit numbers, a[4..1] and b[4..1] to a multiplexer, which sends one of them to a 4-bit output, y[4..1], depending on the status of a channel select input.
- Uses two outputs or a 2-to-4-line decoder to select one of two BCD-to-seven-segment decoders, depending on the status of a digit select input.

2. Assign pin numbers, according to the table below and compile the circuit. Demonstrate the circuit operation to your instructor.

Instructor's Initials ____________

Table 16.1 EPM7128LC84-7 Pin Assignments Altera UP-2 Board

Seven-Segment Digits			
Function	**Pin**	**Function**	**Pin**
oa1	58	oa2	69
ob1	60	ob2	70
oc1	61	oc2	73
od1	63	od2	74
oe1	64	oe2	76
of1	65	of2	75
og1	67	og2	77

DIP Switches			
Function	**Pin**	**Function**	**Pin**
a4	34	digit_select	28
a3	33	channel_select	29
a2	36		
a1	35		
b4	37		
b3	40		
b2	39		
b1	41		

LED Outputs			
Function	**Pin**	**Function**	**Pin**
y4	49		
y3	50		
y2	51		
y1	52		

Table 16.2 EPM7128LC84-7 Pin Assignments RSR PLDT-2 Board

Seven-Segment Digits			
Function	**Pin**	**Function**	**Pin**
oa1	58	oa2	69
ob1	60	ob2	70
oc1	61	oc2	73
od1	63	od2	74
oe1	64	oe2	76
of1	65	of2	75
og1	67	og2	77

DIP Switches			
Function	**Pin**	**Function**	**Pin**
a4	34	digit_select	28
a3	33	channel_select	29
a2	36		
a1	35		
b4	37		
b3	40		
b2	39		
b1	41		

LED Outputs			
Function	**Pin**	**Function**	**Pin**
y4	49		
y3	50		
y2	51		
y1	52		

Table 16.3 EP2C8QZ08C8N Pin Assignments DeVry eSOC Board

Seven-Segment Digits			
Function	**Pin**	**Function**	**Pin**
oa1	137	oa2	169
ob1	138	ob2	170
oc1	139	oc2	171
od1	128	od2	173
oe1	80	oe2	175
of1	81	of2	133
og1	84	og2	135

DIP Switches/Pushbuttons			
Function	**Pin**	**Function**	**Pin**
a4	144	digit_select	131
a3	142	channel_select	130
a2	141		
a1	168		
b4	163		
b3	162		
b2	161		
b1	160		

LED Outputs			
Function	**Pin**	**Function**	**Pin**
y4	112		
y3	113		
y2	114		
y1	115		

Obtain a copy of the datasheets for the following components and add them to your electronic datasheet library:

a. 74LS47/48 BCD-to-seven-segment decoder

b. 74LS139 Dual 2-line-to-4-line decoder

c. 74LS157 Quadruple 2-to-1 multiplexer

3. Refer to the datasheets collected in the previous question and to Figure 16.1 in this lab handout. Answer the following questions:

a. Assume data inputs to the 74157 MUX are as follows: $A_4A_3A_2A_1 = 1001$; $B_4B_3B_2B_1 = 0101$. Write the output value when SEL = 0 and SEL = 1.

i. SEL = 0; $Y_4Y_3Y_2Y_1$ = ____________________

ii. SEL = 1; $Y_4Y_3Y_2Y_1$ = ____________________

b. Complete the truth table for the 74139 decoder, as configured in Figure 16.1.

B_1	A_1	$Y_{10}N$	$Y_{11}N$
0	0		
0	1		

c. Describe the output of the 7448 decoder when its Blanking Input (BIN) is LOW. If a common-cathode seven-segment display was connected to the decoder, what would it display under this condition? What would it display if BIN = 1 and inputs DCBA = 0101?

d. Use the previous answers to complete the following table. (For the last two columns, state the number displayed on the output seven-segment digit, if any. Also indicate if the digit display is blank for a given input condition.)

Channel Select	Digit Select	$A_4A_3A_2A_1$	$B_4B_3B_2B_1$	$Y_4Y_3Y_2Y_1$	Digit 1 Display	Digit 2 Display
0	0	1001	0101			
0	1	1001	0101			
1	0	1001	0101			
1	1	1001	0101			

Lab 17

Magnitude Comparators

Name ______________________________ Class ____________ Date ____________

Objectives Upon completion of this laboratory exercise, you should be able to:

- Create a 3-bit magnitude comparator as a Quartus II Block Diagram File.
- Enter test circuits for the magnitude comparators in the Quartus II graphic editor.

Reference Ken Reid and Robert Dueck, *Introduction to Digital Electronics*
Chapter 5: Combinational Logic Functions

Equipment Required CPLD Trainer:

Altera UP-2 circuit board with ByteBlaster download cable, or
DeVry eSOC board with USB cable, or
RSR PLDT-2 circuit board with straight-through parallel port cable, or
equivalent CPLD trainer board with Altera EPM7128S CPLD

Quartus II Web Edition software
AC adapter, minimum output: 7 VDC, 250 mA DC
Anti-static wrist strap
#22 solid-core wire
Wire strippers

Experimental Notes

In this lab, we will examine the functions of a **magnitude comparator.** This circuit can be implemented in a Block Diagram File. The circuit becomes very much more complex with each additional input bit, so it is often easiest to use an existing magnitude comparator device, rather than its specific Boolean equations.

Magnitude Comparator

A magnitude comparator accepts two binary numbers of equal width and generates outputs that indicate whether the numbers are equal, and if not, which is greater. Figure 17.1 shows the operation of a magnitude comparator that compares two 3-bit numbers, A and B, and determines whether $A = B$, $A > B$, or $A < B$. A true value for any of these conditions makes the appropriate output HIGH.

Figure 17.1a shows $A_2A_1A_0 = 000$ and $B_2B_1B_0 = 001$. Since $0 < 1$, the ALTB (A Less Than B) output goes HIGH and the other two outputs remain LOW. Figure 17.1b indicates that $A_2A_1A_0 = 001$ and $B_2B_1B_0 = 001$. Since $1 = 1$, the *AEQB* (*A* EQuals *B*) output goes HIGH and the other two outputs are LOW. In Figure 17.1c, $A_2A_1A_0 = 101$ and $B_2B_1B_0 = 001$. Since $5 > 1$, the *AGTB* (*A* Greater Than *B*) output goes HIGH and the other two outputs are LOW. Logically, only one output can be HIGH at any time, since *A* can only be less than, equal to, or greater than *B*, but not more than one of these simultaneously.

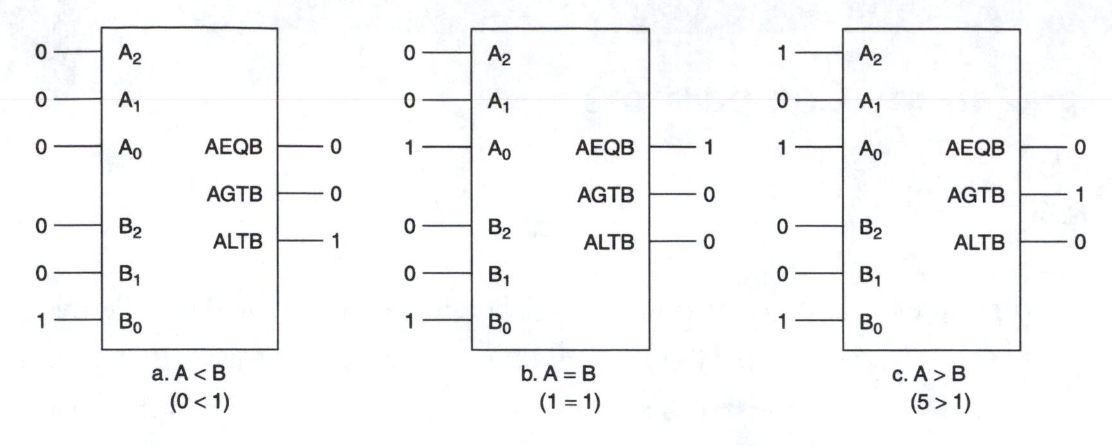

Figure 17.1 3-Bit Magnitude Comparator

Procedure

Magnitude Comparator (Graphic Design File)

1. Figure 17.2 shows the logic diagram of a circuit that implements the 3-bit magnitude comparator of Figure 17.1. Write the Boolean equations for this circuit.

   AEQB = ______________________________

   AGTB = ______________________________

   ALTB = ______________________________

2. Use the Quartus II Block Editor to create the circuit of Figure 17.2. Save the file as ***drive:*\qdesigns\labs\lab17\3bit_cmp\3bit_cmp.bdf.** Assign the device as EPM7128SLC84-7 or EPM7128SLC84-15, depending on which device is in your CPLD board. Compile the design. (The simulation file may contain glitches, or noise spikes, that are generated due to propagation delays in the circuit.)

3. Write a set of simulation criteria to test the correctness of this design.

Simulation Criteria

Instructor's Initials ______________

Use the criteria you wrote to create a Quartus II simulation. Show the simulation to your instructor.

Instructor's Initials ______________

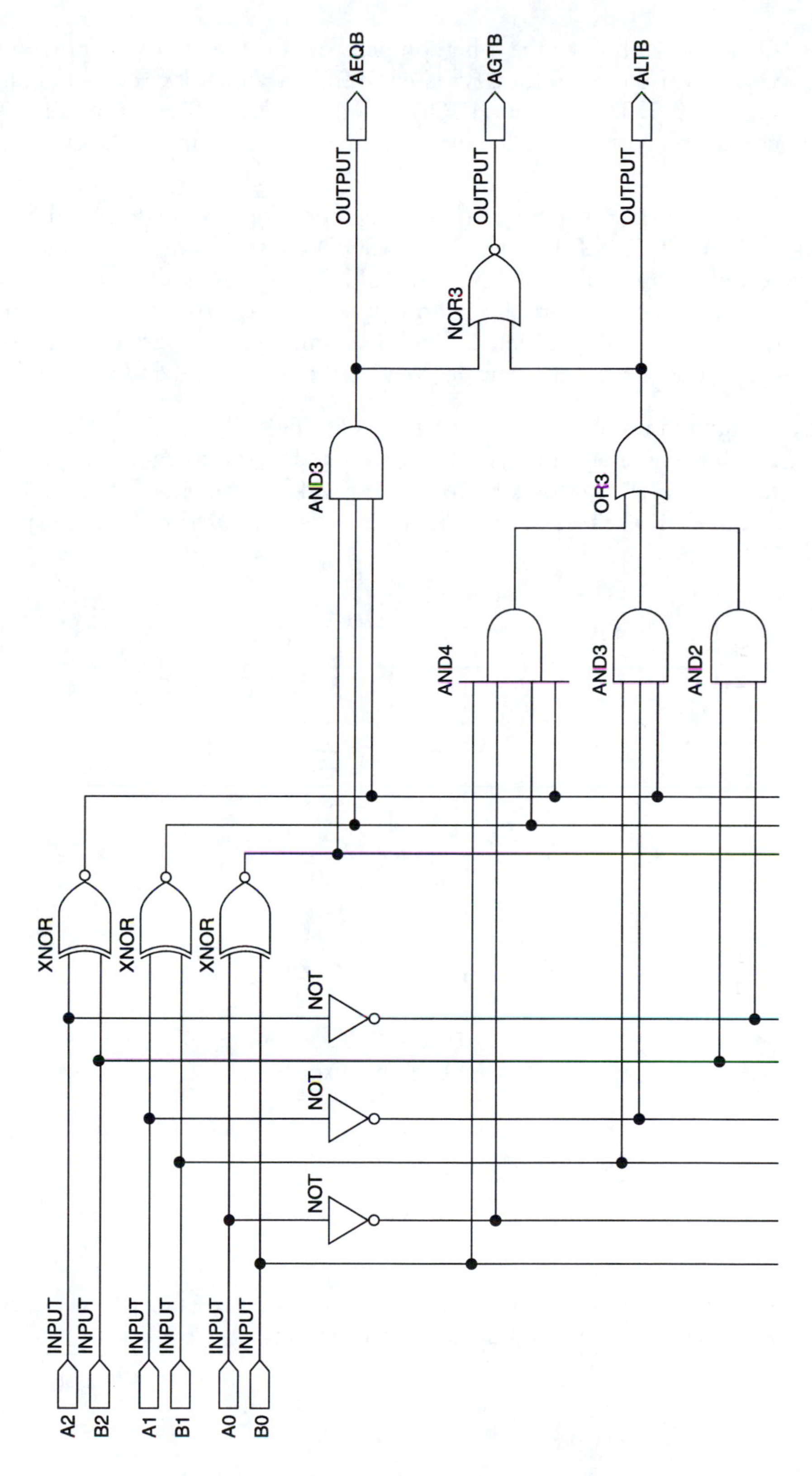

Figure 17.2 3-Bit Magnitude Comparator

4. Create a symbol for the 3-bit comparator you entered in the previous step. Use the Quartus II Block Editor to create the comparator test circuit in Figure 17.4 or Figure 17.5. The outputs *AEQB, AGTB,* and *ALTB* turn on an LED for a true condition. Note that only one LED can be on at a time if the circuit is functioning correctly.

   The components **hi_lo_ca** (common anode for Altera UP-2) and **hi_lo_cc** (common cathode for RSR PLDT-2 and DeVry eSOC) are special decoders that generate an "L" on the seven-segment display if $A < B$, an "H" if $A > B$, and three parallel bars if $A = B$. The segment patterns are shown in Figure 17.3. Any other condition (if it existed) would cause a blank display. Create a Block Diagram File for a circuit that performs this function, create a component (**hi_lo_ca** or **hi_lo_cc**), and add it to the test circuit.

5. Assign pin numbers to the design, according to Table 17.1 or Table 17.2 at the end of this lab exercise. **Outputs[1..8]** correspond to seven-segment digit 2 (**a2** through **dp2** in Table 17.1). **Outputs[9..16]** correspond to seven-segment digit 1 (**a1** through **dp1** in Table 17.1 or Table 17.2), which is disabled. Unused LEDs are also disabled.

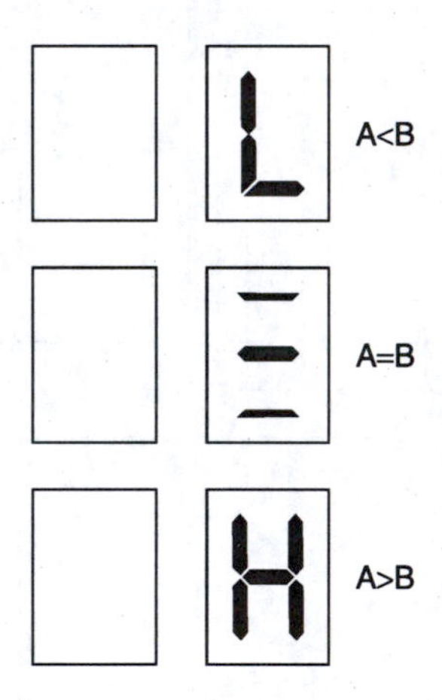

Figure 17.3 Seven-Segment Display Output Generated by Hi_Lo_CC and Hi_Lo_CA

6. Compile the test circuit design and download it to your CPLD board. Demonstrate the operation of the circuit to your instructor.

Instructor's Initials ____________

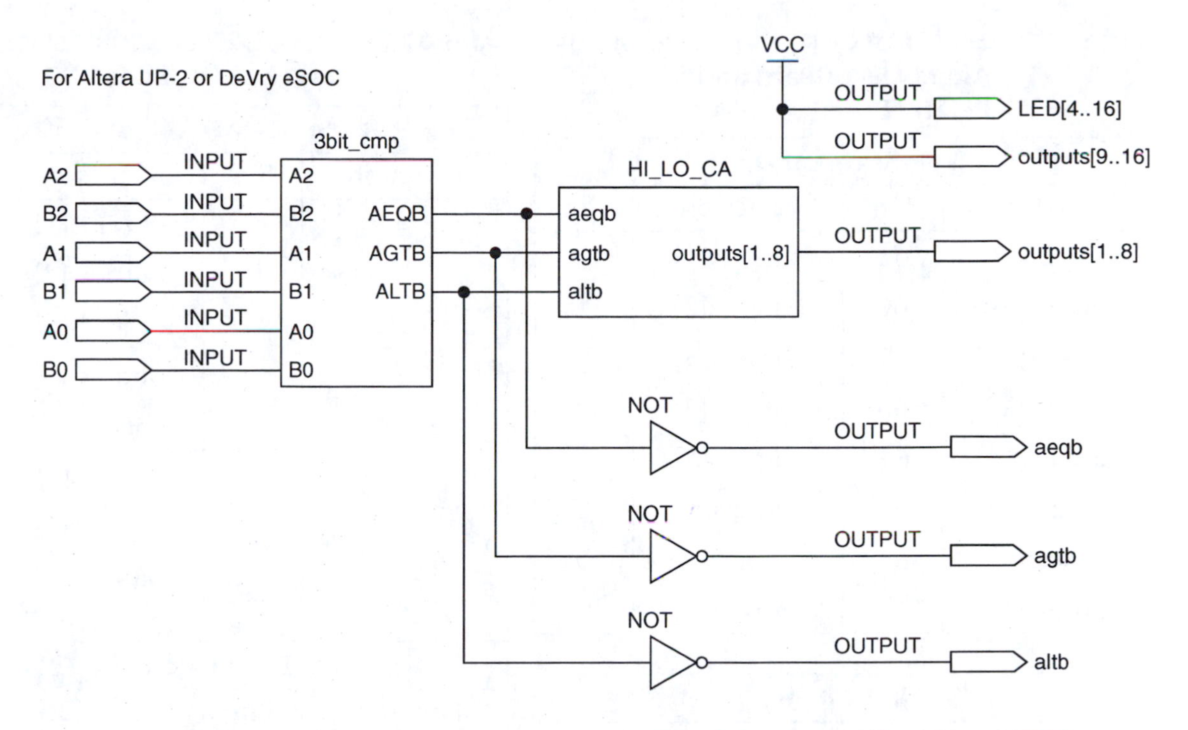

Figure 17.4 3-Bit Comparator Test Circuit (Block Diagram File, Altera UP-2)

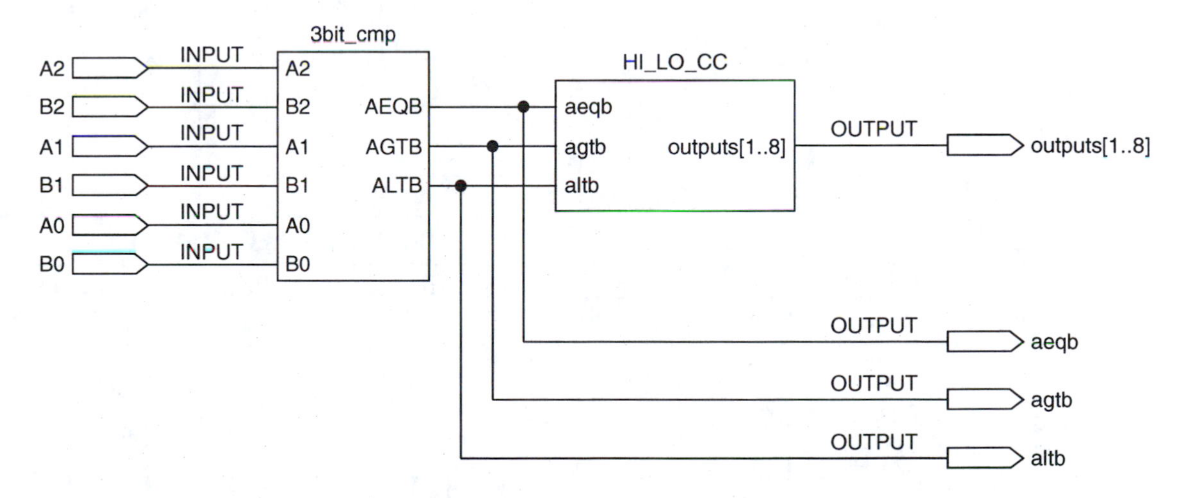

Figure 17.5 3-Bit Comparator Test Circuit (Block Diagram File, RSR PLDT-2 or DeVry eSOC)

Table 17.1 EPM7128LC84-7 Pin Assignments Altera UP-2 Board and RSR PLDT-2 Board

Seven-Segment Digits			
Function	**Pin**	**Function**	**Pin**
outputs[9]	58	outputs[1]	69
outputs[10]	60	outputs[2]	70
outputs[11]	61	outputs[3]	73
outputs[12]	63	outputs[4]	74
outputs[13]	64	outputs[5]	76
outputs[14]	65	outputs[6]	75
outputs[15]	67	outputs[7]	77
outputs[16]	68	outputs[8]	79

DIP Switches			
Function	**Pin**	**Function**	**Pin**
A[2]	34		
A[1]	33		
A[0]	36		
	35		
B[2]	37		
B[1]	40		
B[0]	39		
	41		

LED Outputs			
Function	**Pin**	**Function**	**Pin**
agtb	44	LED [9]	80
aeqb	45	LED [10]	81
altb	46	LED [11]	4
LED [4]	48	LED [12]	5
LED [5]	49	LED [13]	6
LED [6]	50	LED [14]	8
LED [7]	51	LED [15]	9
LED [8]	52	LED [16]	10

Table 17.2 EP2C8Q208L8N Pin Assignments DeVry eSOC Board

Seven-Segment Digits/LEDs			
Function	**Pin**	**Function**	**Pin**
		outputs[1]	137
		outputs[2]	138
		outputs[3]	139
		outputs[4]	128
		outputs[5]	80
		outputs[6]	81
		outputs[7]	84
		outputs[8]	173

DIP Switches			
Function	**Pin**	**Function**	**Pin**
A[2]	160		
A[1]	161		
A[0]	162		
B[2]	141		
B[1]	142		
B[0]	144		

LED Outputs			
Function	**Pin**	**Function**	**Pin**
agtb	115	LED [9]	101
aeqb	114	LED [10]	99
altb	113	LED [11]	97
LED [4]	112	LED [12]	96
LED [5]	110	LED [13]	95
LED [6]	102	LED [14]	92
LED [7]	151	LED [15]	118
LED [8]	150	LED [16]	117

Lab 18

Parity Generators and Checkers

Name ______________________ Class ____________ Date ____________

Objectives Upon completion of this laboratory exercise, you should be able to:

- Implement a 5-bit EVEN parity generator.
- Write a set of simulation criteria to verify the correctness of the parity generator design.
- Create a Quartus II simulation based on your simulation criteria.
- Modify the 5-bit parity generator to implement a 5-bit parity checker.
- Create a test circuit in the Quartus II Block Editor to test the parity generator and checker.

Reference Ken Reid and Robert Dueck, *Introduction to Digital Electronics*

Chapter 5: Combinational Logic Functions

Equipment Required CPLD Trainer:

Altera UP-2 circuit board with ByteBlaster download cable, or
DeVry eSOC board with USB cable, or
RSR PLDT-2 circuit board with straight-through parallel port cable, or
equivalent CPLD trainer board with Altera EPM7128S CPLD
Quartus II Web Edition software
AC adapter, minimum output: 7 VDC, 250 mA DC
Anti-static wrist strap
#22 solid-core wire
Wire strippers

Experimental Notes

Parity Generator

A parity generator can be implemented using a series of XOR functions. For example, a 5-bit EVEN parity generator can be implemented with a design entity having the structure shown in Figure 18.1. In general, each portion of the parity circuit is described by the equation $p(i) = d(i) \oplus p(i-1)$.

Parity Checker

A parity checker compares a parity bit from a parity generator with a parity bit created from the original data applied to the generator. A parity generator can be modified to create a parity checker. Figure 18.2 shows the circuit. In addition to the Exclusive OR equation given for the parity generator, one more equation is required: $p(0) = d(0) \oplus p_{in}$.

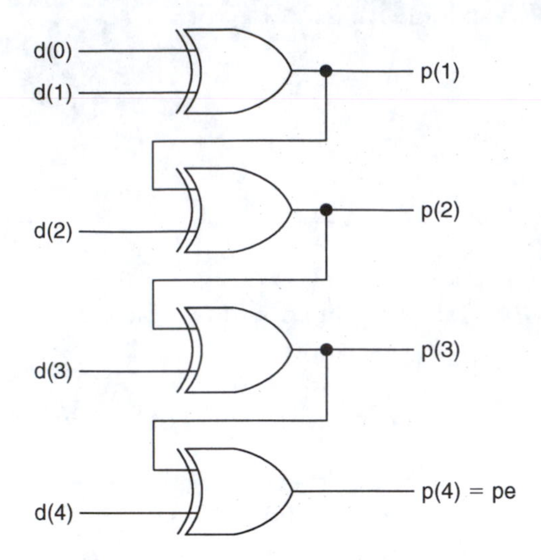

Figure 18.1 5-Bit Parity Generator

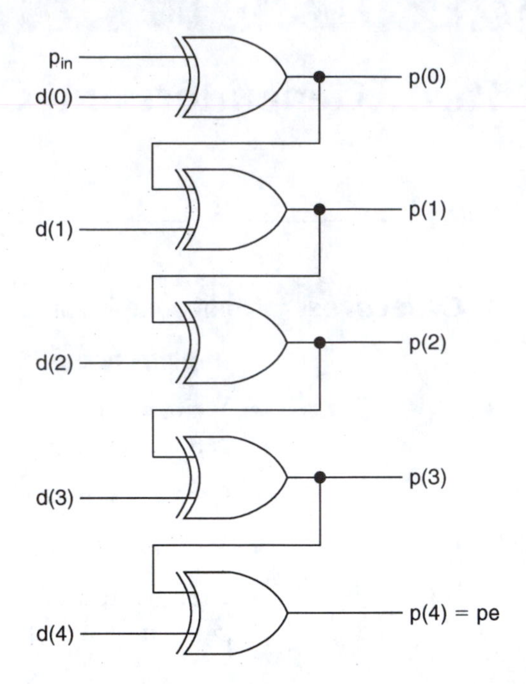

Figure 18.2 5-Bit Parity Checker

Procedure

Design Entry and Simulation (Parity Generator)

1. Enter the circuit shown in Figure 18.1 in a Block Diagram File.
2. Save the .bdf file as ***drive:*\qdesigns\labs\lab18\parity_gen5\parity_gen5.bdf.** Use the file to create a new project.
3. Write a set of simulation criteria that test the correctness of the design. Use these criteria to create a simulation in Quartus II. Show the simulation to your instructor.

Simulation Criteria

Instructor's Initials ______________

Parity Test Circuit

Figure 18.3 shows a normal setup for a 5-bit parity EVEN generator and checker. The same data bits are applied to both the parity generator and checker. The generator determines the EVEN parity bit and applies it to the parity checker, which compares it to a new parity bit determined from the original input data. Separate outputs in the circuit monitor the original parity bit, *PE_GEN*, and the output of the parity checker, P_E.

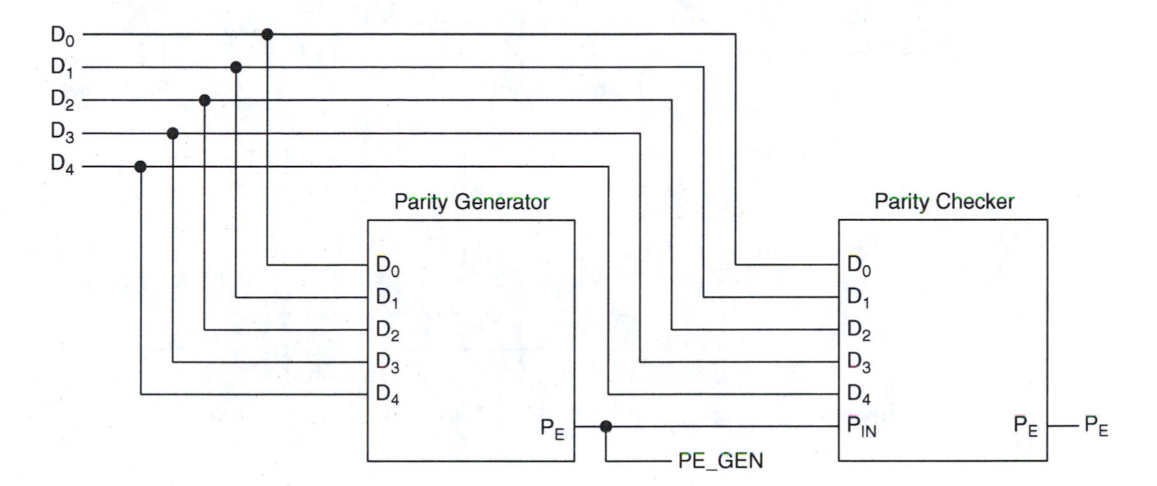

Figure 18.3 Normal Setup for Parity Generation and Checking

This circuit is not suitable for testing how the parity checker reacts to parity errors. The problem is that if the parity checker and parity generator are both working, the P_E light never comes on. We need a way to artificially introduce errors into the system so that we can properly test the response of the parity checker.

Figure 18.4 shows one possibility. Two multiplexers are inserted in the lines connecting D_0 and D_1 to the parity checker. The multiplexers can send normal data or error data to the parity checker. When *Error Select* 0 = 0, parity bit D_0 is the same at both the generator and the checker. When *Error Select* 0 = 1, the parity checker bit D_0 is determined by the status of the logic switch at *Error* 0. This may or may not be the same as the data at parity generator bit D_0. If they are different, an error should be detected by the parity checker. The same arrangement also applies for the multiplexer for D_1.

1. Create a folder called ***drive:*\qdesigns\labs\lab18\parity5_test**. Use the Quartus II New Project Wizard (**File** menu) to create a new project in the folder.

2. Copy the file **parity5_gen.bdf** to the new folder and use it to create a symbol. Add the file to the active project.

3. Modify the 5-bit parity generator to make a 5-bit parity checker. Save the file as **parity_chk5.bdf.** (Make sure to add the file to the project.) Create a symbol for the file.

4. Create a circuit for a 2-to-1 multiplexer and save it in the project folder, making sure to add the file to the project. (The example in this laboratory exercise uses a multiplexer created from a Block Diagram File called **2to1mux.bdf.**) Create a symbol for the multiplexer.

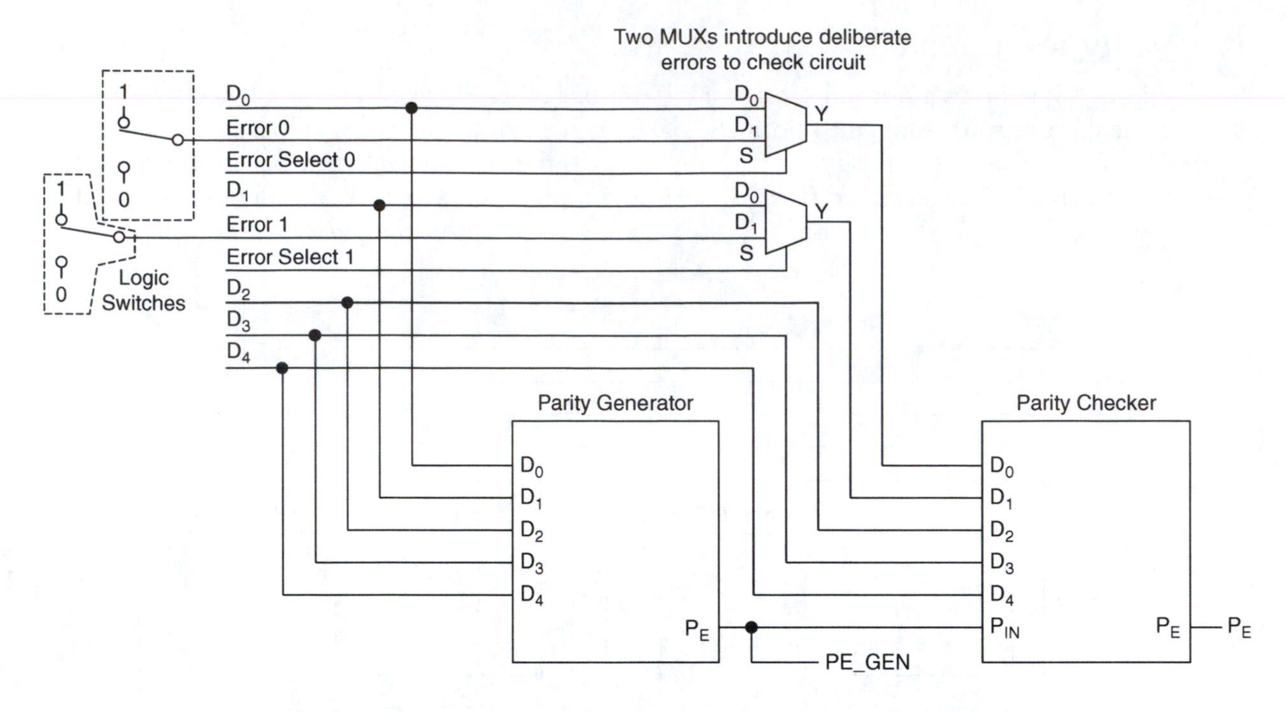

Figure 18.4 Test Circuit for Parity Generation and Checking

5. Create a Block Diagram File, as in Figure 18.5 (Altera UP-2 or DeVry eSOC) or Figure 18.6 (RSR PLDT-2). Save the file as:

   ***drive*:\qdesigns\labs\lab18\parity 5_test\parity5_test.bdf.**

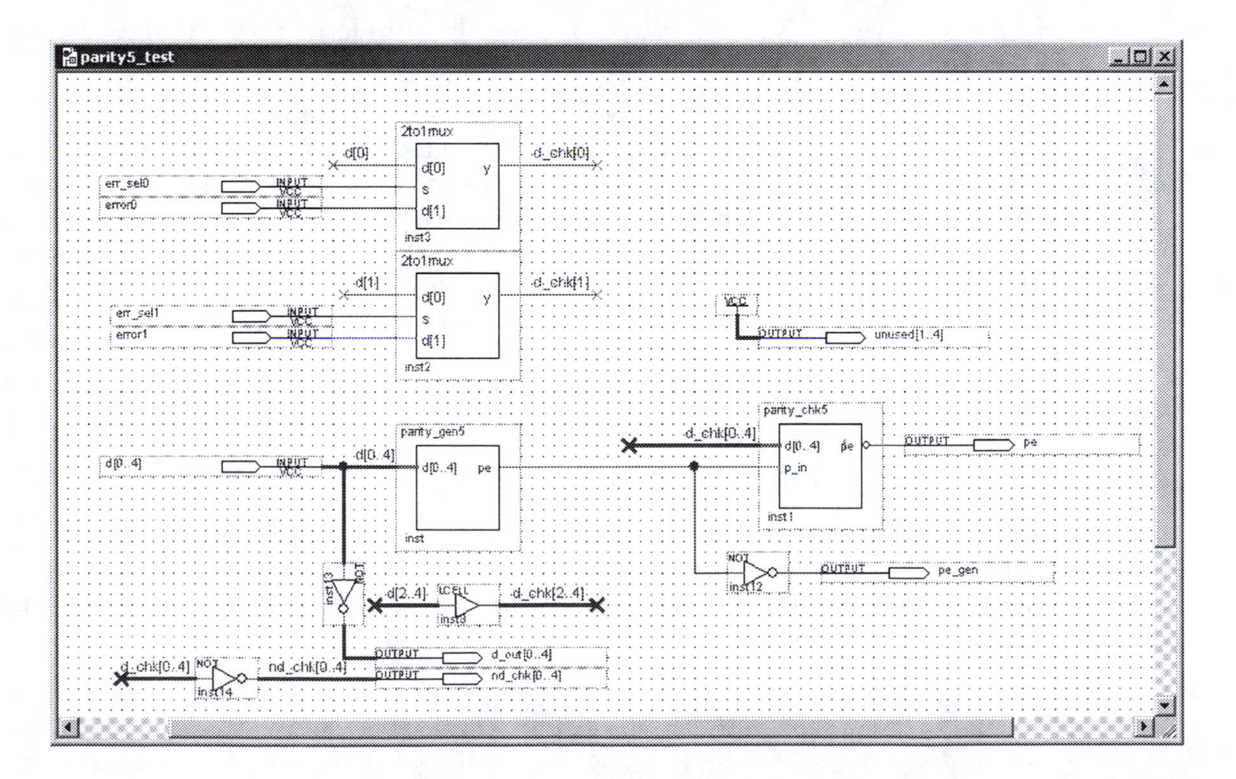

Figure 18.5 Parity Test Circuit (Altera UP-2)

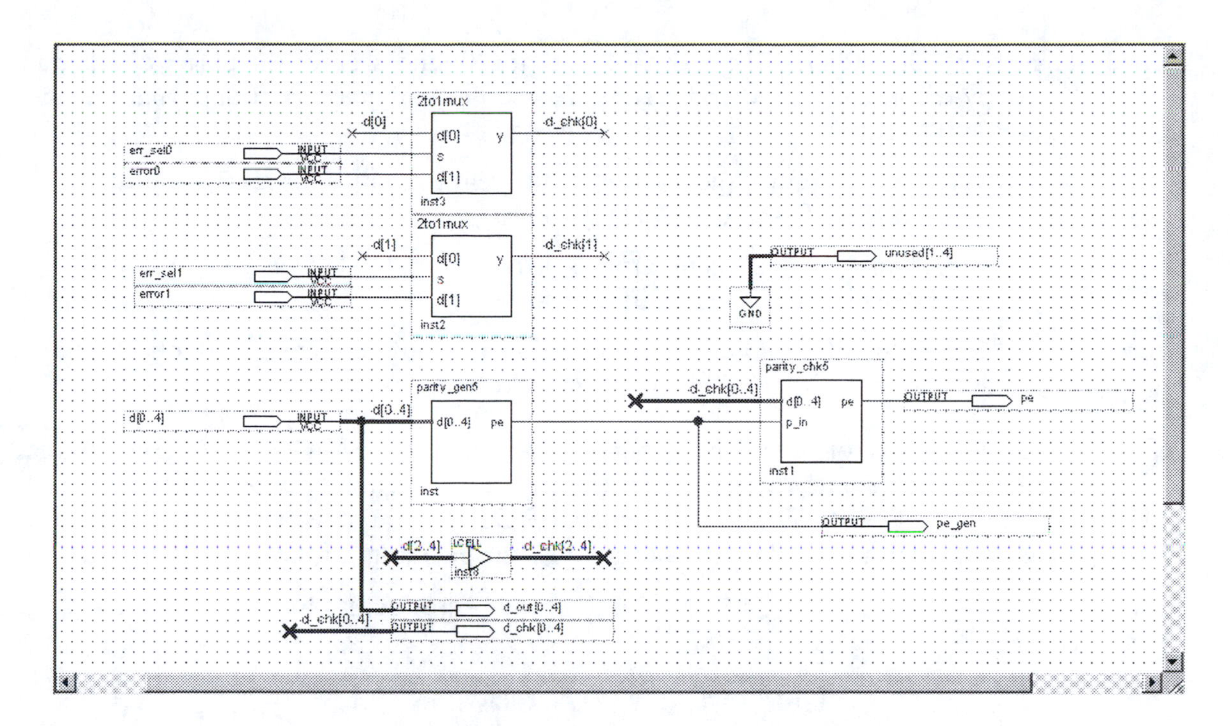

Figure 18.6 Parity Test Circuit (RSR PLDT-2 and DeVry eSOC)

Note The **pe** output in the parity checker in Figure 18.5 is inverted. (This can be done by highlighting the block, then right-clicking and selecting **Properties.** In the **Ports** tab, select **pe,** and click **All** in the **Inversion** box.) This and the other NOT gates are required for the Altera UP-2 and DeVry eSOC boards to make the board LEDs active-HIGH.

Note Both Figure 18.5 and Figure 18.6 make use of an **LCELL buffer.** This component, shown in detail in Figure 18.7, allows us to direct three of the five elements of the input bus **d[0..4]** to the parity checker and the other two elements to the multiplexers. The component can be inserted in the Block Diagram File by typing **LCELL** in the **Symbol** dialog box.

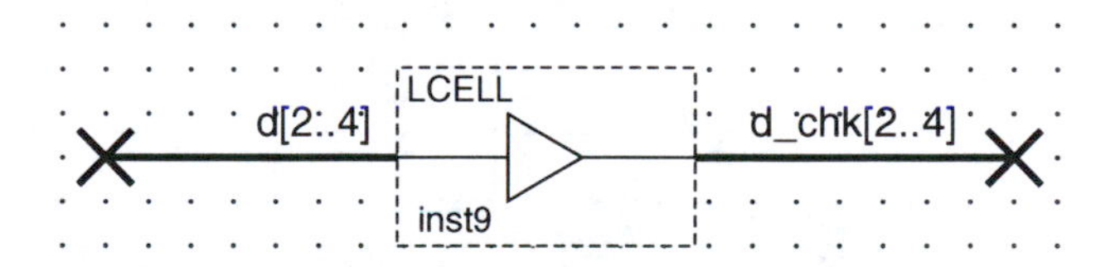

Figure 18.7 LCELL buffer

6. Compile the parity generator/checker test file. Add pin numbers, as listed in Tables 18.1 through 18.3 at the end of this laboratory exercise. Compile the file again and download it to the CPLD test board.

7. Set **err_sel0** and **err_sel1** both to 0 so that the input data is the same for both the parity generator and parity checker. Test some combinations of data input. For each combination, note whether the number of 1s at the data inputs is ODD or EVEN. Which condition causes the **pe_gen** light to come on? The **pe** light should not come on for any combination. Why? Show your instructor the operation of the circuit and explain the behavior of **pe** and **pe_gen.**

Instructor's Initials ________________

8. Set **err_sel0** and **err_sel1** both to 1 so that **d[0]** and **d[1]** for the parity checker are now set by the inputs **error0** and **error1.** Test some combinations of data that include the following cases:

   - **"Error"on d[0]: d[0]** (generator) is different from **d[0]** (checker), but **d[1]** (generator) is the same as **d[1]** (checker)
   - **"Error"on d[1]: d[1]** (generator) is different from **d[1]** (checker), but **d[0]** (generator) is the same as **d[0]** (checker)
   - **"Errors"on d[0] and d[1]: d[0]** (generator) is different from **d[0]** (checker) and **d[1]** (generator) is different from **d[1]** (checker)

How is an error on one bit shown on the **pe** output? How are errors on two bits shown on the **pe** output? What does this say about the ability of a parity checker to detect an ODD or EVEN number of errors? Demonstrate the operation of the circuit to your instructor and explain the behavior of the **pe** output of the parity checker.

Instructor's Initials ____________________

Table 18.1 EPM7128LC84-7 Pin Assignments Altera UP-2 Board

DIP Switches			
Function	**Pin**	**Function**	**Pin**
d[0]	34	err_sel0	28
d[1]	33	error0	29
d[2]	36		30
d[3]	35		31
d[4]	37		57
	40		55
	39	err_sel1	56
	41	error1	54

LED Outputs			
Function	**Pin**	**Function**	**Pin**
d_out[0]	44	nd_chk[0]	80
d_out[1]	45	nd_chk[1]	81
d_out[2]	46	nd_chk[2]	4
d_out[3]	48	nd_chk[3]	5
d_out[4]	49	nd_chk[4]	6
unused[1]	50	unused[3]	8
unused[2]	51	unused[4]	9
pe_gen	52	pe	10

Table 18.2 EPM7128LC84-7 Pin Assignments RSR PLDT-2 Board

DIP Switches			
Function	**Pin**	**Function**	**Pin**
d[0]	34	err_sel0	28
d[1]	33	error0	29
d[2]	36		30
d[3]	35		31
d[4]	37		57
	40		55
	39	err_sel1	56
	41	error1	54

LED Outputs			
Function	**Pin**	**Function**	**Pin**
d_out[0]	44	d_chk[0]	80
d_out[1]	45	d_chk[1]	81
d_out[2]	46	d_chk[2]	4
d_out[3]	48	d_chk[3]	5
d_out[4]	49	d_chk[4]	6
unused[1]	50	unused[3]	8
unused[2]	51	unused[4]	9
pe_gen	52	pe	10

Table 18.3 EP2C8Q208C8N Pin Assignments DeVry eSOC Board

DIP Switches/Pushbuttons			
Function	**Pin**	**Function**	**Pin**
d[0]	160	err_sel0	142
d[1]	161	error0	144
d[2]	162		
d[3]	163		
d[4]	168		
		(pushbutton)	
		err_sel1	131
		error1	130

LED Outputs			
Function	**Pin**	**Function**	**Pin**
d_out[0]	115	nd_chk[0]	101
d_out[1]	114	nd_chk[1]	99
d_out[2]	113	nd_chk[2]	97
d_out[3]	112	nd_chk[3]	96
d_out[4]	110	nd_chk[4]	95
unused[1]	102	unused[3]	92
unused[2]	151	unused[4]	118
pe_gen	150	pe	117

Lab 19

Full Adder and Parallel Binary Adder

Name ______________________ Class ____________ Date ____________

Objectives Upon completion of this laboratory exercise, you should be able to:

- Create and simulate a full adder, assign pins to the design, and test it on a CPLD circuit board.
- Use the full adder as a component in an 8-bit parallel binary adder.
- Create a hierarchical design, including components for full adders and seven-segment decoders.
- Design an overflow detector for use in a two's complement adder/subtractor.

Reference Ken Reid and Robert Dueck, *Introduction to Digital Electronics*
Chapter 6: Digital Arithmetic and Arithmetic Circuits

Equipment Required CPLD Trainer:
Altera UP-2 circuit board with ByteBlaster download cable, or
DeVry eSOC board with USB cable, or
RSR PLDT-2 circuit board with straight-through parallel port cable, or
equivalent CPLD trainer board with Altera EPM7128S CPLD
Quartus II Web Edition software
AC adapter, minimum output: 7 VDC, 250 mA DC
Anti-static wrist strap
#22 solid-core wire
Wire strippers

Experimental Notes

Arithmetic Circuits

Circuits for performing binary arithmetic are based on half adders, which add two bits and produce a sum and carry, and full adders, which also account for a carry added from a less-significant bit. Full adders can be grouped together to make a parallel binary adder, with *n* full adders allowing two *n*-bit numbers to be added, generating an *n*-bit sum and a carry output.

Procedure

Note LEDs on the Altera UP-2 are active-LOW. LEDs on the RSR PLDT-2 and DeVry eSOC are active-HIGH. This must be accounted for in all your designs.

Full Adders

1. The logic diagram for a full adder is shown in Figure 19.1. Use this diagram for a file called *drive:\qdesigns\labs\lab19\full_add\full add.bdf.*

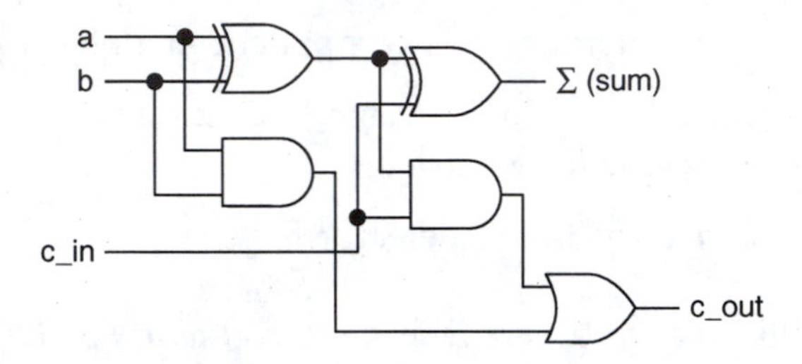

Figure 19.1 Full Adder Circuit

2. Save the full adder design file and use it to create a project in Quartus II. Compile the project.
3. Write a set of simulation criteria to verify the correctness of your design. Use the criteria to create a simulation for the full adder. Show the criteria and simulation to your instructor.

Simulation Criteria

Instructor's Initials 

4. Assign pin numbers as shown in Table 19.1. Compile the project again.

Table 19.1 Pin Assignments for a Full Adder

	Pin Numbers		
Function	**UP-2**	**PLDT-2**	**eSOC**
a	34	34	160
b	33	33	161
c_in	36	36	162
c_out	51	51	151
sum	52	52	150

5. Download the full adder design to your CPLD board. Take the truth table of the full adder to verify its operation. Show the result to your instructor.

Instructor's Initials ____________

Parallel Adder

1. Create a new folder for an 8-bit parallel adder. Use the Quartus II New Project Wizard to make a project for the parallel adder.
2. Create an 8-bit parallel adder using the component **full_add.bdf** from the previous section. Permanently assign the carry input (C_0) to a logic LOW. Use C_0 as the carry input to the first full adder.
3. Calculate the sums shown in Table 19.2 for the 8-bit parallel adder.
4. Use the sums calculated in Table 19.2 to create a simulation of the 8-bit parallel adder you created in step 1 of this section. The sums in the simulation must match those in Table 19.2. (This looks harder than it really is. Figure 19.2 shows the input waveforms on a Quartus II Vector Waveform File. The patterns allows easy testing of the circuit on the board. All you have to do is set all DIP switches LOW, then make them HIGH in sequence, from b[1] to a[8], then go back and make them LOW in the same sequence. This should be very easy to test.)

   Show the simulation to your instructor.

Instructor's Initials ____________

Table 19.2 Sample Sums for an 8-Bit Parallel Adder

A	*B*	Carry	Sum
00000000	00000000		
00000000	00000001		
00000000	00000011		
00000000	00000111		
00000000	00001111		
00000000	00011111		
00000000	00111111		
00000000	01111111		
00000000	11111111		
00000001	11111111		
00000011	11111111		
00000111	11111111		
00001111	11111111		
00011111	11111111		
00111111	11111111		
01111111	11111111		
11111111	11111111		
11111111	11111110		
11111111	11111100		
11111111	11111000		
11111111	11110000		
11111111	11100000		
11111111	11000000		
11111111	10000000		
11111111	00000000		
11111110	00000000		
11111100	00000000		
11111000	00000000		
11110000	00000000		
11100000	00000000		
11000000	00000000		
10000000	00000000		
00000000	00000000		

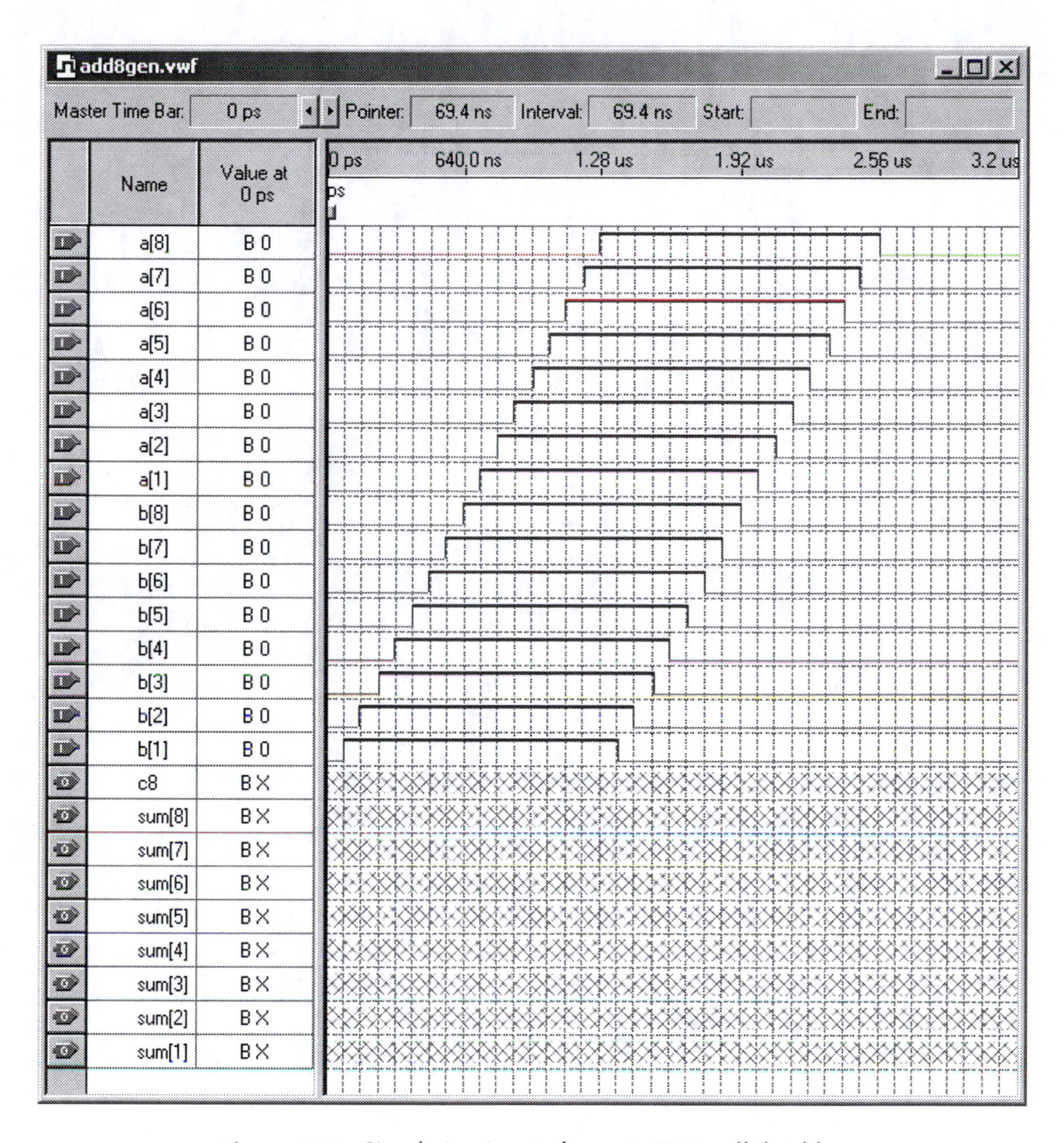

Figure 19.2 Simulation Inputs for an 8-Bit Parallel Adder

5. Assign pins to the 8-bit adder as shown in Table 19.3.

Note This design requires two 8-bit toggle switches; the DeVry eSOC board has only one. The design can't be implemented on this board.

Table 19.3 Pin Assignments for Parallel Adder

	Pin Numbers	
Function	**UP-2**	**PLDT-2**
a[8]	34	34
a[7]	33	33
a[6]	36	36
a[5]	35	35
a[4]	37	37
a[3]	40	40
a[2]	39	39
a[1]	41	41
b[8]	28	28
b[7]	29	29
b[6]	30	30
b[5]	31	31
b[4]	57	57
b[3]	55	55
b[2]	56	56
b[1]	54	54
unused[1]	44	44
unused[2]	45	45
unused[3]	46	46
unused[4]	48	48
unused[5]	49	49
unused[6]	50	50
unused[7]	51	51
c8	52	52
sum[8]	80	80
sum[7]	81	81
sum[6]	4	4
sum[5]	5	5
sum[4]	6	6
sum[3]	8	8
sum[2]	9	9
sum[1]	10	10

6. Compile the file and download the design for the 8-bit parallel adder to your CPLD board. Test the operation of the 8-bit parallel adder by applying the combinations of inputs *A* and *B* listed in Table 19.2. Show the results to your instructor.

Instructor's Initials ____________________

Lab 20

Two's Complement Adders/Subtractor

Name ________________________ Class ____________ Date ____________

Objectives Upon completion of this laboratory exercise, you should be able to:

- Create a hierarchical design, including components for full adders and seven-segment decoders, using the Quartus II Block Editor.
- Use a full adder in an 8-bit two's complement adder/subtractor.
- Design an overflow detector for use in a two's complement adder/subtractor.

Reference Ken Reid and Robert Dueck, *Introduction to Digital Electronics*

Chapter 6: Digital Arithmetic and Arithmetic Circuits

Equipment Required CPLD Trainer:

Altera UP-2 circuit board with ByteBlaster download cable, or
RSR PLDT-2 circuit board with straight-through parallel port cable, or
equivalent CPLD trainer board with Altera EPM7128s CPLD

Quartus II Web Edition software
AC adapter, minimum output: 7 VDC, 250 mA DC
Anti-static wrist strap
#22 solid-core wire
Wire strippers

Note This design requires 16 toggle inputs; it can't be physically implemented on the DeVry eSOC board.

Experimental Notes

A parallel adder, such as the one constructed in Lab 19, can be converted to a two's complement adder/subtractor by including XOR functions on the inputs of one set of operand bits, say input *B*, allowing the operations $A + B$ or $A - B$ to be performed. A control input, *SUB* (for SUBtract), causes the XORs to invert the *B* bits if HIGH, producing the one's complement of *B*. *SUB* will not invert *B* if LOW, transferring *B* to the parallel adder without modification. If *SUB* is also tied to the carry input of the parallel adder, the output is $(A + B + 0 = A + B)$ when $SUB = 0$ and $(A + \overline{B} + 1 = A - B)$ when $SUB = 1$, where $\overline{B}$ is the one's complement of *B* and $(\overline{B} + 1)$ is its two's complement.

Sign-bit overflow occurs when a two's complement sum or difference exceeds the permissible range of numbers for a given bit size. This can be detected by an SOP circuit that compares the operand and result sign bits of a parallel adder, or by an XOR gate that compares carry into and out of the MSB.

Procedure

Note LEDs on the Altera UP-2 and DeVry eSOC are active-LOW. LEDs on the RSR PLDT-2 are active-HIGH. This must be accounted for in all your designs.

Two's Complement Adder/Subtractor

1. Modify the 8-bit adder you created in Lab 19 to make an 8-bit two's complement adder/subtractor. Include an overflow detector that will turn on an LED when the output of the adder subtractor overflows beyond the permissible range of values for an 8-bit signed number. Display the sum outputs on LEDs, as with the previous 8-bit adder, but also add a pair of seven-segment decoders to display the result numerically on the board's seven-segment display.

   Note

   1. If you are using the Altera UP-2 board, recall that the LEDs and numerical displays are active-LOW. Binary sum outputs must be active-LOW to display the binary sum properly on the LEDs, but the internal signals connecting the adder/subtractor sum outputs to the seven-segment decoder inputs must be active-HIGH.
   2. If you are using the RSR PLDT-2 board, recall that the LEDs and numerical displays are active-HIGH. Internal signals for LED outputs and seven-segment decoder inputs can be the same since they are both active-HIGH.

2. Create a simulation based on the test data in Table 20.1. (Refer to Figure 19.2 in Lab 19.) The simulation should contain all combinations in the table with the *SUB* input LOW, then again with the *SUB* input HIGH. The simulation should also account for the overflow output.

3. Assign pins as shown in Tables 20.2 and 20.3 at the end of this lab exercise.

4. Compile the project and download it to the CPLD board. Test the operation of the circuit by applying the *A* and *B* inputs from simulation test data. Show the results to your instructor.

Instructor's Initials ___________________

Table 20.1 Test Data for Adder/Subtractor

A	*B*	Carry	Sum
00000000	00000000		
00000000	00000001		
00000000	00000011		
00000000	00000111		
00000000	00001111		
00000000	00011111		
00000000	00111111		
00000000	01111111		
00000000	11111111		
00000001	11111111		
00000011	11111111		
00000111	11111111		
00001111	11111111		
00011111	11111111		
00111111	11111111		
01111111	11111111		
11111111	11111111		
11111111	11111110		
11111111	11111100		
11111111	11111000		
11111111	11110000		
11111111	11100000		
11111111	11000000		
11111111	10000000		
11111111	00000000		
11111110	00000000		
11111100	00000000		
11111000	00000000		
11110000	00000000		
11100000	00000000		
11000000	00000000		
10000000	00000000		
00000000	00000000		

Table 20.2 EPM7128LC84-7 Pin Assignments Altera UP-2 Board and RSR PLDT-2 Board

Seven-Segment Digits			
Function	**Pin**	**Function**	**Pin**
a1	58	a2	69
b1	60	b2	70
c1	61	c2	73
d1	63	d2	74
e1	64	e2	76
f1	65	f2	75
g1	67	g2	77
dp1	68	dp2	79

DIP Switches			
Function	**Pin**	**Function**	**Pin**
a[8]	34	b[8]	28
a[7]	33	b[7]	29
a[6]	36	b[6]	30
a[5]	35	b[5]	31
a[4]	37	b[4]	57
a[3]	40	b[3]	55
a[2]	39	b[2]	56
a[1]	41	b[1]	54

LED Outputs			
Function	**Pin**	**Function**	**Pin**
unused[1]	44	sum[8]	80
unused[2]	45	sum[7]	81
unused[3]	46	sum[6]	4
unused[4]	48	sum[5]	5
unused[5]	49	sum[4]	6
unused[6]	50	sum[3]	8
overflow	51	sum[2]	9
c8	52	sum[1]	10

Other functions:

sub: Altera UP-2 MAX_PB1 (pin 11) RSR PLDT-2: s5-1 (pin 12)

Lab 21

Combinational Project—Building a Calculator

Name ______________________ Class __________ Date __________

Objectives Upon completion of this laboratory exercise, you should be able to:

- Create each component of a small calculator individually.
- Use the Block Diagram Editor to combine individual components forming a system.
- Simulate the calculator to verify its operation.
- Download and test the calculator on a development board.

Reference Ken Reid and Robert Dueck, *Introduction to Digital Electronics*
Chapter 7: Digital System Application

Equipment Required CPLD Trainer:
Altera UP-2 circuit board with ByteBlaster download cable, or
DeVry eSOC board with USB cable, or
RSR PLDT-2 circuit board with straight-through parallel cable

Experimental Notes

This experiment involves the design of a digital system in the form of a small calculator. This example follows Chapter 7 of Reid & Dueck, *Introduction to Digital Electronics*. Following effective problem solving techniques as specified in the text will ensure a smooth design implementation.

Problem Statement

Design a small calculator that takes two 4-bit binary numbers as inputs and outputs results as shown in Table 21.1 according to the "choice" inputs. The final design should have two 4-bit inputs (*A* and *B*), two "choice" inputs (*C*), and eight outputs (*Z*).

Table 21.1 Description of C Inputs to Calculator Problem

Choice	Function	Output
00	Add	Z = A + B
01	Subtract	Z = A − B (negative numbers in 2's complement form)
10	Multiply	Z = A * B
11	Min/Max	Upper 4 bits of Z = maximum of A or B Lower 4 bits of Z = minimum of A or B

This circuit has two inputs that are each 4-bit binary numbers. The output will be one of four results: first, the output could be $A + B$. Since these are both 4-bit numbers, the highest value we could expect would be $1111 + 1111 = 11110$ binary, which easily fits into the 8 bits allowed for the output. The output could be $A - B$, for which we may need to deal with complementing one of the variables. While there is a possibility that overflow could occur if there are not enough output bits to store our subtraction result, overflow should not occur since we are using 4-bit input values and 8 bits for the output. A third possibility is multiplication. Since we have two 4-bit inputs, the highest value we can expect is $1111 \times 1111 = 11100001$ binary, which fits into our 8-bit output. Finally, we have a requirement to compare A and B, sending the maximum of the two 4-bit inputs to the upper four bits of the output and the minimum of A and B to the lower four bits. What do we do if $A = B$? It doesn't matter in this case if we send A or B to the upper or lower four bits: they would be identical.

We also have inputs that will allow the user to choose their desired function. These inputs may be considered one 2-bit value or two individual inputs. We will consider this a 2-bit input value.

Inputs and Outputs

The following **inputs** are described:

Inputs A and B will form the operands for our calculator.

A_3 A_2 A_1 and A_0: the 4-bit value A

B_3 B_2 B_1 and B_0: the 4-bit value B

We also have a 2-bit input variable called C that allows us to select which operation will be done:

C_1 and C_0: two bits used to select the operation.

Notice that A and B will usually be treated as a 4-bit binary number instead of four individual bits. There is no reason to consider the value of C to be considered a single value rather than two individual bits.

The following **outputs** are given:

Z_7 Z_6 Z_5 Z_4 Z_3 Z_2 Z_1 Z_0

These form the output value from the inputs specified by the user. Note that for addition, subtraction, and multiplication, the output value Z is treated like one 8-bit binary number. For the Max/Min function, it is essentially thought of as two 4-bit numbers.

Block Diagram

The overall block diagram is shown in Figure 21.1.

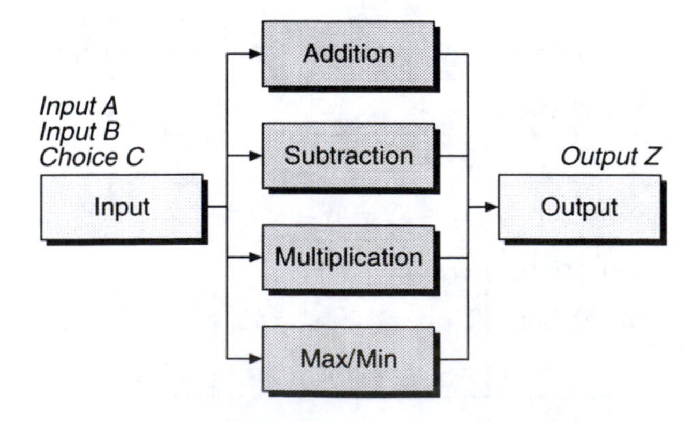

Figure 21.1 Block Diagram for Calculator

Procedure

Input Block

1. Start the Quartus II software, begin with the New Project Wizard and name your design **calculator.** It is recommended that you store your design in the directory ***drive*:\qdesigns\textbook\ch07\calculator,** where drive is the drive letter where you store data files (usually, this is c: for a stand-alone computer system, or another letter such as **f:** or **g:** for a networked PC).

2. Start a new **block diagram file** in Quartus II and save it with the name **calculator.** This will be our overall design when the calculator is finished. In the block diagram file, add the component named **input** a total of 10 times. Change the name of each input to the actual input signal name. Your block diagram file should be similar to Figure 21.2, which shows the **input** components with a small signal line drawn at the end of each input. Signals will be named without using subscripts in the Quartus II software: for example, signal A_2 will named $A2$ throughout our drawing.

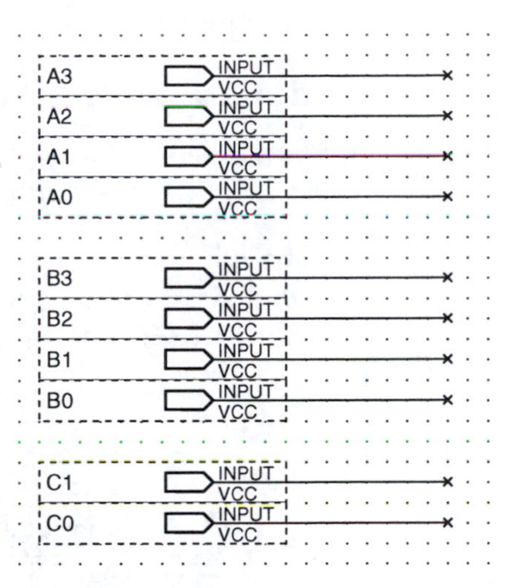

Figure 21.2 Input Block

Addition Block

1. Open a new Block Diagram File: choose **File, New, Block Diagram/Schematic File.** Add a 7483 component, draw inputs to all of the inputs of the 7483, and draw outputs from the outputs. This is shown in Figure 7.9. Save this as **adder.bdf** and create a symbol for this adder component. The 7483 is a TTL part that functions as a 4-bit parallel adder; using this device allows the adder block to be drawn with only one device. This is an effective use of a device that has previously been developed. This solution still does not take the select inputs (C_1 and C_0) into account; these signals do not affect the sum, but there must be additional circuitry within our final calculator to choose the output of the addition block as the output of the calculator. Note that the 7483 symbol has a C_0 input that will be used later for this purpose.

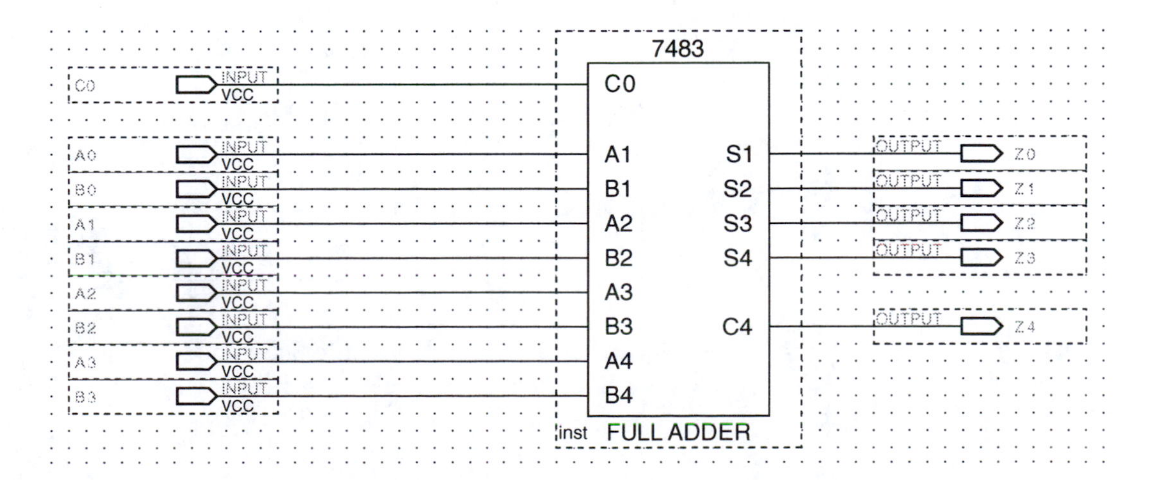

Figure 21.3 Addition Block Using a 7483 Device

2. Create a symbol; choose **File, Create/Update, Create Symbol Files for Current File.** This creates a symbol file we can add to our top level (calculator). Once the symbol is created, you can close the window with the 7483 inside, return to **calculator** and add the **adder** symbol.

Adder/Subtractor Module

While the adder module performs addition, we need a module to perform addition and subtraction. We will modify the **adder** to build the **AddSub** component.

1. The schematic for a 4-bit adder/subtractor is shown in Figure 21.4. Draw this schematic, create a symbol and call it **AddSub.** The next step is to add this component to the overall file, **calculator.** This is shown in Figure 21.5.

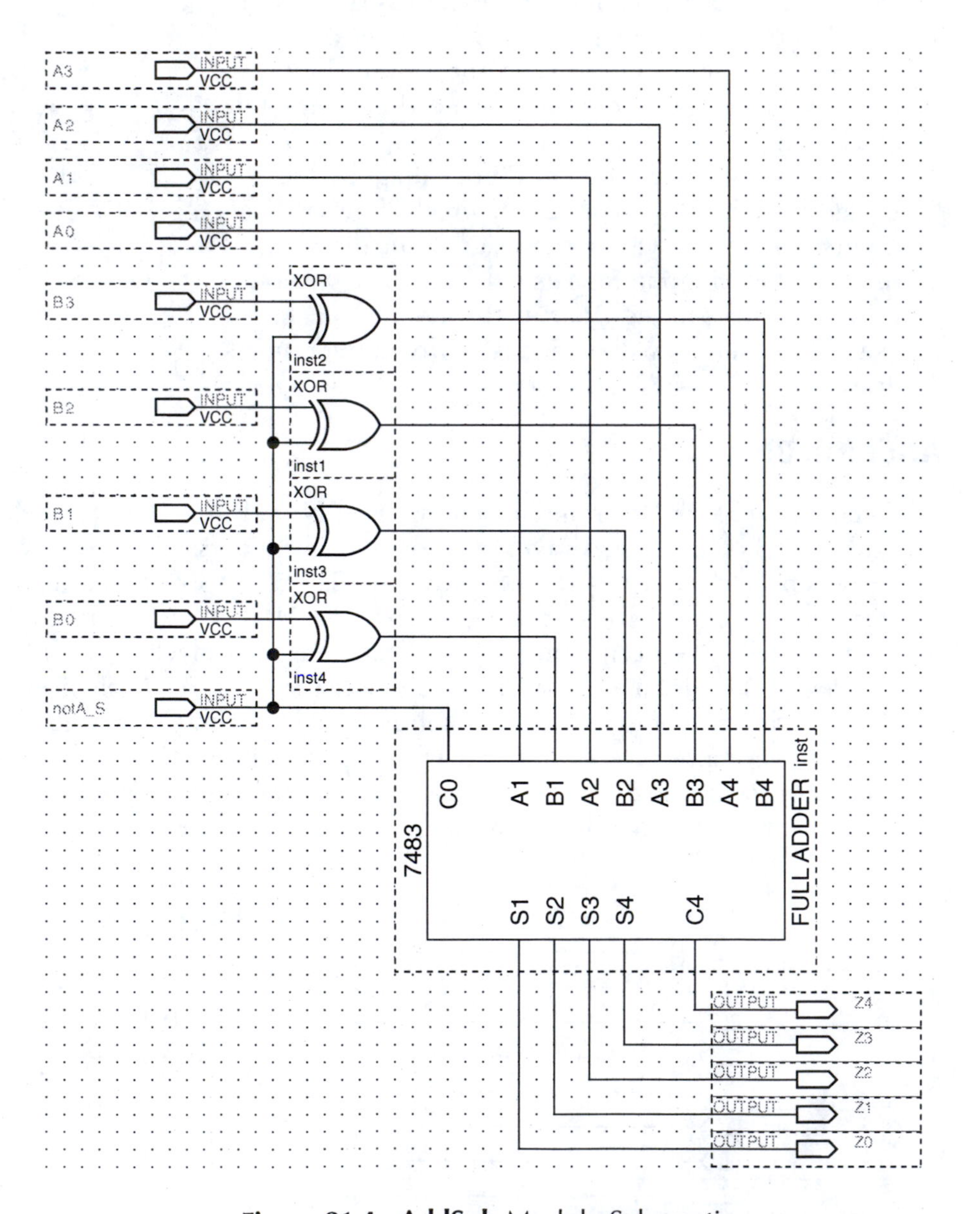

Figure 21.4 **AddSub** Module Schematic

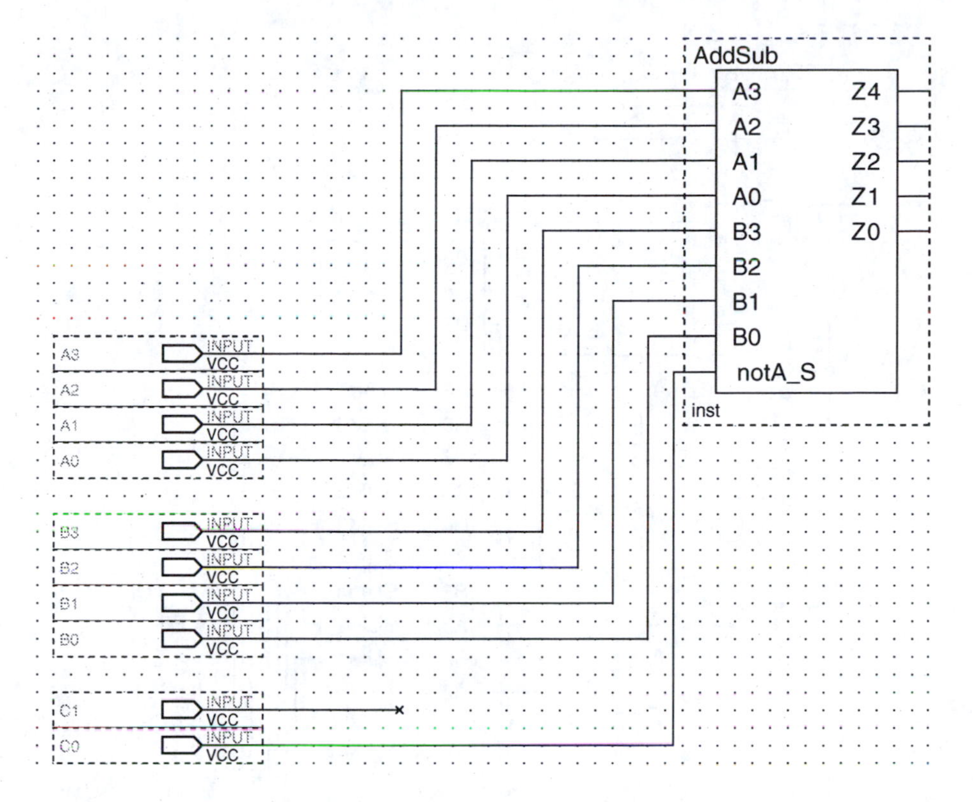

Figure 21.5 **AddSub** Module in Our Overall Calculator Design

Multiplication

We will design a solution using our observation that we *multiply* using *repeated* addition as described in Chapter 7 of *Introduction to Digital Electronics*. We start with the multiplicand (since the LSB of the multiplier is 1). Next, we shift to the left and add zero (since the next bit in the multiplier is 0). We continue with this pattern: if the next bit of the multiplier is zero, shift one bit to the left and add zero. If the next multiplier bit is 1, shift one bit to the left and add the multiplicand.

Since we are adding numbers repeatedly, adders can be used instead of a gate level solution designed from scratch. Our solution will quickly exceed 5 bits since the multiplier and multiplicand are both four bits, so the solution requires an adder capable of adding more than two 4-bit numbers. Quartus II has an 8-bit adder available named **8fadd** that could form the basic building block of our solution. This component adds two 8-bit numbers and finds a 9-bit result.

1. Draw the circuit schematic as shown in Figure 21.6, and add it to our overall design as shown in Figure 21.7.

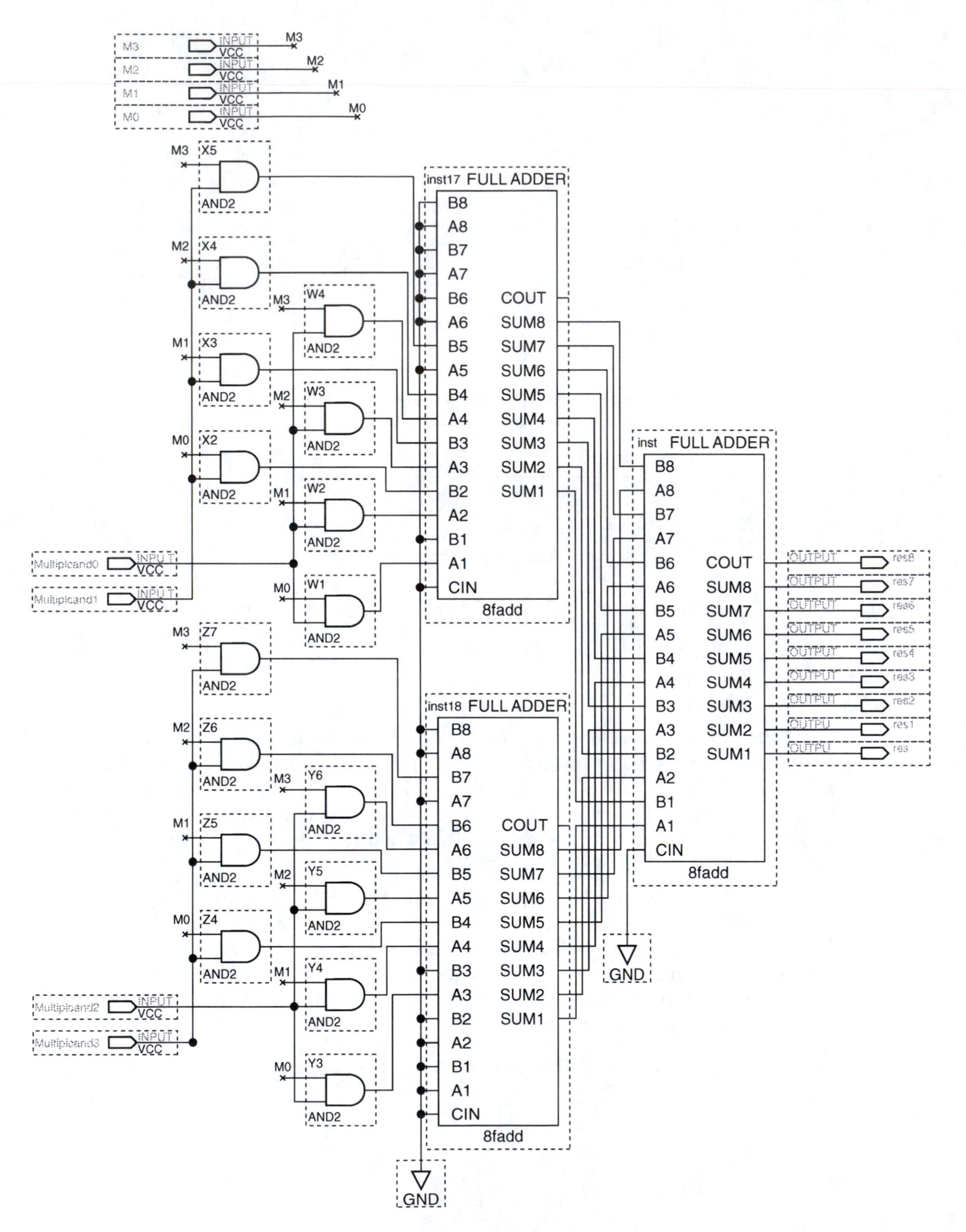

Figure 21.6 Schematic for **Multiplier** Using Quartus Components **8fadd**

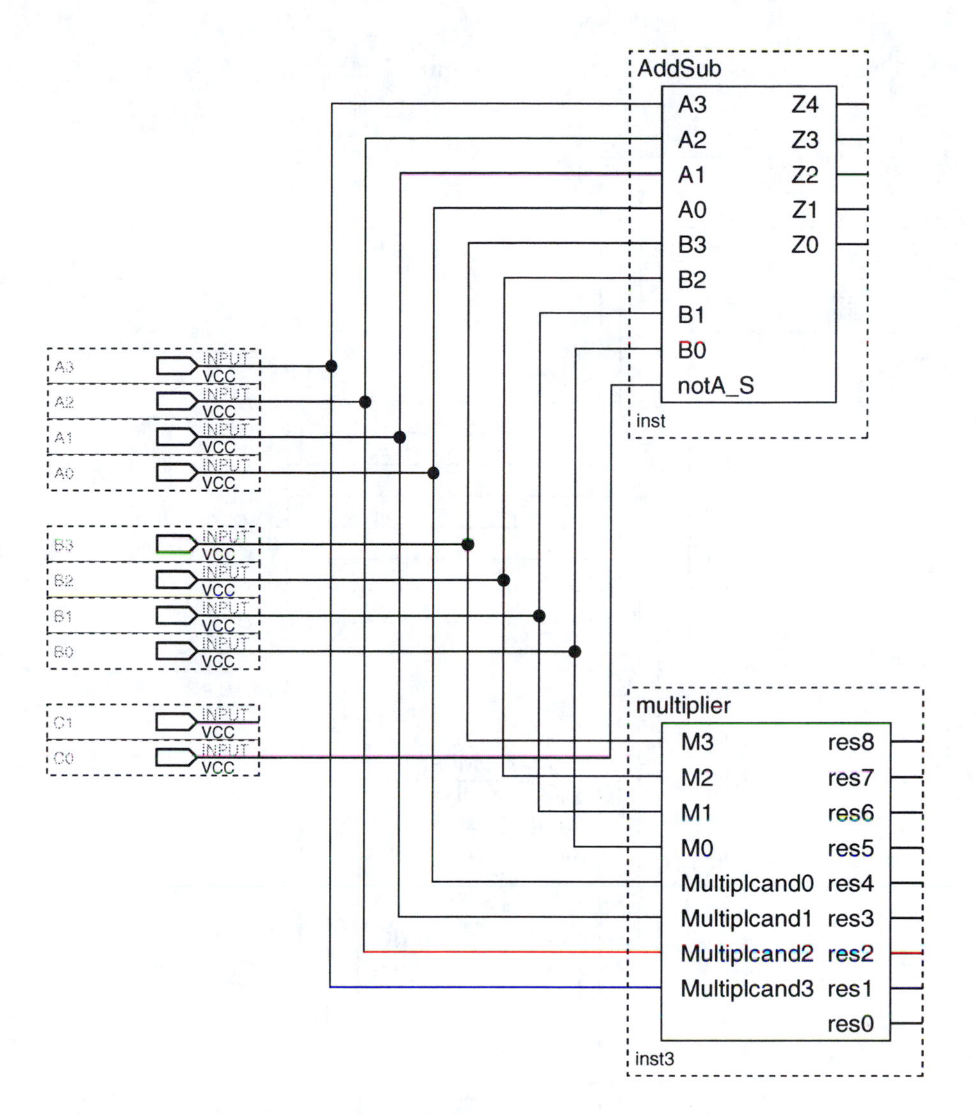

Figure 21.7 **Multiplier** Component Added to the Overall **Calculator** Schematic

Max/Min Block

This circuit is based on a comparison of the two inputs *A* and *B*, directing the highest value to the upper four bits of the output and the lesser to the lower four bits. Figure 21.8 shows this pictorially for the case where $B > A$; our task is to design the circuitry in the center block.

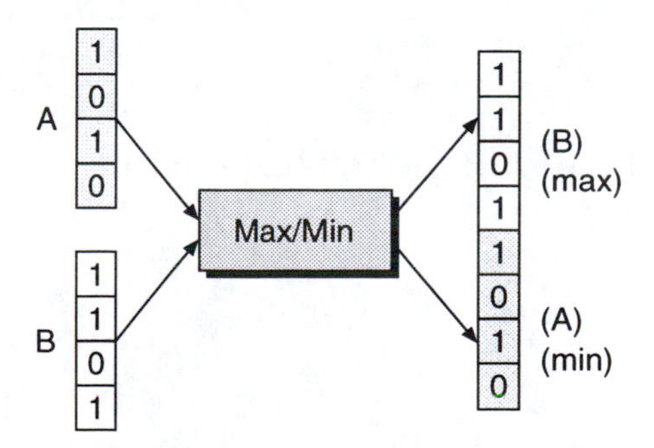

Figure 21.8 Block Diagram for Max/Min Block

1. Draw the schematic as shown in Figure 21.9

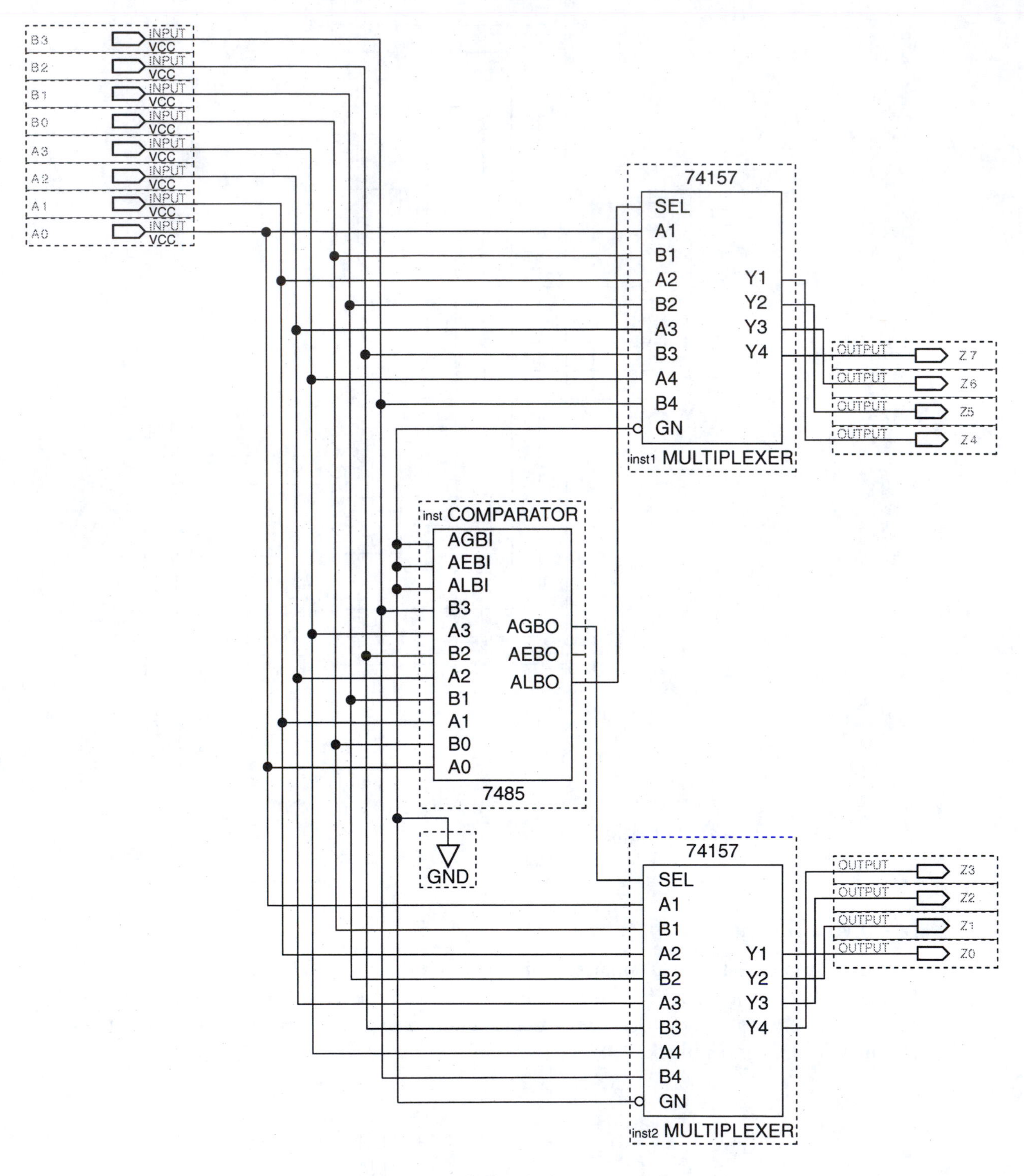

Figure 21.9 **Max/Min** Component Schematic

2. Add the Max/Min component to the overall design as shown in Figure 21.10.

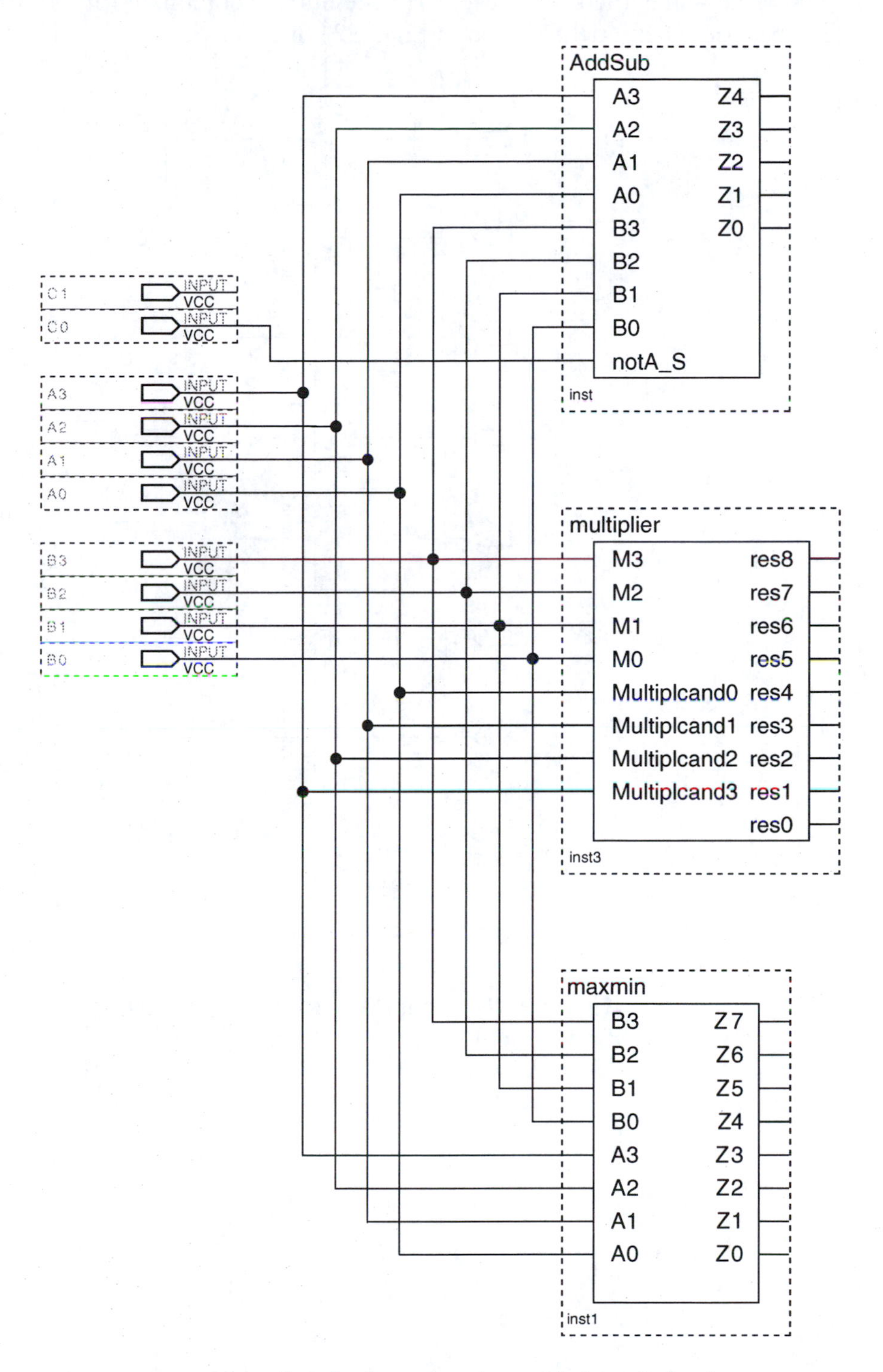

Figure 21.10 Overall **Calculator** Design with **Max/Min** Component

Output Block

1. Draw the schematic for the 8-bit tri-state buffer component for use in the output block. The schematic is shown in Figure 21.11.

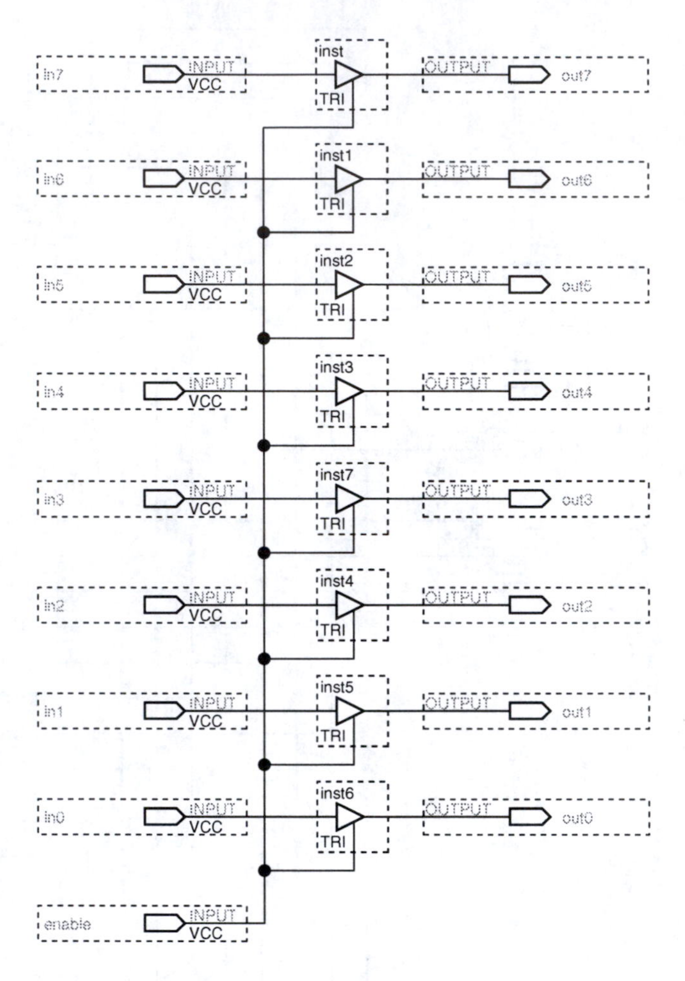

Figure 21.11 Schematic for **Octaltristate** Component

2. Using the **octaltristate** component, build the **calcout** schematic as shown in Figure 21.12.

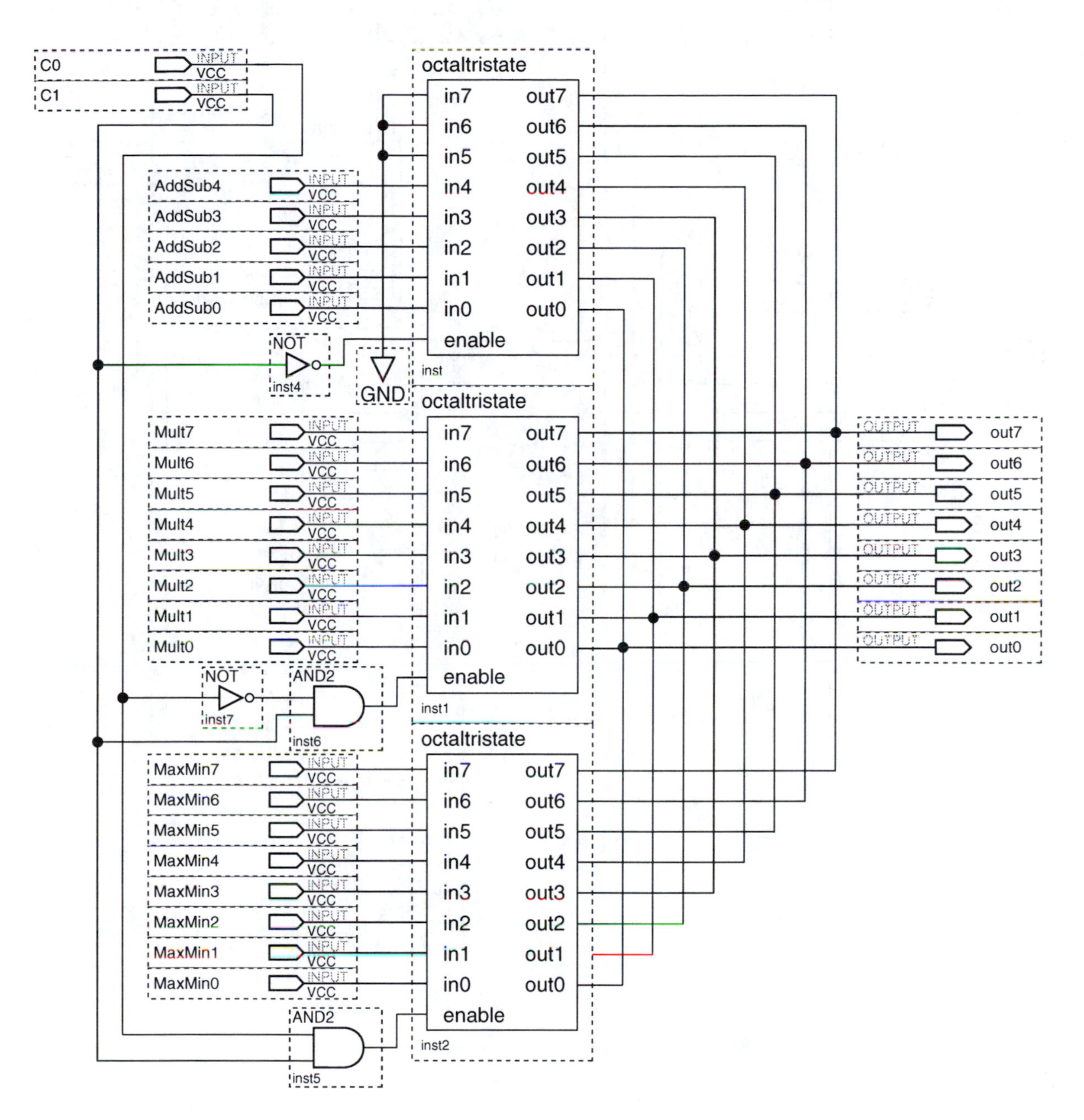

Figure 21.12 Calcout Component Schematic

Final Diagram

1. Complete the final block diagram for our calculator as shown in Figure 21.13:

Simulation

1. Design a set of simulation criteria to verify the proper operation of the **calculator**. Recall procedures studied to **group** like signals together into one group of signals—this makes the output z much easier to interpret.

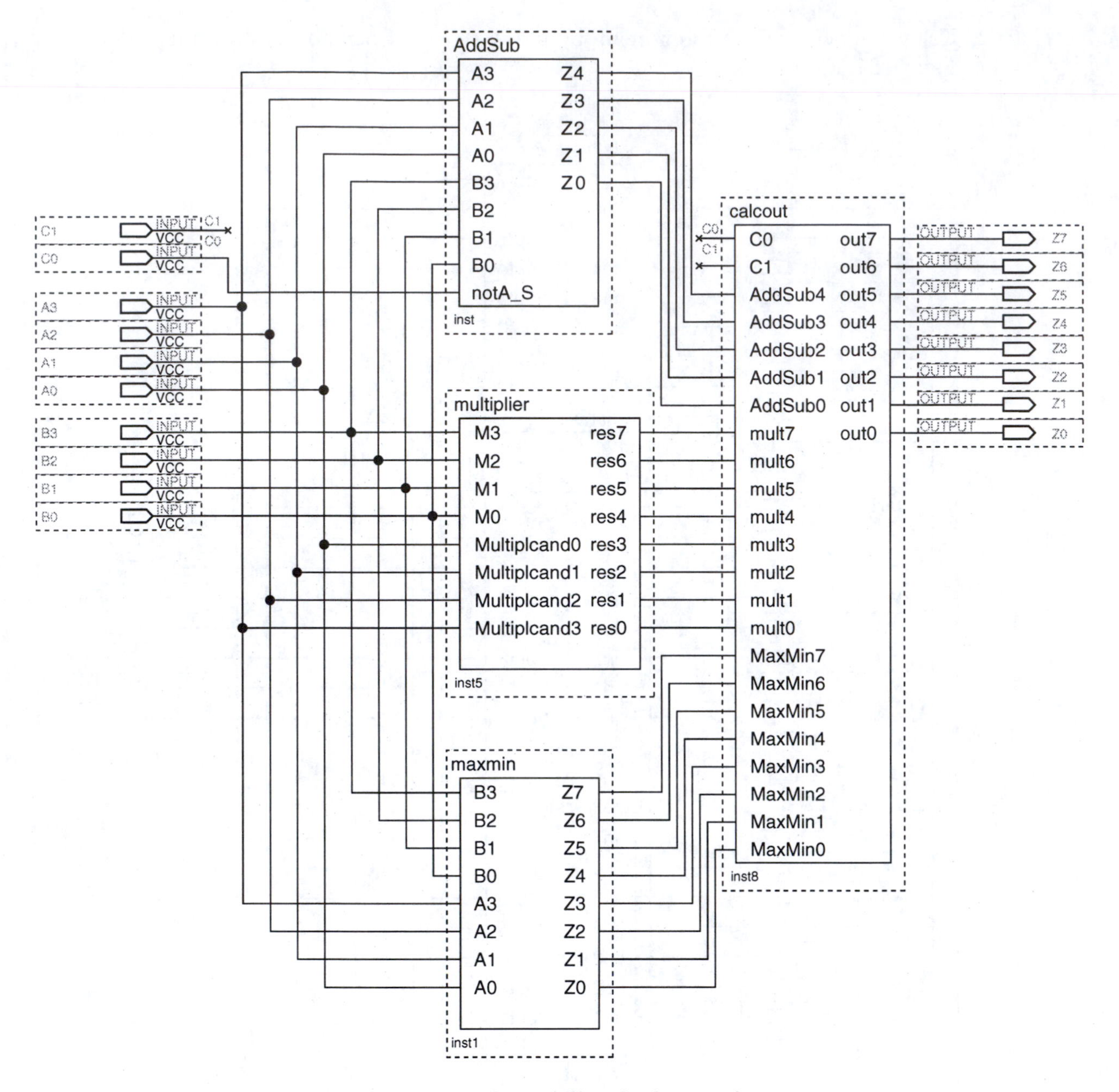

Figure 21.13 Final Overall **Calculator** Schematic

2. Notice that there are far too many combinations of inputs to test. Your test plan should include many examples, including operations that result in positive, negative, and zero answers as well as numbers in extreme ranges.

Download and Test

1. Assign proper pins to all inputs and outputs according to previous labs, recompile, and download your design to a development board. Demonstrate your circuit's operation to your instructor.

Lab 22

Combinational Project—Cola Machine Logic

Name ______________________ Class __________ Date __________

Objectives Upon completion of this laboratory exercise, you should be able to:

- Use effective problem-solving techniques to design a combinational logic system.
- Design a circuit from a word description of a given problem.

Reference Ken Reid and Robert Dueck, *Introduction to Digital Electronics*

Chapters 3–7: Note that this experiment follows the design process as described in Chapter 7 and uses the techniques and circuits found throughout the text to this point.

Equipment Required CPLD Trainer:

Altera UP-2 circuit board with ByteBlaster download cable or
DeVry eSOC board with USB or
RSR PLDT-2 circuit board with straight-through parallel cable

Experimental Notes

The Student Association at your college has decided to subsidize cola products in all on-campus vending machines. Students can now buy a cola for 20 cents! You have been commissioned to build the coin- and product-handling hardware for the new cola vending machine. Welcome to the team!

College Cola

.20

Design Specifications

This vending machine requires 20 cents to release a cola. However, ***only*** the ***minimum*** change is acceptable. Thus, the customer will receive a cola if they deposit the following sets of coins:

–	4 nickels	(N1,N2,N3,N4)
–	2 dimes	(D1,D2)
–	2 nickels and 1 dime	(N1,N2,D1)
–	One quarter.	(Q1)

When a quarter is deposited, the machine must also check whether it has nickels to return. If so, it delivers a nickel and a cola. If not, it returns the patron's coin. In all cases, if the vending machine is out of cola, it must return the patron's money.

To simplify the design's complexity, this machine only accepts the change combinations explicitly stated. Other possible change combinations will not be considered.

Control Signals

As the hardware developer, the following ***input*** signals are available to you:

- Active-HIGH coin signals (N1,N2,N3,N4,D1,D2,Q1)
  - Each of these signals is HIGH when the corresponding coin is inserted into the machine.
- Active-HIGH Nickel sense (Ns)
  - This signal is HIGH when nickels are available and LOW if there are no nickels.
- Active-HIGH Cola sense (Cs)
  - This signal is HIGH when colas are available.

The inputs will be simulated with logic switches.

You must provide the following ***output*** signals

- Active-HIGH Nickel Return (NR)
  - Dispenses one nickel
- Active-LOW Coin Return (/CR)
  - Returns all the patron's change
- Active-HIGH Dispense Cola (DC)
  - Opens a pneumatic valve allowing the cola to pass

The outputs will be simulated with LEDs.

Note /CR = (NOT CR)

Procedure

1. Use the Block Diagram Editor to draw the logic diagram for the circuit in sum-of-products form. The simplest implementation requires only the following types and quantities of gates: 2 4-input ANDs, 2 3-input ANDs, 2 2-input ANDs; 1 4-input OR; 1 2-input NOR, and 2 inverters.
2. Find the Boolean expressions for each of the DC, NR, and /CR outputs.
3. Build the circuitry and download to a development board for testing. Use switches as inputs and LED's as outputs.
4. Test the circuit. Verify:
   - The correct coin combinations activate the dispense coke output.
   - The effect of Nickel sense (Ns) and Cola sense (Cs) on the circuit.
   - The correct activation of the Dispense Cola (DC), Nickel Return (NR) and Coin Return (/CR).

Lab 23

NAND Latches, D Latches, and D Flip-Flops

Name ______________________ Class ____________ Date ____________

Objectives Upon completion of this laboratory exercise, you should be able to:

- Use the Quartus II Block Editor to create a NAND latch.
- Create multiple-bit latches and flip-flops.
- Create simulations for the latches and flip-flops.
- Test the latches and flip-flops on a CPLD test board.

Reference Ken Reid and Robert Dueck, *Introduction to Digital Electronics*

Chapter 8: Introduction to Sequential Logic

Equipment Required

CPLD Trainer:

Altera UP-2 circuit board with ByteBlaster download cable, or
DeVry eSOC board with USB cable, or
RSR PLDT-2 circuit board with straight-through parallel port cable, or
equivalent CPLD trainer board with Altera EPM7128S CPLD

Quartus II Web Edition software
AC adapter, minimum output: 7 VDC, 250 mA DC
Anti-static wrist strap
#22 solid-core wire
Wire strippers

Experimental Notes

NAND Latch

Figure 23.1 shows a NAND latch created using the Quartus II Block Editor.

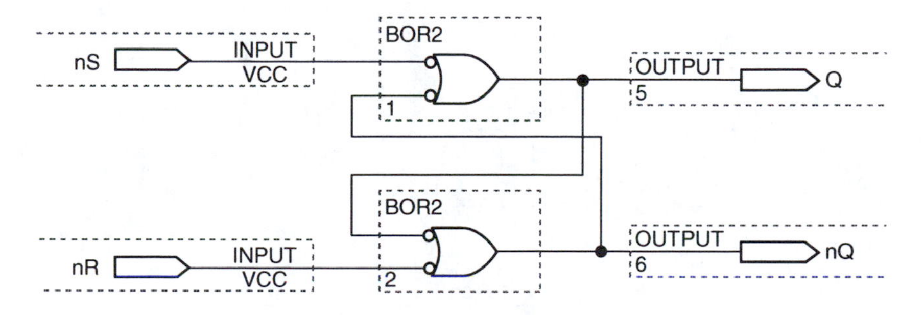

Figure 23.1 NAND Latch

The latch gates in Figure 23.1 are negative NAND gates. The component name for this gate in Quartus II is BOR2, meaning, somewhat inaccurately, "bubbled-OR, 2-input." We label the inputs *nS*, *nR* and the outputs *Q*, *nQ*, where the "*n*" indicates logical negation (NOT function), thus telling us that *nS*, *nR*, and *nQ* are all active-LOW terminals.

D Latches and D Flip-Flops

Two important digital circuits are the D latch and the D flip-flop (Figure 23.2), both of which are used to store data. The main difference between the circuits is the conditions under which the data are stored.

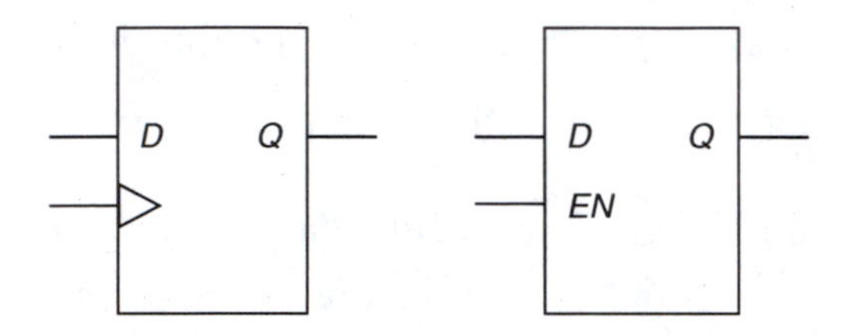

Figure 23.2 D Flip-Flop and D Latch

In a D latch (or "transparent latch"), the output *Q* takes on the value of the input *D* only when an enable input *(EN)* is HIGH. (We say, "*Q* follows *D*.") If *D* changes while $EN = 1$, *Q* will follow immediately. If $EN = 0$, *Q* retains its previous value and does not change with *D*.

In a flip-flop, *Q* also follows *D*, but only when there is a transition on an enabling input called the clock *(CLK)*. Typically, the clock input is shown as a triangle, indicating that it is a dynamic input. For a positive edge-triggered D flip-flop, *Q* follows *D* when the clock makes a transition from LOW to HIGH. Otherwise, *Q* will not change.

Latches and flip-flops can also have multiple inputs and outputs, as shown in Figure 23.3. For these devices, all *Q*s follow all *D*s on the appropriate enable condition. For example, if $D_3D_2D_1D_0 = 0101$, then $Q_3Q_2Q_1Q_0$ will equal 0101 after the first positive clock edge for the flip-flop or as soon as $EN = 1$ for the latch.

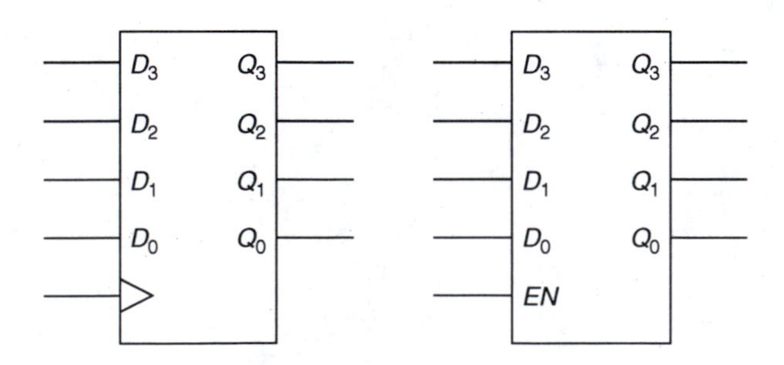

Figure 23.3 4-Bit D Flip-Flop and Latch

Procedure

NAND Latch

1. Use the Quartus II Block Editor to draw the NAND latch shown in Figure 23.1. Save the file as ***drive:*\qdesigns\labs\lab23\ndlatch\nand_latch.bdf** and use it to create a Quartus II project. Compile the project.

2. Create a simulation that shows the Set\Reset behavior of the latch. Start by making *nS* and *nR* both HIGH for the whole simulation period. Then add a LOW-going pulse on an input by dragging the mouse cursor across a portion of one of the input waveforms and clicking the "0" toolbar button. Repeat in several non-overlapping places for both input waveforms. Run the simulation and show the results to your instructor.

   Instructor's Initials ________________

3. Add pin numbers to the NAND latch, as indicated in Table 23.1. Recompile the circuit and download it to your CPLD board. The inputs *nS* and *nR* should be connected to the two pushbuttons on the board. The output *Q* and *nQ* should be connected via jumpers to two LEDs. Show the operation of the latch to your instructor.

   Instructor's Initials ________________

Table 23.1 NAND Latch Pin Assignments

	Pin Numbers		
Function	**UP-2**	**PLDT-2**	**eSOC**
nS	11	11	145
nR	1	1	146
Q	51	51	151
nQ	52	52	150

4-Bit D Latch

1. Use the Quartus II Block Editor to create the latch circuit in Figure 23.4 with 4 *D* latches and a common enable input. Save the file in a new folder as **g:\qdesigns\lab23\dlatch1\dlatch1.bdf** and use it to create a Quartus II project. (Don't name or compile the project.)

2. Compile the project and create a simulation. In the simulation, show the *D* inputs as a group and the *Q* outputs as another group. The simulation should show that the *Q* outputs do not change if ENABLE = 0, even if the *D* inputs change and also that *Q* follows *D* immediately when ENABLE = 1.

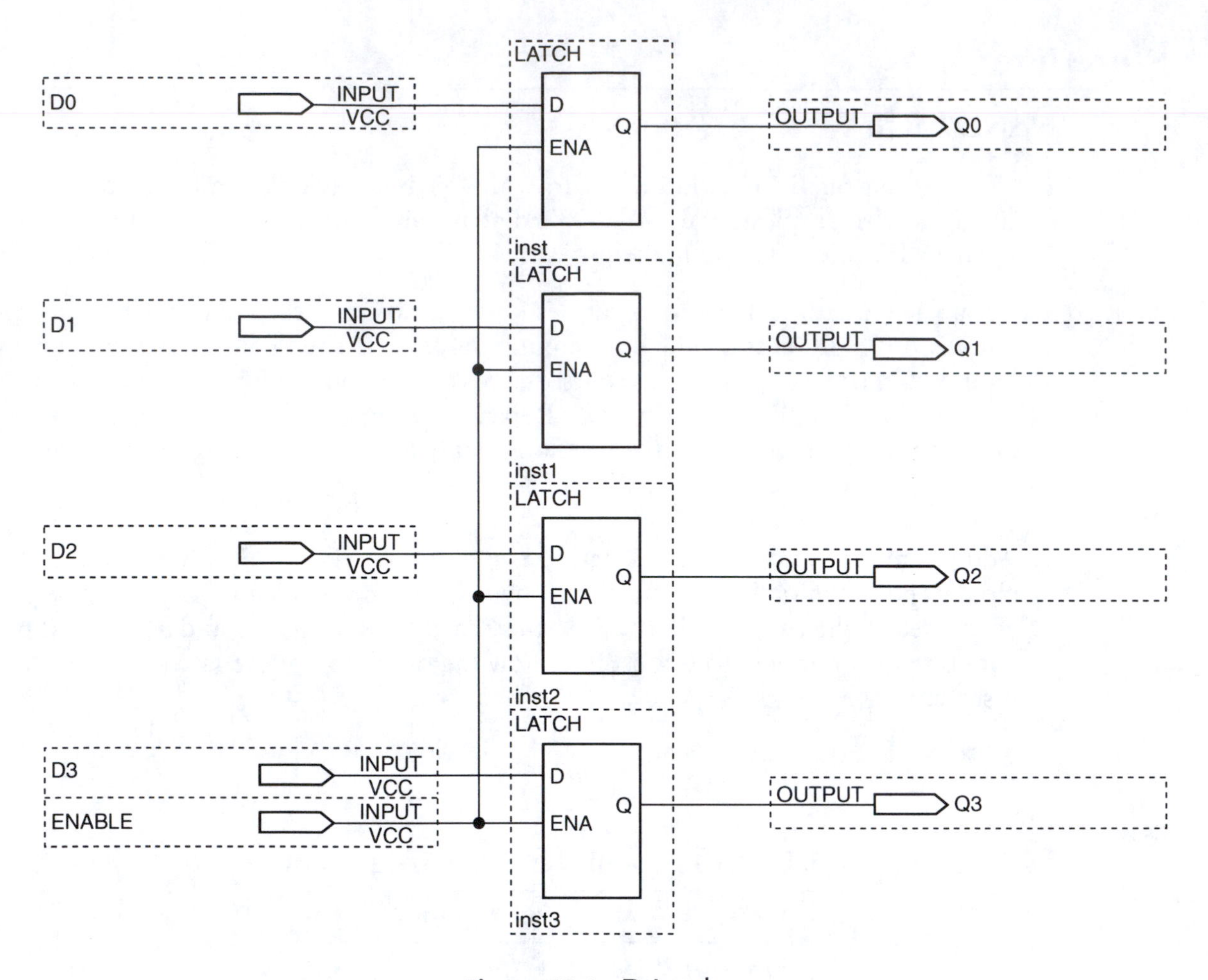

Figure 23.4 D Latches

Suggested Input

- Overwrite the **d** input with a **group value** (any single-digit hexadecimal number). Do this by highlighting the **d** waveform and clicking the "Arbitrary Value" toolbar button on the Waveform Editor toolbar.
- Highlight the middle half of the **d** waveform by dragging the mouse cursor across it. Invert the value in the portion of **d** by clicking the "INV" toolbar button on the Waveform Editor toolbar.

- Create a HIGH-going pulse on **enable:** highlight a small portion of the **enable** input close to the beginning of the waveform and click the "1" toolbar button.
- Highlight a second portion of the **enable** input near the change on the **d** input and click the "1" toolbar button to create a second pulse.

Run the simulation and show the result to your instructor.

Instructor's Initials ________________

3. Create a graphic symbol from the Block Diagram file for the latch by selecting **Create/Update, Create Symbol Files for Current File** from the Quartus II **File** menu.

4. Make a new block diagram file and save it as **g:\qdesigns\lab23\dlatch1\dlatch_tester.bdf.** Use the Quartus II Block Editor to create the circuit shown in Figure 23.5.

Important Step: *Make this new file the top level of the project by selecting **Set as Top-Level Entity** from the **Project** menu.*

Compile the project.

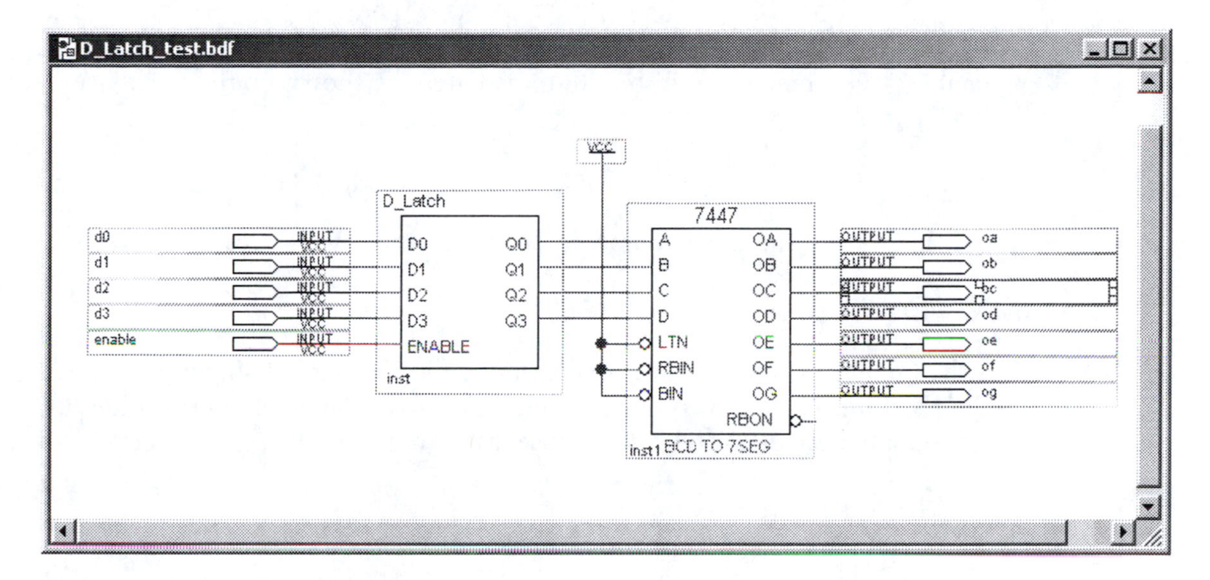

Figure 23.5 D Latch and Seven-Segment Decoder

5. Assign pins to the design, as shown below (Table 23.2). Compile the project and download it to the CPLD circuit board.

Table 23.2 Pin Assignments

	Pin	
Function	**UP-2/PLDT-2**	**DeVry eSOC**
d3	34	160
d2	33	161
d1	36	162
d0	35	163
enable	41	144
oa	69	137
ob	70	138
oc	73	139
od	74	128
oe	76	80
of	75	81
og	77	84

6. Connect the DIP switches for **d[3..0]** and for **enable.** (The switch connections are already jumpered on the RSR PLDT-2 board.) The seven-segment display is already connected, so nothing further needs to be done with these outputs.

7. Test the 4-bit latch for the following conditions:
   - **enable** = 0. Changes on the **d** inputs should not affect the seven-segment display.
   - **enable** = 1. Changes on **d** should immediately be shown on the seven-segment display.

Instructor's Initials ______________

8. **Close the project and create a new folder for the next section of this lab.**

4-Bit D Flip-Flop

1. Use the Quartus II Block Editor to create the flip-flop circuit shown in Figure 23.6 with 4 D flip-flops and a common clock input. Save the file in a new folder as **g:\qdesigns\lab23\dff1\dff1.bdf** and use it to create a Quartus II project.

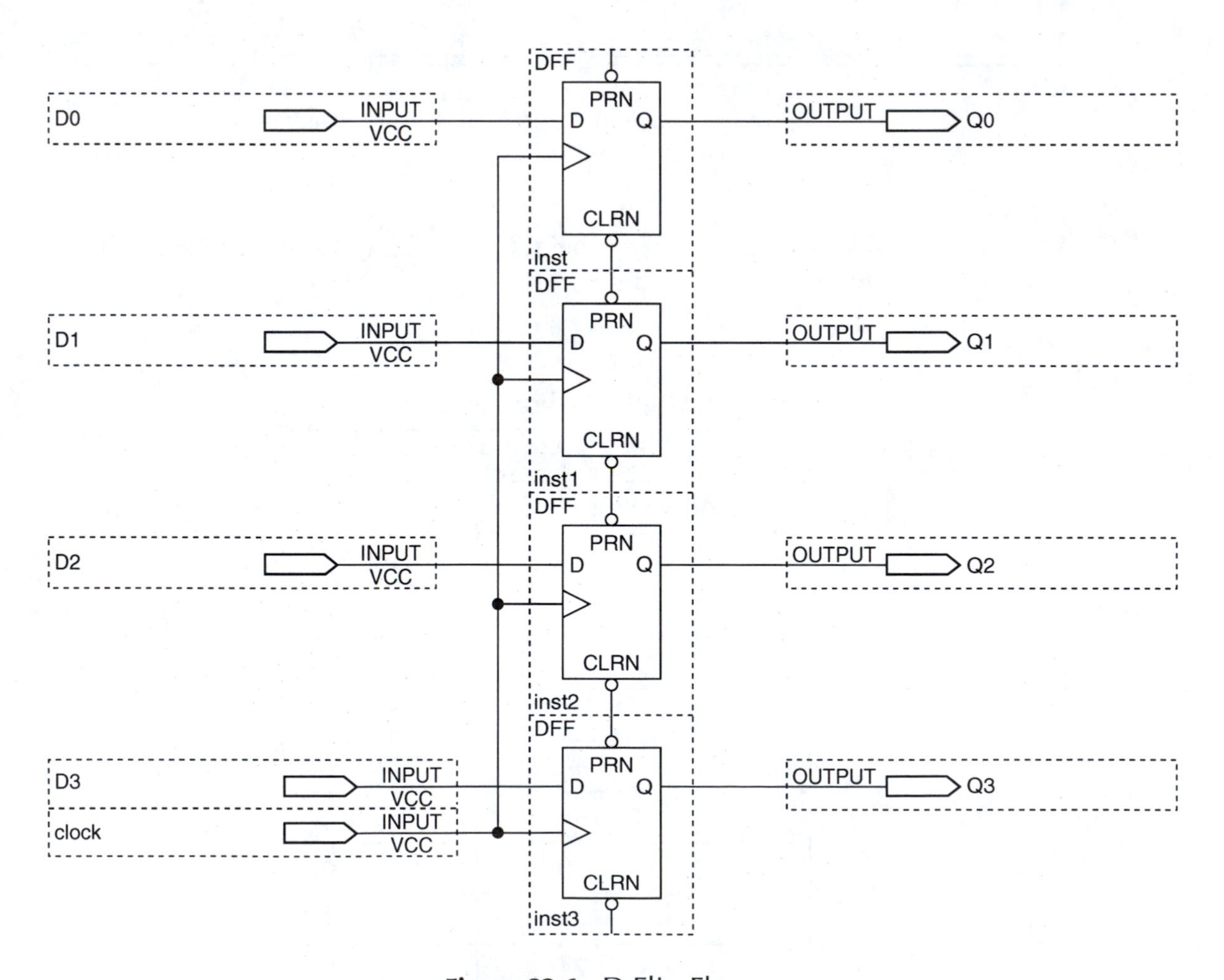

Figure 23.6 D Flip-Flops

2. Compile the project and create a simulation. In the simulation, show the D inputs as a group and the Q outputs as another group. The simulation should show that the outputs do not change unless there is a positive edge in the clock, even if the D inputs change. Write a set of simulation criteria that will test this function.

Simulation Criteria

Run the simulation and show the result to your instructor.

Instructor's Initials ________________

3. Create a graphic symbol from the Block Diagram file for the flip-flops by selecting **Create/Update, Create Symbol Files for Current File** from the Quartus II **File** menu. Also create a symbol for the NAND latch you created earlier in this lab exercise.

4. Make a new block diagram file and save it as g:**\qdesigns\lab23\dff1\dff_tester.bdf** Use the Quartus II Block Editor to create the circuit shown in Figure 23.7. (Note that the NAND latch from the earlier part of the lab is used to debounce a pair of pushbutton inputs.)

Important Step: *Make this new file the top level of the project by selecting* ***Set as Top-Level Entity*** *from the* ***Project*** *menu.*

Compile the project.

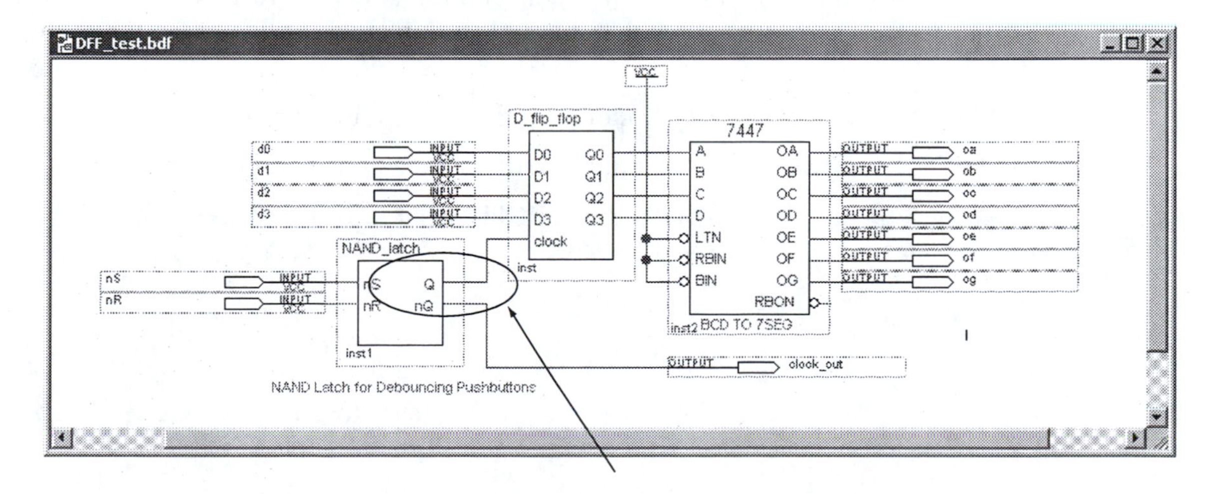

Figure 23.7 4-Bit Flip-Flop and Seven-Segment Decoder (UP-2 Shown; for RSR PLDT-2 or DeVry eSOC, Connect clock_out to Q of NAND_latch rather than nQ)

5. Assign pin numbers to the design, as shown below (Table 23.3). Recompile the project and download the design to the CPLD board.

Table 23.3 Pin Assignments

	Pin	
Function	**UP-2/PLDT-2**	**DeVry eSOC**
d3	34	160
d2	33	161
d1	36	162
d0	35	163
nS	11	131
nR	1	130
clock_out	52	150
oa	69	137
ob	70	138
oc	73	139
od	74	128
oe	76	80
of	75	81
og	77	84

6. The *D* input connections are the same as for the 4-bit latch and flip-flop. Connect pushbuttons to pins 11 and 1 (NAND latch inputs) and an LED to pin 52. (This LED monitors the logic level of the debounced flip-flop clock.) Test for the following conditions:
   - **clock** = 0. Changes on **d** should not affect the circuit output.
   - **clock** = 1. Changes on **d** should not affect the circuit output.
   - **clock** = positive edge (LOW then HIGH) should cause **q** to follow **d.**

Instructor's Initials ____________________

Lab 24

JK and T Flip-Flops

Name ______________________ Class ____________ Date ____________

Objectives Upon completion of this laboratory exercise, you should be able to:

- Use the Quartus II Block Editor to create a circuit for an asynchronous binary counter, using JK or T flip-flops.
- Create a simulation that verifies the operation of the counter made with JK flip-flops.
- Test the counters on a CPLD test board.

Reference Ken Reid and Robert Dueck, *Introduction to Digital Electronics*

Chapter 8: Introduction to Sequential Logic

Equipment Required CPLD Trainer:

Altera UP-2 circuit board with ByteBlaster download cable, or
DeVry eSOC board with USB cable, or
RSR PLDT-2 circuit board with straight-through parallel port cable, or
equivalent CPLD trainer board with Altera EPM7128S CPLD

Quartus II Web Edition software
AC adapter, minimum output: 7 VDC, 250 mA DC
Anti-static wrist strap
#22 solid-core wire
Wire strippers

Experimental Notes

The JK flip-flop can operate in a number of synchronous modes, which include no change ($JK = 00$), reset ($JK = 01$), set ($JK = 10$), and toggle ($JK = 11$). One of the more useful modes of the JK flip-flop is the toggle mode, which allows the device to be used as an element in a binary counter. If several flip-flops are all configured to toggle, they can be arranged so that the Q output on one flip-flop clocks the next one. If the flip-flops are negative-edge triggered, the effect of this arrangement is to generate a binary count sequence.

The T ("toggle") flip-flop can fulfill the same function as a JK flip-flop. It has a synchronous input called *T*, which switches the flip-flop between a toggle mode (when $T = 1$) and a no change mode (when $T = 0$). If a T flip-flop is configured to toggle on each clock pulse, it can be directly substituted into a circuit that uses JK flip-flops for the same function.

Both types of flip-flops have asynchronous clear and preset functions in addition to the synchronous JK and T inputs. These functions act immediately when made LOW to set the *Q* output of the flip-flop to 0 (when preset = 0) or to 1 (when clear = 0). The primary use of these functions is to set the flip-flop to a known initial state, from which point functions are usually determined by the state of the synchronous inputs and the clock.

Procedure

Asynchronous Counter (JK Flip-Flops): Design Entry and Simulation

1. Create a 4-bit asynchronous counter in the Quartus II Block Editor, as shown in Figure 24.1, and save it as ***drive:*\qdesigns\labs\lab24\asynch_ctr_JK\ asynch_ctr_JK.bdf.** Use the file to create a new project.

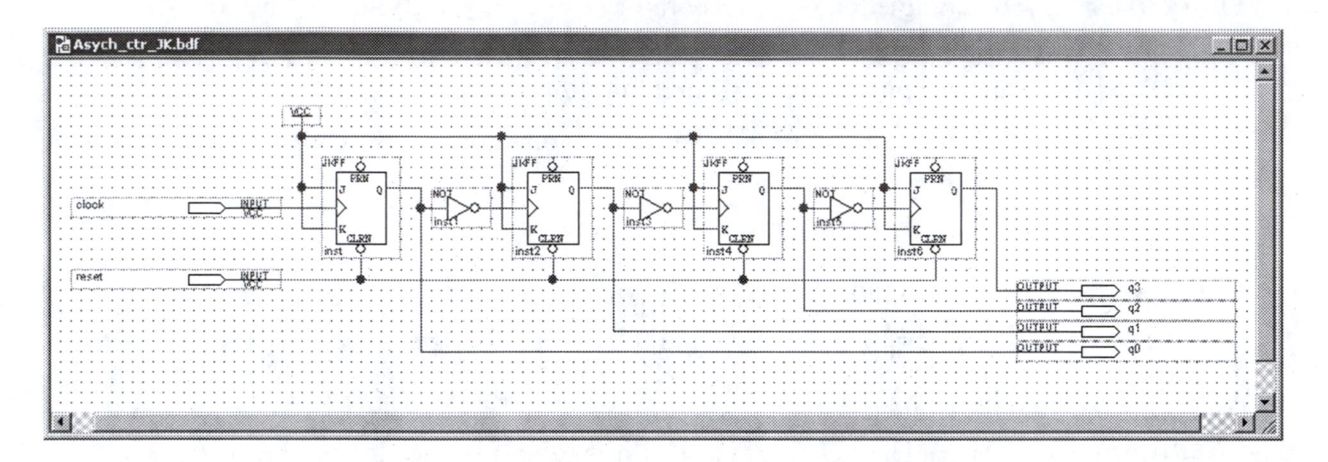

Figure 24.1 4-Bit Asynchronous Binary Counter

2. Write a set of simulation criteria to verify the correctness of the circuit design. Use these criteria to make a Quartus II simulation. Show the criteria and the simulation to your instructor.

Simulation Criteria

Instructor's Initials 

Test Circuit

If we are to download the circuit of Figure 24.1 directly to a CPLD test board, we would need to apply a clock pulse to the clock input of the circuit, either from the on-board oscillator or from a manual pushbutton. If we did this one of two things would happen. If we used the oscillator, which runs at several megahertz, the counter outputs would run so fast that we would not be able to observe them change on the board LEDs. If we used the pushbutton, the counter would not progress in a predictable fashion, due to the mechanical bounce of the pushbutton switch. This requires us to use a clock divider or a switch debouncer for reliable operation that is slow enough to observe visually.

Our CPLD test boards have pushbuttons, each of which has only a single normally open contact. This type of switch can be debounced using a clocked component called **debouncer.vhd,** whose use is shown in Figure 24.2 (for the RSR PLDT-2 and DeVry eSOC boards) and Figure 24.3 (for the Altera UP-2 board). The VHDL code for this debouncer is available on the CD that accompanies *Introduction to Digital Electronics,* in a folder called **Student_Lab_Files.** This VHDL file can be copied to the working folder for a project and used to make a debouncer component, as shown in Figures 24.2 and 24.3.

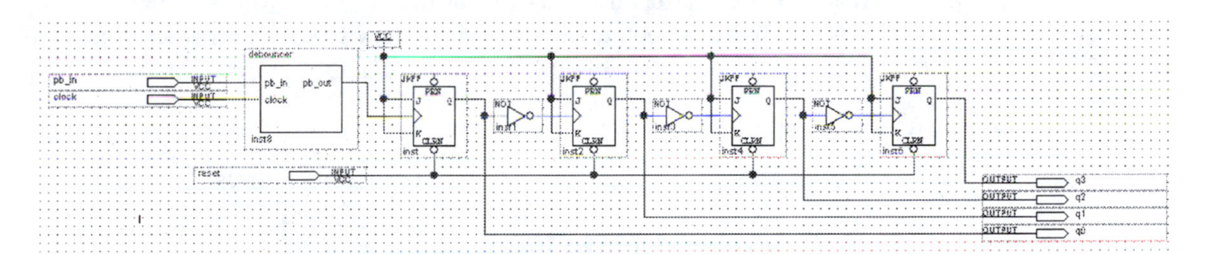

Figure 24.2 4-Bit Counter with Switch Debouncer (RSR PLDT-2)

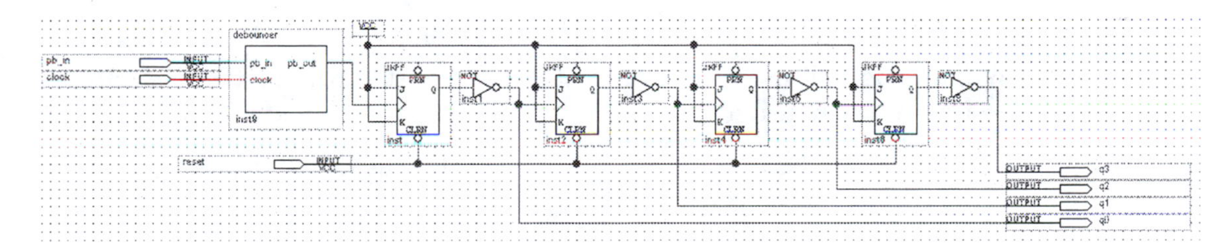

Figure 24.3 4-Bit Counter with Switch Debouncer (Altera UP-2 and DeVry eSOC)

1. Modify the counter circuit shown in Figure 24.1 to make either the circuit in Figure 24.2 (for the RSR PLDT-2 or DeVry eSOC board) or Figure 24.3 (for the Altera UP-2 board). Note that the Q outputs in the circuit in Figure 24.3 are inverted to adapt them to the active-LOW LEDs of the Altera UP-2 board.

2. Assign pin numbers to the design as follows and compile the project.

	Pin Numbers		
Function	**UP-2**	**PLDT-2**	**eSOC**
pb_in	11	11	145
clock	83	83	24
reset	1	1	131
q3	49	49	110
q2	50	50	102
q1	51	51	151
q0	52	52	150

3. Program your CPLD test board with the asynchronous counter design. Connect **pb_in** to a pushbutton on the board and **reset** to another pushbutton. Connect the **q** outputs to LEDs, if not already connected. It is not necessary to make a direct connection to pin 83 (clock), as this connection is hardwired to the CPLD on your test board.

4. Demonstrate the operation of the counter to your instructor.

Instructor's Initials ____________________

Asynchronous Counter (T Flip-Flops)

1. Replace the JK flip-flops in Figure 24.2 or 24.3 with T flip-flops. The symbol name for a T flip-flop is **TFF.**

2. Compile the project and use it to program your board.

3. Show the modified Block Diagram File to your instructor and demonstrate the operation of the new circuit.

Instructor's Initials ____________________

Lab 25

Binary Counters (Block Diagram File)

Name ______________________ Class ____________ Date ____________

Objectives Upon completion of this laboratory exercise, you should be able to:

- Enter the design for a binary counter with synchronous load using the Quartus II Block Editor.
- Simulate the functions of the counter in this laboratory exercise.
- Download and test the output of the counter in this exercise.
- Display counter outputs as binary values on LEDs.

Reference Ken Reid and Robert Dueck, *Introduction to Digital Electronics*
Chapter 9: Counters and Shift Registers

Equipment Required
CPLD Trainer:
Altera UP-2 circuit board with ByteBlaster download cable, or
DeVry eSOC board with USB cable, or
RSR PLDT-2 circuit board with straight-through parallel port cable, or
equivalent CPLD trainer board with Altera EPM7128S CPLD
Quartus II Web Edition software
AC adapter, minimum output: 7 VDC, 250 mA DC
Anti-static wrist strap
#22 solid-core wire
Wire strippers

Experimental Notes

Binary Counters from Flip-Flops

Though binary counters can be designed using any type of flip-flop, JK and D flip-flops are commonly used for this function. The synchronous input equations of the JK flip-flops of a binary counter follow a predictable sequence:

$$J_0 = K_0 = 1$$
$$J_1 = K_1 = Q_0$$
$$J_2 = K_2 = Q_1Q_0$$
$$J_3 = K_3 = Q_2Q_1Q_0$$

D flip-flops can also be used for binary counter. Their synchronous input equations also follow a predictable sequence:

$$D_0 = \bar{Q}_0$$
$$D_1 = Q_1 \oplus Q_0$$
$$D_2 = Q_2 \oplus Q_1Q_0$$
$$D_3 = Q_3 \oplus Q_2Q_1Q_0$$

These equations determine the next state of a counter by generating a toggle condition or a no change condition for each flip-flop throughout the count sequence. Typically, designing a counter with JK flip-flops yields simpler equations, but requires more work than D flip-flops, since the synchronous input of a D flip-flop is the same as its required next state, whereas the inputs of a JK flip-flop must be derived from an excitation table. For more detail, see Section 9.3 in *Introduction to Digital Electronics.*

The logic diagram of a 3-bit binary counter based on D flip-flops is shown in Figure 25.1. Figure 25.2 shows a simulation of the 3-bit counter.

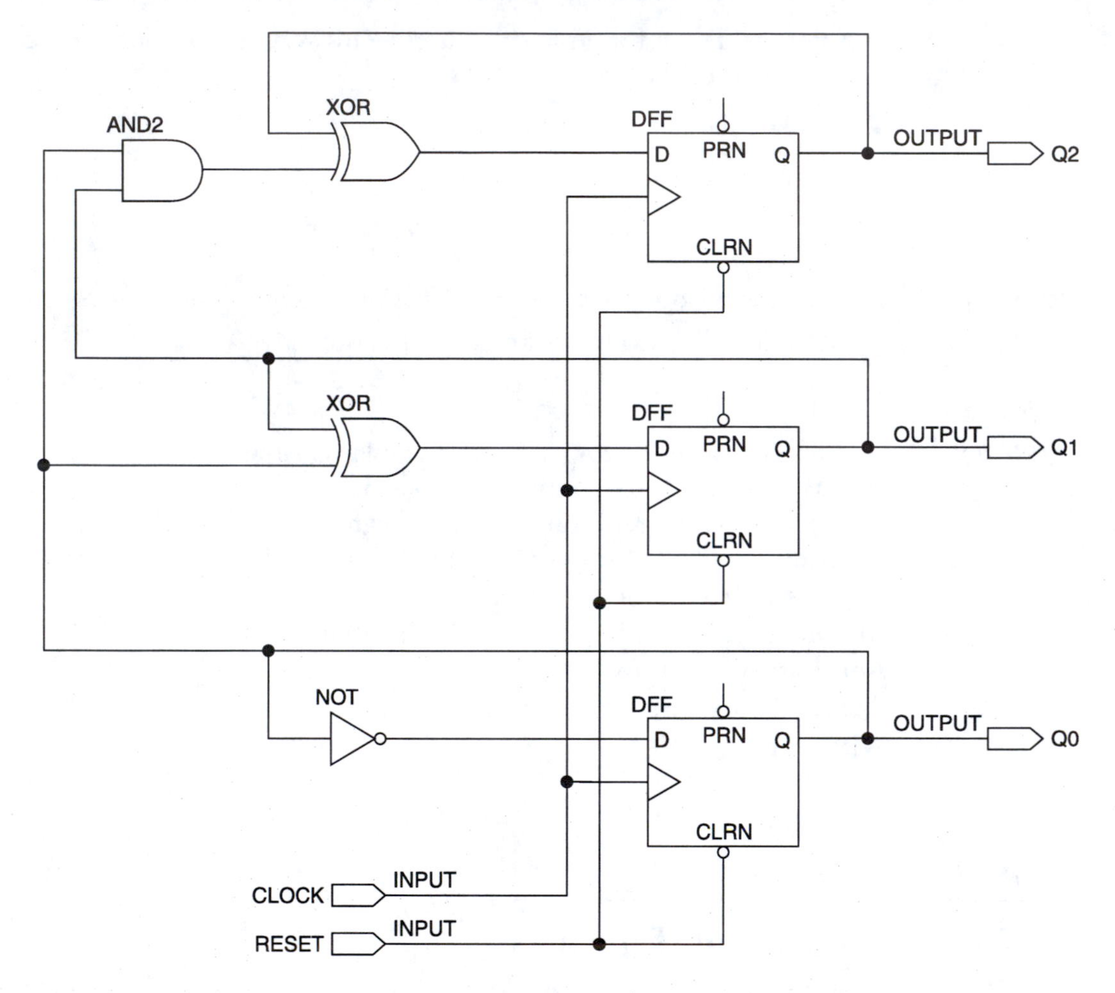

Figure 25.1 3-Bit Binary Counter

3bit_dct Simulation Report
Simulation Waveforms
Master Time Bar: 0 ps Pointer: 0 ps Interval: 0 ps Start: End:

Name	Value at 0 ps
RESET	B 1
CLOCK	B 0
Q[2..0]	H 0
Q2	B 0
Q1	B 0
Q0	B 0

0 ps 160.0 ns 320.0 ns 480.0 ns 640.0 ns 800.0 ns 960.0 ns

Figure 25.2 Simulation of a 3-Bit Counter

The 3-bit counter can be modified as shown in Figure 25.3 to include a synchronous load function. Each flip-flop is fed by an AND-OR network (essentially a 2-to-1 multiplexer) that directs either the count logic or a parallel input to the synchronous input of the flip-flop. A simulation of this counter is shown in Figure 25.4. We can see that the load function is really synchronous because the first load pulse does not overlap or immediately precede a positive clock edge and is therefore ignored.

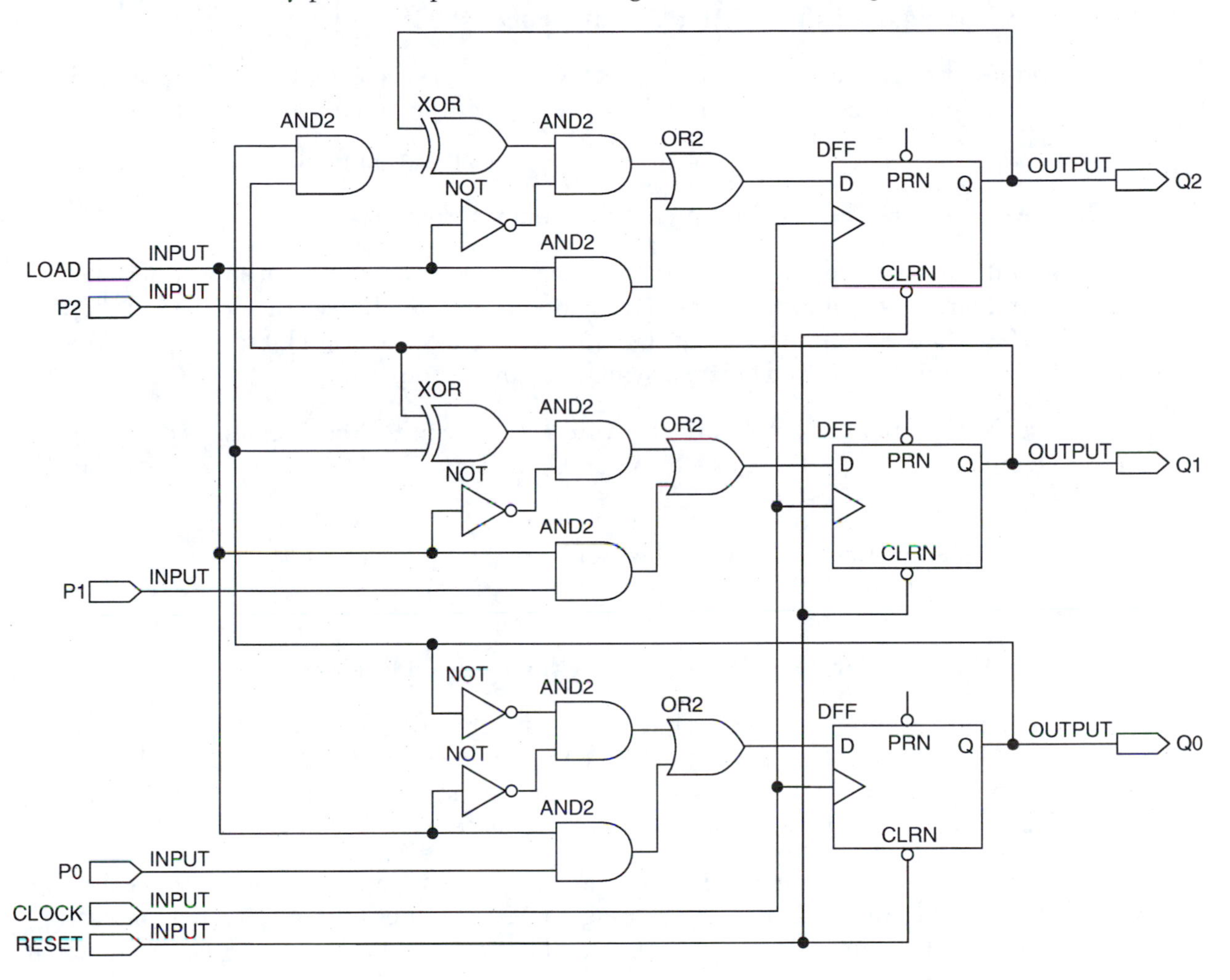

Figure 25.3 3-Bit Binary Counter with Synchronous Load

Figure 25.4 Simulation of a 3-Bit Binary Counter with Synchronous Load

Procedure

3-Bit Presettable Counter from Flip-Flops

1. Use the Quartus II Block Editor to enter the presettable 3-bit counter shown in Figure 25.3. Save the file as:

   ***drive*:\qdesigns\labs\lab25\3bit_dct_sl\3bit_dct_sl.bdf**

   and use it to create a Quartus II project. Compile the project and create a simulation to verify its operation. Show the simulation to your instructor.

   Instructor's Initials ________________

2. Create a symbol for the 3-bit presettable counter from procedure 1.

3. In order to manually clock the counter from procedure 1 with a pushbutton switch, we require a switch debouncer, which is provided in the file **debouncer.vhd** in the **Student_Lab_Files** folder for this lab. Copy the file **debouncer.vhd** to your working folder, open the file, and create a symbol for the debouncer.

4. Use the Quartus II Block Editor to create a test circuit for the counter, as shown in Figure 25.5 (RSR PLDT-2 or DeVry eSOC) or Figure 25.6 (Altera UP-2). Save the file as:

   ***drive*:\qdesigns\labs\lab25\3bit_load\3bit_load.bdf** (Figure 25.5)

   or

   ***drive*:\qdesigns\labs\lab25\3bit_load_a\3bit_load_a.bdf** (Figure 25.6).

   The *CLK* input to the debouncer is the system clock from the on-board oscillator (4 MHz for the RSR PLDT-2, 25.175 MHz for the Altera UP-2, and 24 MHz for the DeVry eSOC board). PB_IN is an input from an on-board pushbutton. PB_OUT is a debounced output used as the clock for the counter.

   Note that the counter in Figure 25.6 has inverted outputs, so that it will drive the LEDs on the Altera UP-2 board causing a HIGH counter output to turn an indicator LED on. The counter symbol outputs can be inverted by using the Properties function of the Quartus II block.

   To invoke the function, click on the counter symbol to highlight it, then right-click to get a pop-up menu. Select **Properties.** Select **Q0** from the **Name** box and click **All** in the **Inversion** box. Do not click **OK** just yet. Select **Q1** and invert it. Select and invert **Q2.** Click **OK.** Figures 25.7 and 25.8 show the counter symbol before and after the outputs are inverted.

   Note *Unused* outputs are not necessary and may be left off the design.

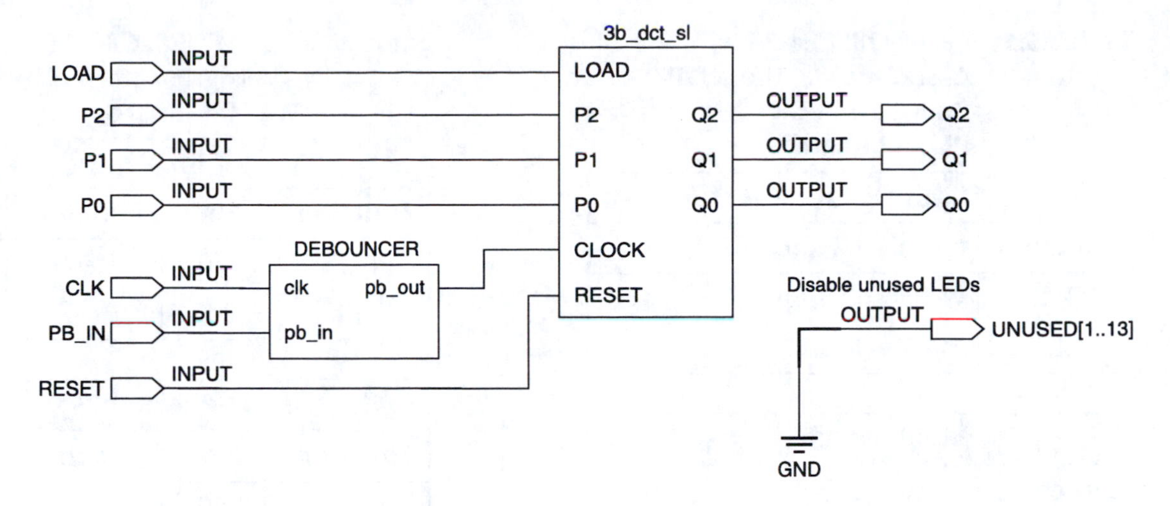

Figure 25.5 3-Bit Counter Test Circuit (RSR PLDT-2 or DeVry eSOC)

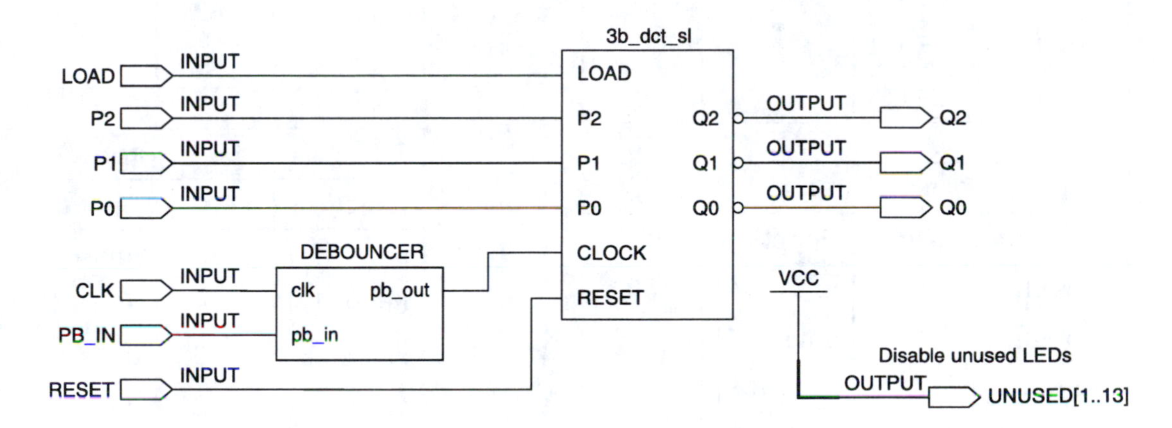

Figure 25.6 3-Bit Counter Test Circuit (Altera UP-2)

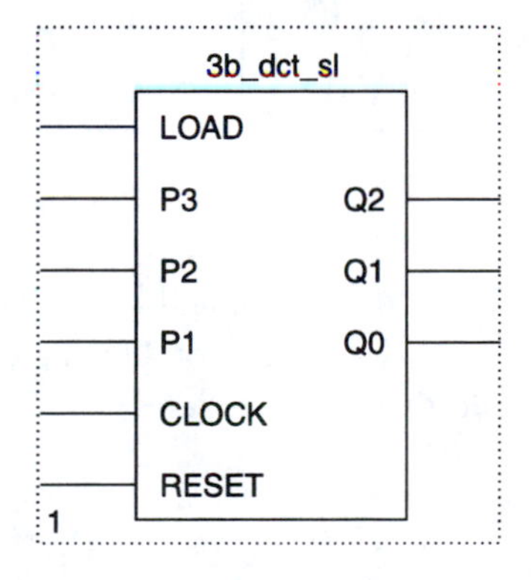

Figure 25.7 Symbol for a 3-Bit Presettable Counter

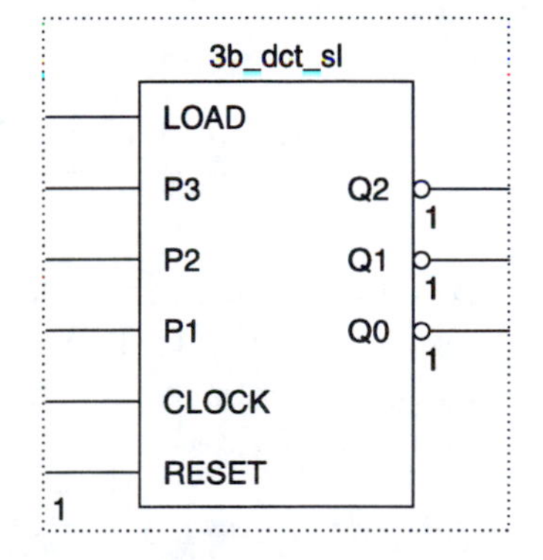

Figure 25.8 Symbol for a 3-Bit Presettable Counter with Inverted Outputs

Table 25.1 EPM7128LC84-7 Pin Assignments Altera UP-2 Board and RSR PLDT-2 Board

Pushbuttons and Clock			
Function	**Pin**	**Function**	**Pin**
PB_IN	11	Reset	1
CLK	83		

DIP Switches			
Function	**Pin**	**Function**	**Pin**
P2	34		
P1	33		
P0	36		
Load	41		

LED Outputs			
Function	**Pin**	**Function**	**Pin**
unused[1]	44	unused[6]	80
unused[2]	45	unused[7]	81
unused[3]	46	unused[8]	4
unused[4]	48	unused[9]	5
unused[5]	49	unused[10]	6
Q2	50	unused[11]	8
Q1	51	unused[12]	9
Q0	52	unused[13]	10

Table 25.2 EPM7128LC84-7 Pin Assignments DeVry eSOC Board

Pushbuttons and Clock			
Function	**Pin**	**Function**	**Pin**
PB_IN	130	Reset	131
CLK	24		

DIP Switches			
Function	**Pin**	**Function**	**Pin**
P2	160		
P1	161		
P0	162		
Load	144		

LED Outputs			
Function	**Pin**	**Function**	**Pin**
unused[1]	115	unused[6]	101
unused[2]	114	unused[7]	99
unused[3]	113	unused[8]	97
unused[4]	112	unused[9]	96
unused[5]	110	unused[10]	95
Q2	102	unused[11]	92
Q1	151	unused[12]	118
Q0	150	unused[13]	117

5. Assign pin numbers as shown in Tables 25.1 and 25.2. Disable unused LEDs by setting them HIGH (Altera UP-2) or LOW (RSR PLDT-2 or DeVry eSOC) and assigning appropriate pin numbers. Compile the file and download it to the CPLD board.

6. Test the counter by setting the **LOAD** input LOW and pressing the **CLK** pushbutton several times. Set the **LOAD** input HIGH, set the **P[2..0]** inputs to 101, and press the **CLK** pushbutton. Set **LOAD** LOW and clock a few times. Press the **RESET** pushbutton to clear the counter outputs. Show the operation of the circuit to your instructor.

Instructor's Initials ____________________

Lab 26

Parameterized Counters

Name ______________________ Class ____________ Date ____________

Objectives Upon completion of this laboratory exercise, you should be able to:

- Create a Block Diagram File in Quartus II that contains a counter from the Altera Library of Parameterized Modules (LPM).
- Customize the LPM counter to select required ports and parameters of the correct active level and value.
- Create on LPM counter in VHDL.
- Create Quartus II simulation files to verify the operation of the circuits designed for this laboratory exercise.
- Download the counter and shift register designs to a CPLD test board and demonstrate their operation.

Reference Ken Reid and Robert Dueck, *Introduction to Digital Electronics*

Chapter 9: Counters and Shift Registers

Altera, *LPM Quick-Reference Guide* (Available on the CD in the back of *Introduction to Digital Electronics* in the file **lpm.pdf**

Equipment Required

CPLD Trainer:

DeVry eSOC board with USB cable, or
Altera UP-2 circuit board with ByteBlaster download cable, or
RSR PLDT-2 circuit board with straight-through parallel port cable, or
equivalent CPLD trainer board with Altera EPM7128S CPLD

Quartus II Web Edition software
AC adapter, minimum output: 7 VDC, 250 mA DC
Anti-static wrist strap
#22 solid-core wire
Wire strippers

Experimental Notes

Altera offers a library of components that can be used in Quartus II as part of a VHDL or graphic design file. These **LPM (Library of Parameterized Modules)** components can be easily modified to create designs of any required size. For example, a parameter called LPM_WIDTH can be set to a given value to make a counter or shift register from 1 to 256 bits wide, subject to the number of logic cells available in a given CPLD.

A list of LPM components and a summary of their functions are given in the document LPM_Reference_Guide.pdf, available on the CD in the back of *Introduction to Digital Electronics* in the folder called "Lab Manual Files." We will examine the use of a parameterized counter (**lpm_counter**). A parameterized shift register will be examined in another lab.

An LPM module is specified by **ports** and **parameters.** A **port** is an input or output of the device, with a function such as clock, clear, or load. A **parameter** is a property of the block, such as **LPM_WIDTH,** a parameter that specifies how many bits its parallel input or output has. Port names are written in lower case; parameter names are in upper case, in the form LPM_*parameter.* Some ports and parameters, such as **clock** and **LPM_WIDTH,** must be used in all instances of **lpm_counter.** Others, such as **aclr** and **LPM_DIRECTION,** are optional.

Figure 26.1 shows the symbol for the **lpm_counter** component with its complete set of ports and parameters. The function of each port and permissible values of the parameters are listed in the **Help** for each component, accessible from the **Properties** dialog box or the Quartus II **Help** menu.

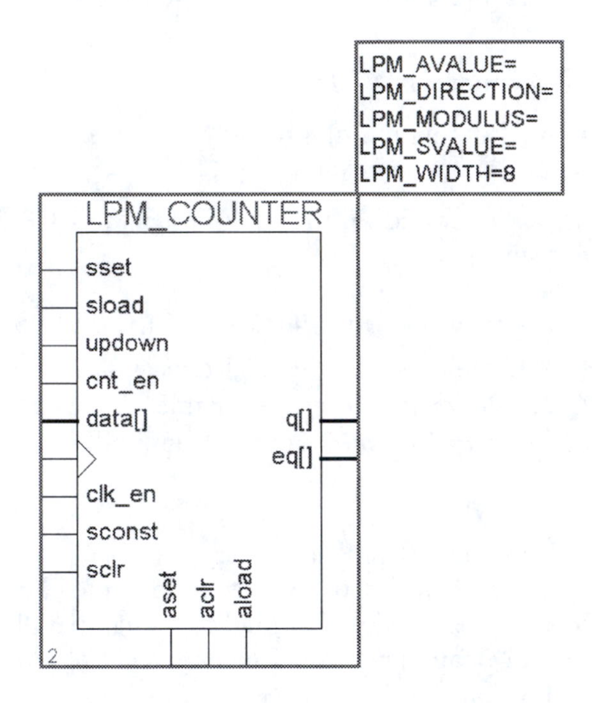

Figure 26.1 Available Port and Parameters for an LPM Counter

Procedure

LPM Counter (Block Diagram File)

1. Use the Quartus II Block Editor to create an LPM counter with the ports and parameters shown in Figure 26.2. Save the file as ***drive:*\qdesigns\labs\lab26\ct8lpm\ct8lpm.bdf** and use it to create a project in Quartus II. Compile the project.

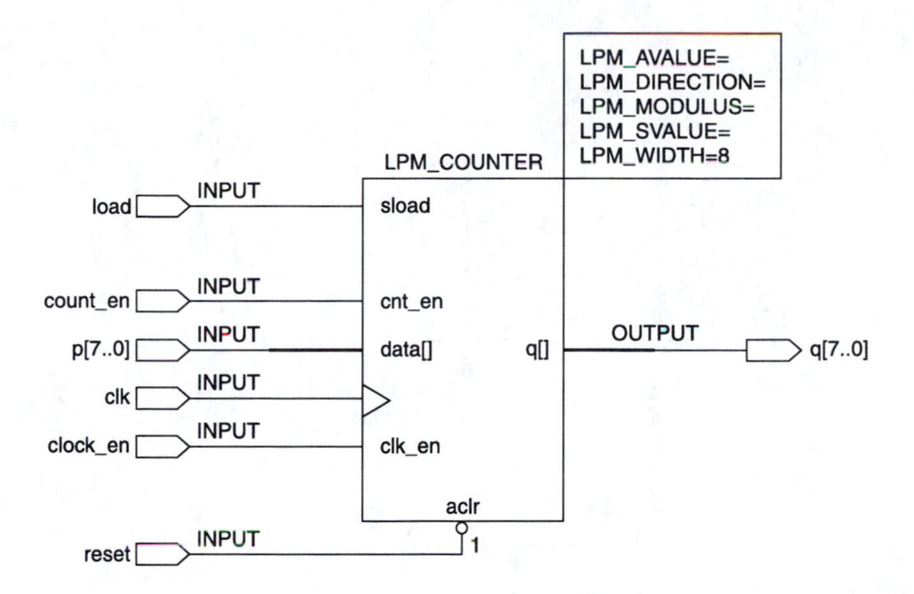

Figure 26.2 8-Bit LPM Counter

2. Create a simulation of the counter that tests all of the functions according to the following checklist:

   ____ Tests one full cycle of the count sequence (requires a modified end time)

   ____ Shows that load is really synchronous

   ____ Shows the difference between count enable and clock enable functions

   ____ Tests asynchronous reset

   Show the simulation to your instructor.

   Instructor's Initials ________________

3. Create a test circuit, shown in Figure 26.3 (RSR PLDT-2) or 26.4 (Altera UP-2 or DeVry eSOC). Assign pin numbers as shown in Table 26.1 (PLDT-2 or UP-2) or in Table 26.2 (DeVry eSOC). Add a debouncer to your design. Remember to invert your counter outputs if you are using the Altera UP-2 board so that the LEDs on the board will sequence correctly. Assign pins as needed. Compile the design and download it to the CPLD test board. Demonstrate the operation of the counter to your instructor. Compile the file and download it to the CPLD test board.

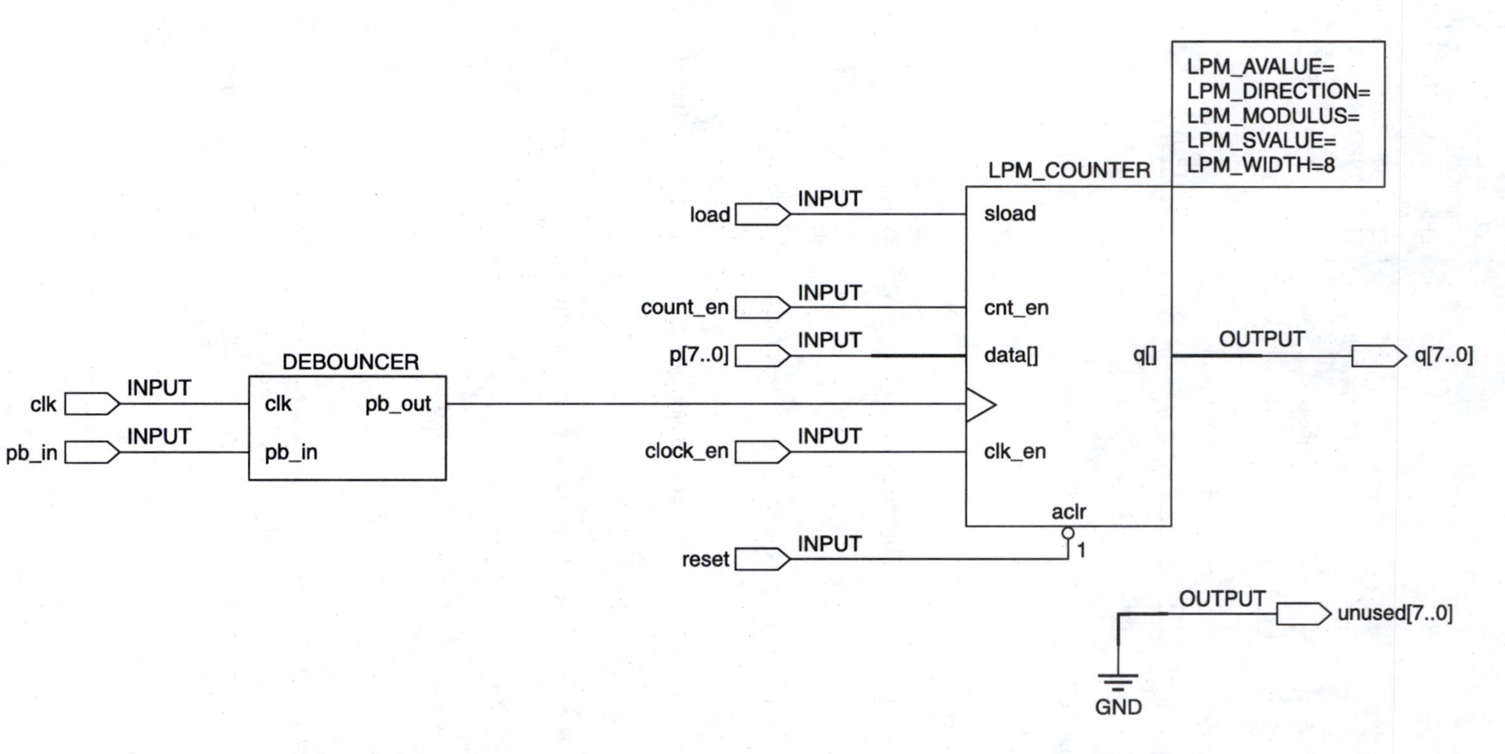

Figure 26.3 LPM Counter Test Circuit (RSR PLDT-2 or DeVry eSOC)

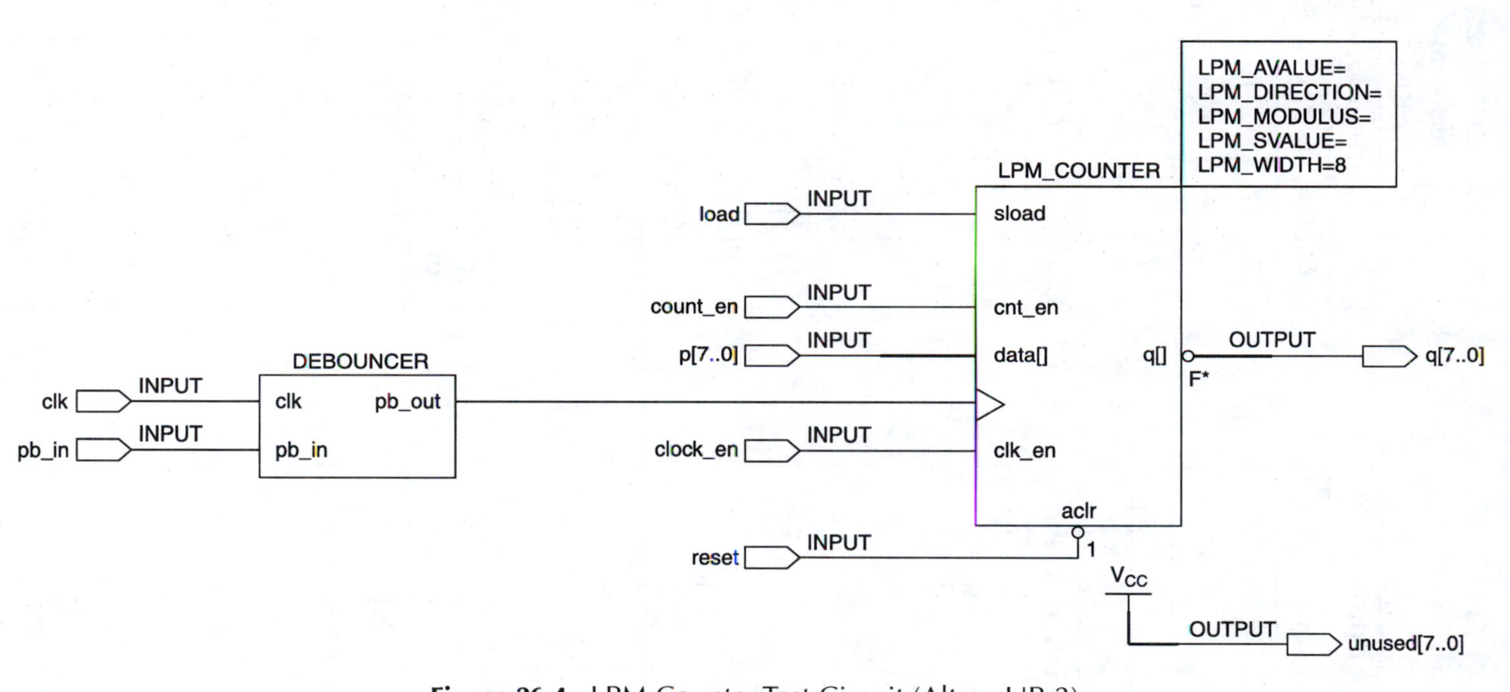

Figure 26.4 LPM Counter Test Circuit (Altera UP-2)

Table 26.1 EPM712BSLC84-7 Pin Assignments (UP-2 or PLDT-2)

Pushbuttons and Clock			
Function	**Pin**	**Function**	**Pin**
pb_in	11	Reset	1
clk	83		

DIP Switches			
Function	**Pin**	**Function**	**Pin**
p[7]	34		28
p[6]	33		29
p[5]	36		30
p[4]	35		31
p[3]	37		57
p[2]	40	clock_en	55
p[1]	39	count_en	56
p[0]	41	load	54

LED Outputs			
Function	**Pin**	**Function**	**Pin**
q[7]	44	unused[7]	80
q[6]	45	unused[6]	81
q[5]	46	unused[5]	4
q[4]	48	unused[4]	5
q[3]	49	unused[3]	6
q[2]	50	unused[2]	8
q[1]	51	unused[1]	9
q[0]	52	unused[0]	10

Table 26.2 EP2C8Q208C8N Pin Assignments DeVry eSOC Board

Pushbuttons and Clock			
Function	**Pin**	**Function**	**Pin**
PB_IN	131	clock_en	145
Reset	130	count_en	146
		load	147

DIP Switches			
Function	**Pin**	**Function**	**Pin**
p[7]	160		60
p[6]	161		61
p[5]	162		63
p[4]	163		64
p[3]	168		65
p[2]	141		
p[1]	142		
p[0]	144		

LED Outputs			
Function	**Pin**	**Function**	**Pin**
q[7]	115	unused[7]	101
q[6]	114	unused[6]	99
q[5]	113	unused[5]	97
q[4]	112	unused[4]	96
q[3]	110	unused[3]	95
q[2]	102	unused[2]	92
q[1]	151	unused[1]	118
q[0]	150	unused[0]	117

Special Function: Pin 83 (CLK; hardwired)

4. Test the functions of the LPM counter, including count, load, and reset. Show how count enable differs from clock enable. Demonstrate the circuit to your instructor.

Instructor's Initials ________________

Lab 27

Sequential Project: Digital Stopwatch

Name ______________________________ Class ______________ Date ______________

Objectives Upon completion of this laboratory exercise, you should be able to:

- Create a binary counter.
- Create a BCD counter using a counter from the Altera Library of Parameterized Modules (LPM) in a Block Diagram File.
- Combine components to make a digital stopwatch.

Reference Ken Reid and Robert Dueck, *Introduction to Digital Electronics*

Chapter 9: Counters and Shift Registers

Equipment Required CPLD Trainer:

Altera UP-2 circuit board with ByteBlaster download cable, or
DeVry eSOC board with USB cable, or
RSR PLDT-2 circuit board with straight-through parallel port cable, or
equivalent CPLD trainer board with Altera EPM7128S CPLD

Quartus II Student Web Edition software
AC adapter, minimum output: 7 VDC, 250 mA DC
Anti-static wrist strap
#22 solid-core wire
Wire strippers

Experimental Notes

LPM Counters

Figure 27.1 shows the symbol for the **lpm_counter** component with a minimal set of ports and a modulus of 10. The function of each port and permissible values of all parameters for this component are listed in the Help for the component, accessible from the Quartus II **Help** menu.

The function of one port is worthy of special mention. The multibit port called **eq[]** is an output decoder, consisting of 16 outputs: **eq0, eq1, eq2, . . . , eq15.** One of these bits goes HIGH when the internal flip-flops of the counter hold a particular value. For example, **eq0** goes HIGH when **q** *equals* 0, **eq1** goes HIGH when **q** *equals* 1, and so on, until finally **eq15** goes HIGH when **q** *equals* 15. We are using **eq9** (which goes HIGH when **q** *equals* 9) in the BCD counters used later in the lab to tell us when a BCD digit has reached its highest value (9). This allows us to enable the next counter digit by activating a **count enable** input, so it can count up in its turn.

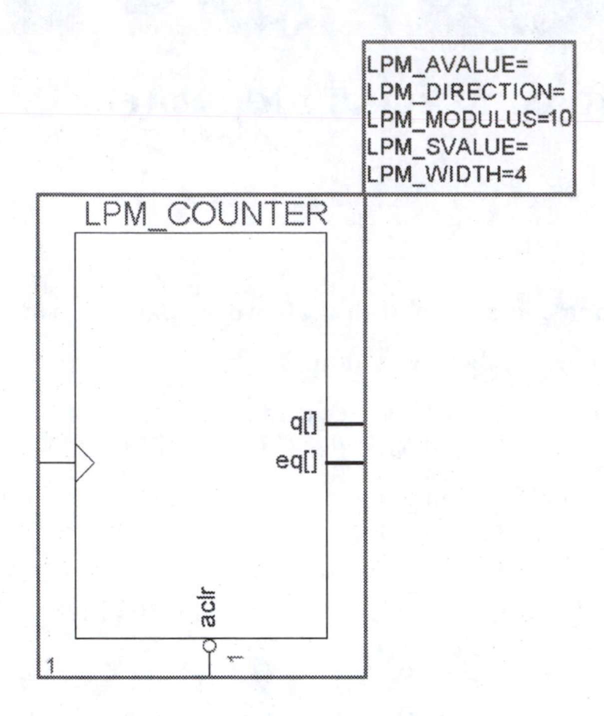

Figure 27.1 LPM_COUNTER Symbol

Digital Stopwatch

A digital stopwatch circuit is the major portion of this lab. It brings together the concepts of counters, decoders, logic gates, and flip-flops. We will build the counter in three stages:

1. We will build a two-digit binary-coded decimal (BCD) counter that counts at intervals of one second. This counter can be reset at any time by a pushbutton.
2. We will add a JK flip-flop that toggles an enable input on each counter to alternately start and stop the counter.
3. We will add two more counters and some logic gates to measure the tenths and hundredths of a second.

Two-Digit BCD Counter

Figure 27.2 shows a 2-digit BCD counter that counts from 00 to 99 then starts over. The module labeled **CLOCK_DIV_4M** is a counter that, for the RSR PLDT-2, divides the 4 MHz on-board clock to a value of 1 Hz. This can be created as a separate LPM counter with a suitable modulus. Both counters are **reset** when the **reset** input goes LOW by pressing a pushbutton switch.

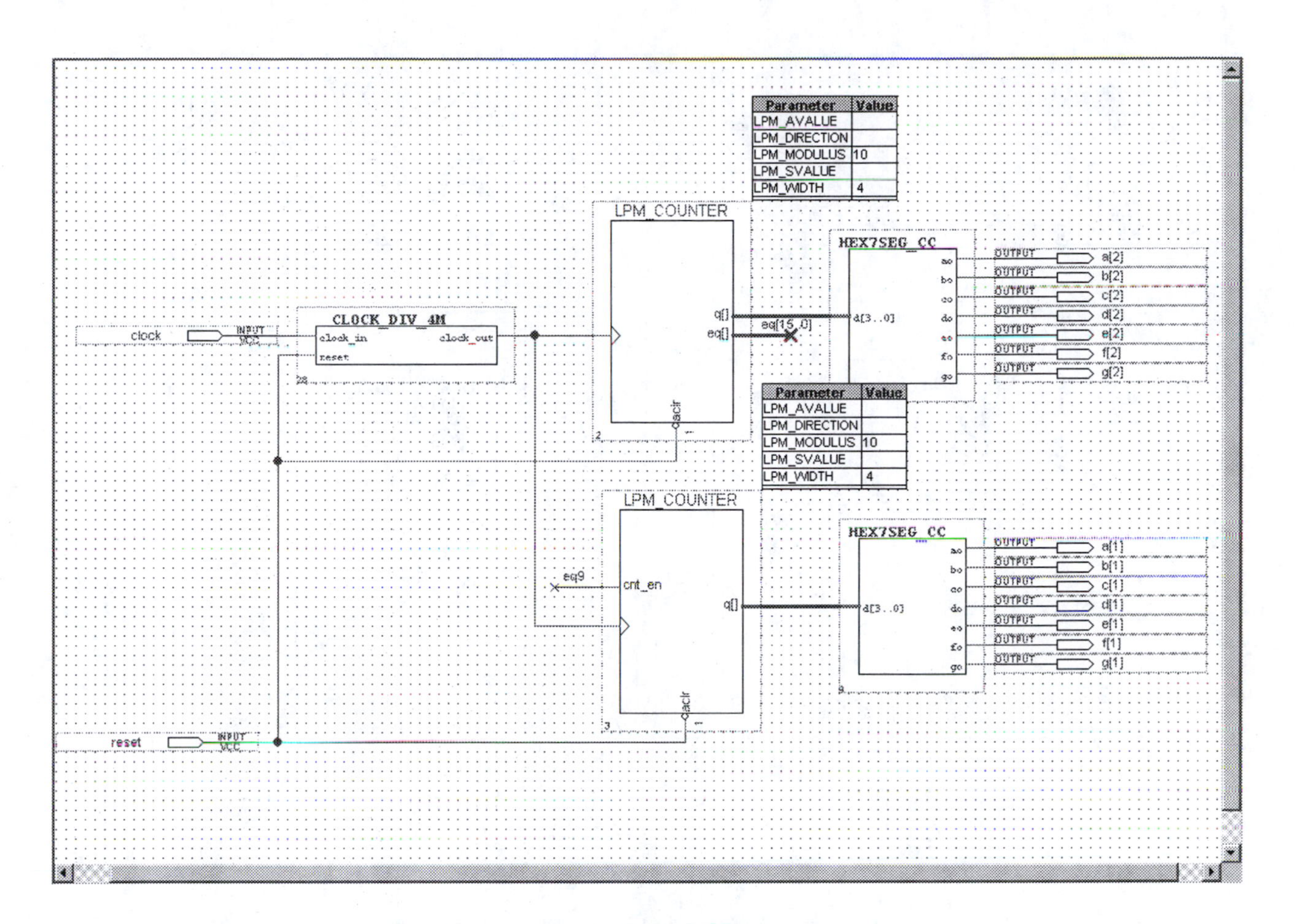

Figure 27.2 2-Digit BCD Counter

Note A similar counter having a different modulus must be used to divide the clock for the Altera UP-2 board from 25.175 MHz down to 1 Hz. If you are using the DeVry eSOC board, your counts must divide 24 MHz down to 1 MHz.

The **cnt_en** (count enable) port on the high-order counter prevents it from proceeding in its count sequence except when the low-order digit has a value of 9. When the low-order counter is at 9, then **eq9** goes HIGH, enabling the high-order counter.

Start/Stop Circuit

Figure 27.3 shows a modified version of the BCD counter. A new input has been added to each counter. The **clk_en** (clock enable) input allows the count to proceed when it is HIGH and stops the count when it is LOW. This state is toggled by the **start/stop** input, connected, via a switch debouncer circuit, to an input pushbutton.

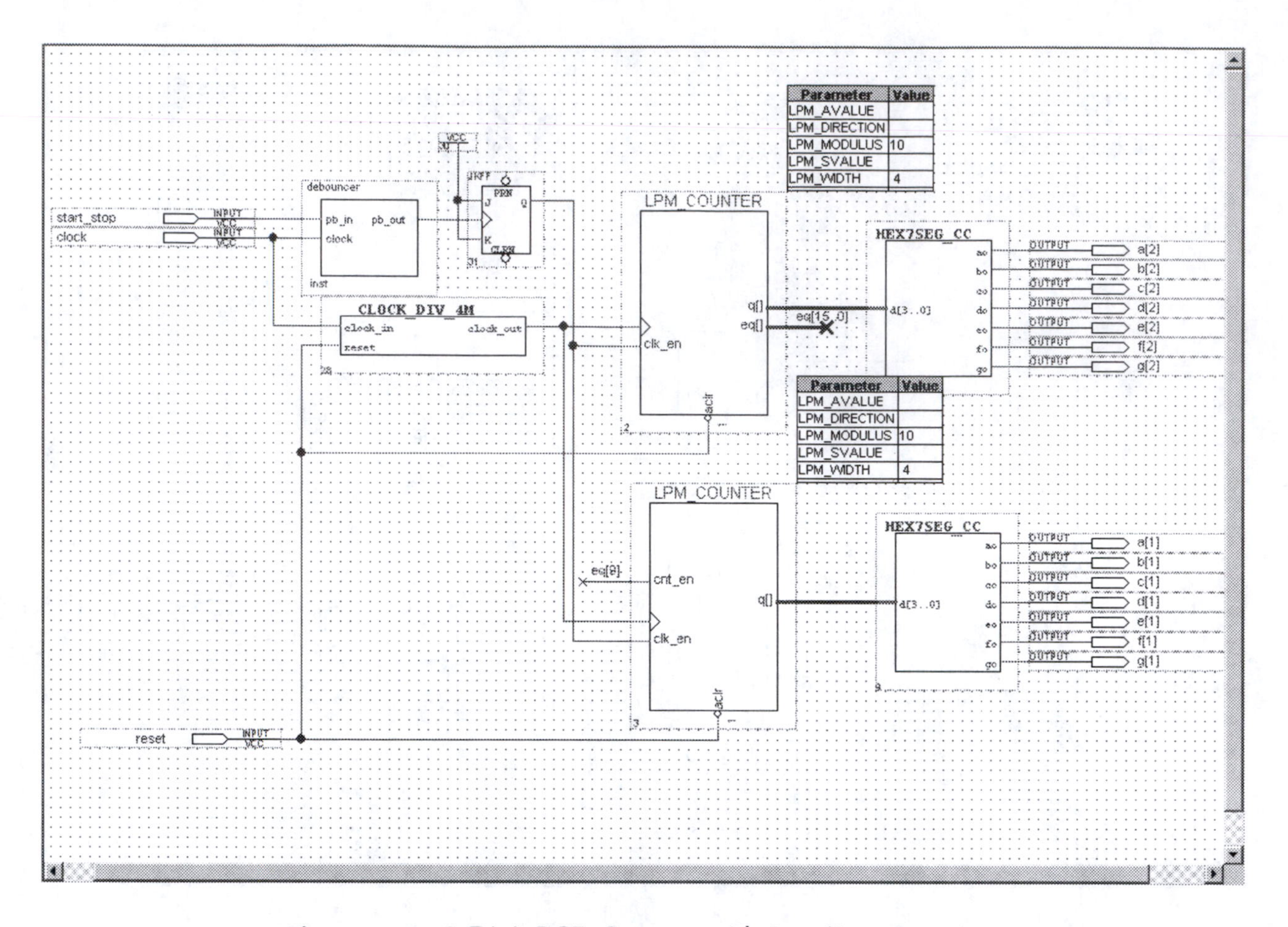

Figure 27.3 2-Digit BCD Counter with Start/Stop Function

The debouncer circuit is a VHDL file called **debouncer.vhd.** To use this file, copy it from the CD accompanying *Introduction to Digital Electronics* to the folder that holds the files for this project. Create a symbol for the file and insert it into the block diagram file for the counter.

Tenths and Hundredths

The tenths and hundredths digits can be added by modifying the circuit in Figure 27.4. (Figure 27.4 can be viewed on the CD at the back of *Introduction to Digital Electronics* and printed in landscape format on 11 × 17 paper.) The clock divider is changed so that its output is 100 Hz, rather than 1 Hz. Also the count enable circuits require logic gates at each stage so that the counter changes only when all previous digits are at 9 (e.g., 19.98, 19.99, 20.00, etc.) Tenths and hundredths of a second are shown on the LED displays of the CPLD test board. Units and tens are shown on the seven-segment displays.

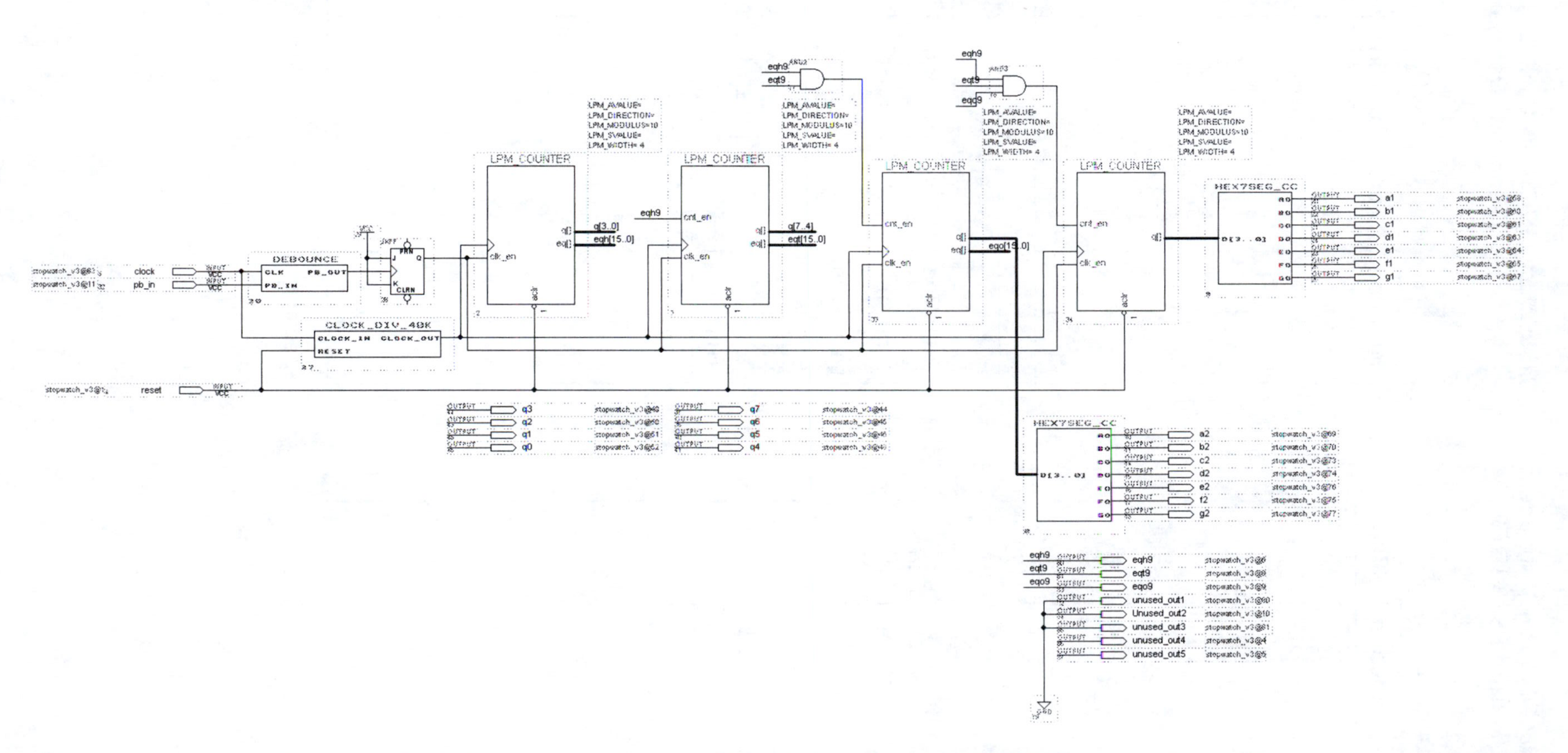

Figure 27.4 Tenths and Hundredths (Full-size image found on CD-ROM enclosed with *Introduction to Digital Electronics,* Reid/Dueck.)

Table 27.1 EPM7128LC84-7 Pin Assignments Altera UP-2 and RSR PLDT-2 Board

Seven-Segment Digits			
Function	**Pin**	**Function**	**Pin**
a1	58	a2	69
b1	60	b2	70
c1	61	c2	73
d1	63	d2	74
e1	64	e2	76
f1	65	f2	75
g1	67	g2	77
dp1	68	dp2	79

Pushbuttons			
Function	**Pin**	**Function**	**Pin**
pb_in (start/stop)	11	reset	1

LED Outputs			
Function	**Pin**		
q7	44		
q6	45		
q5	46		
q4	48		
q3	49		
q2	50		
q1	51		
q0	52		

Unassigned: Pins 12, 15, 16, 17, 18, 20, 21, 22, 24, 25, 27

Special Function: Pin 83–clock

Table 27.2 EP2C8Q208C8N Pin Assignments DeVry eSOC Board

Seven-Segment Digits			
Function	**Pin**	**Function**	**Pin**
a1	169	a2	137
b1	170	b2	138
c1	171	c2	139
d1	173	d2	128
e1	175	e2	80
f1	133	f2	81
g1	135	g2	84

Pushbuttons			
Function	**Pin**	**Function**	**Pin**
pb_in (start/stop)	131	reset	130

LED Outputs			
Function	**Pin**		
q7	115		
q6	114		
q5	113		
q4	112		
q3	110		
q2	102		
q1	151		
q0	150		

Special Function: Pin 24–clock

Procedure

1. Make a graphic file for a 2-digit binary-coded decimal (BCD) counter, with a 1-Hz clock, as shown in Figure 27.2. Save the file as ***drive:*\qdesigns\labs\lab27\stopwatch_v1\stopwatch_v1.bdf** and use it to create a project in Quartus II. Compile the project. Download the project to your board. Show the operation of the circuit to your instructor.

   Instructor's Initials ____________

2. Save the counter file as ***drive:*\qdesigns\labs\lab27\stopwatch_v2\stopwatch_v2.bdf** and make the modifications as shown in Figure 27.3. Use the file to create a new project and compile. Download the project to your board. Show the operation of the circuit to your instructor.

   Instructor's Initials ____________

3. Save the counter file as ***drive:*\qdesigns\labs\lab27\stopwatch_v3\stopwatch_v3.bdf.** Add the tenths and hundredths display by creating the circuit shown in Figure 27.4. Use the file to make a new project and compile. Download the project to your board. Show the operation of the circuit to your instructor.

   Instructor's Initials ____________

Lab 28

Shift Registers

Name ______________________ Class ____________ Date ____________

Objectives Upon completion of this laboratory exercise, you should be able to:

- Enter a 4-bit serial shift register circuit, using the Quartus II Block Editor.
- Create simulations of the shift register designs in this laboratory exercise.
- Download and test the shift register designs on a CPLD board.

Reference Ken Reid and Robert Dueck, *Introduction to Digital Electronics*

Chapter 9: Counters and Shift Registers

Equipment Required

CPLD Trainer:

Altera UP-2 circuit board with ByteBlaster download cable, or
DeVry eSOC board with USB cable, or
RSR PLDT-2 circuit board with straight-through parallel port cable, or
equivalent CPLD trainer board with Altera EPM7128S CPLD

Quartus II Web Edition software
AC adapter, minimum output: 7 VDC, 250 mA DC
Anti-static wrist strap
#22 solid-core wire
Wire strippers

Experimental Notes

A shift register is a synchronous sequential circuit that can move and store digital data. A serial shift register, which moves data in a straight line from one flip-flop to another, can be easily constructed with D flip-flops, as shown in Figure 28.1.

For this circuit, the Q output of one flip-flop connects directly to the D input of the adjacent one. This defines the direction of data movement in the register. Conventions vary, but we will state that movement from the most significant bit (MSB) to the least significant bit (LSB), as in Figure 28.1, constitutes a shift to the right.

Data enter these circuits at *serial_in* and move in sequence through the flip-flops.

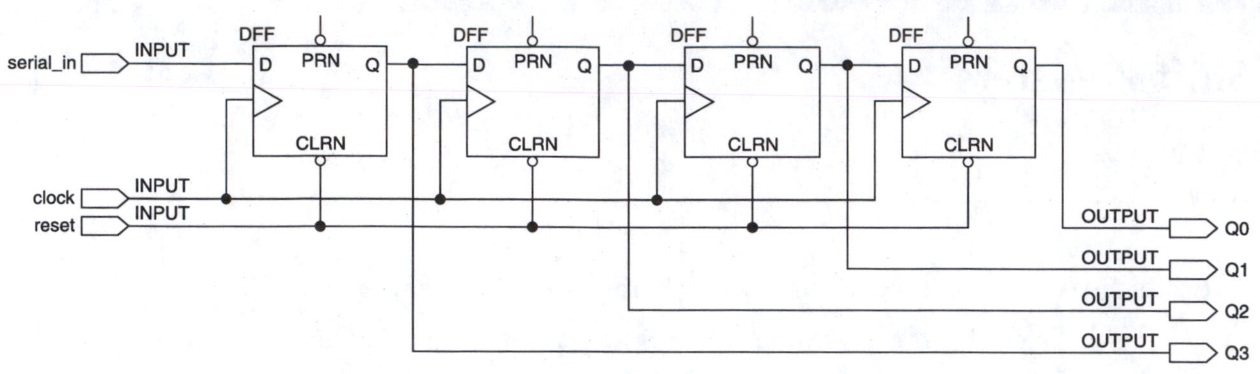

Figure 28.1 4-Bit Serial Shift Register (Shift Right)

Universal Shift Register

A universal shift register is one that can be configured in any combination of serial or parallel input or output. Many universal shift registers also provide the option of shift-left or shift-right.

The shift operation is controlled by a set of control inputs. A common configuration is as follows:

S_1S_0	Function
0 0	Hold (no change of output)
0 1	Shift right (fill with zeros)
1 0	Shift left (fill with zeros)
1 1	Parallel load

Procedure

Serial Shift Register from Flip-Flops

1. Use the Quartus II Block Editor to enter the circuit for a 4-bit serial shift register (shift right), as shown in Figure 28.1. Save the file as:

   ***drive:*\qdesigns\labs\lab28\srg4_sr\srg4_bdf.**

   Use the file to create a new project. Compile the project.

2. Create a simulation to verify the operation of the right-shift register. Show the simulation to your instructor.

   Instructor's Initials ____________________

3. Create a test circuit for the right-shift register, as shown in Figure 28.2 (RSR PLDT-2 or DeVry eSOC) or Figure 28.3 (Altera UP-2). The switch debouncer component, **debouncer.vhd,** is found in the student files found on the accompanying CD. Assign pin numbers as shown in Table 28.1. Also, assign pin numbers to disable any unused LEDs, as required. Recompile the file and download it to the CPLD test board.

Table 28.1 Pin Assignments for Serial Shift Register

	Pin Number		
Pin Name	**UP-2**	**PLDT-2**	**eSOC**
clock	83	83	124
pb_in	11	11	130
reset	1	1	131
serial_in	34	34	160
Q3	44	44	115
Q2	45	45	114
Q1	46	46	113
Q0	48	48	112

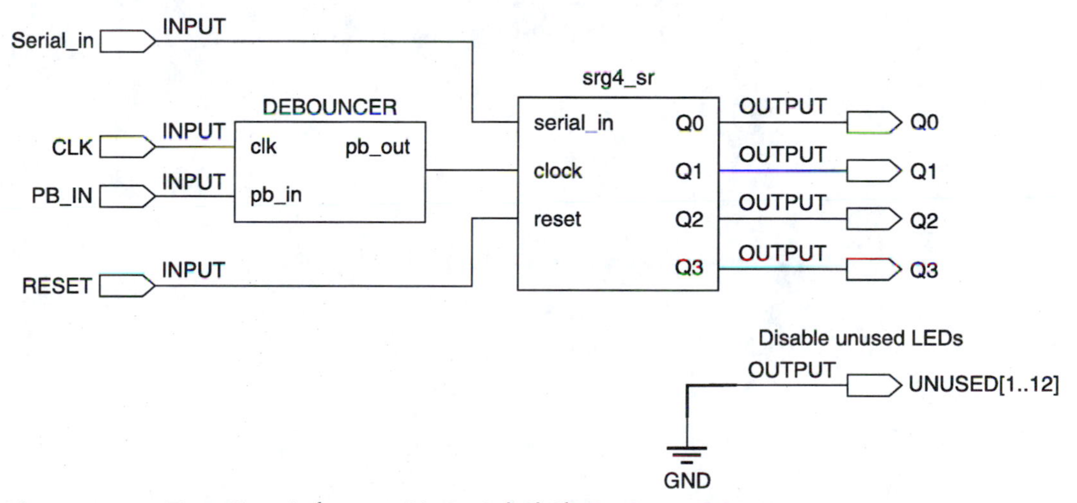

Figure 28.2 Test Circuit for a 4-Bit Serial Shift Register (RSR PLDT-2 or DeVry eSOC)

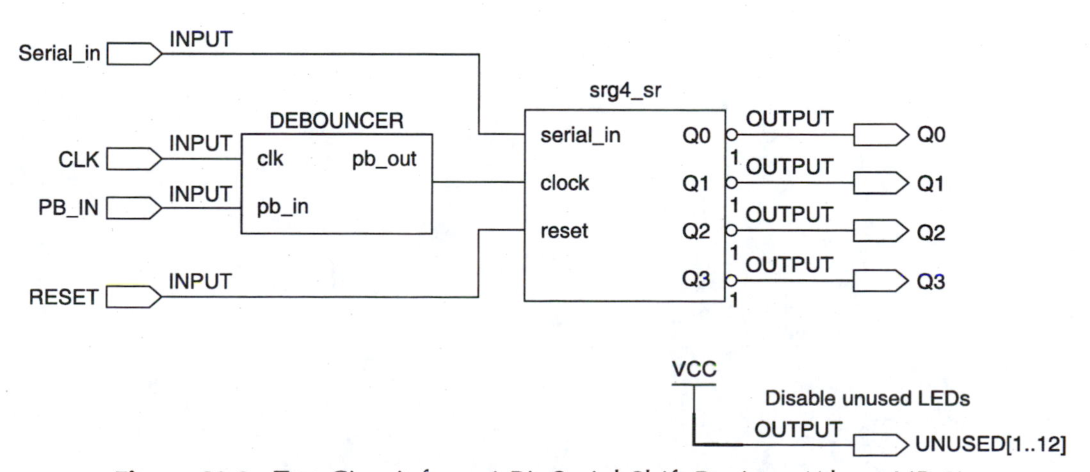

Figure 28.3 Test Circuit for a 4-Bit Serial Shift Register (Altera UP-2)

4. Wire a DIP switch to the serial input of the shift register. Connect PB_IN and RESET to pushbuttons and the Q outputs to LEDs as follows: Q3 to LED1, Q2 to LED2, Q1 to LED3, and Q0 to LED4.

5. Push the RESET pushbutton once to clear the shift register. Set the serial input HIGH, then press PB_IN once to clock the circuit once. Set the serial input LOW and press PB_IN several times. What do you observe?

   ____

6. Set the serial input HIGH. Clock the circuit several times. What do you observe?

   ____

7. Demonstrate the operation of the circuit to your instructor.

   Instructor's Initials ____________________

Lab

29

Parameterized Shift Registers

Name ______________________ Class ______________ Date ______________

Objectives

Upon completion of this laboratory exercise, you should be able to:

- Instantiate LPM shift registers as components, customized with the desired ports and parameters.
- Create Quartus II simulation files to verify the operation of the circuits designed for this laboratory exercise.
- Download the shift register designs to a CPLD test board and demonstrate their operation.

Reference

Ken Reid and Robert Dueck, *Introduction to Digital Electronics*

Chapter 9: Counters and Shift Registers

Altera, *LPM Quick-Reference Guide* (Available on the CD at the back of *Introduction to Digital Electronics* in the file **lpm.pdf.**)

Equipment Required

CPLD Trainer:

DeVry eSOC board with USB cable, or
Altera UP-2 circuit board with ByteBlaster download cable, or
RSR PLDT-2 circuit board with straight-through parallel port cable, or
equivalent CPLD trainer board with Altera EPM7128S CPLD

Quartus II Web Edition software
AC adapter, minimum output: 7 VDC, 250 mA DC
Anti-static wrist strap
#22 solid-core wire
Wire strippers

Experimental Notes

Altera offers a library of components that can be used in Quartus II as part of a graphic design file. These **LPM (Library of Parameterized Modules)** components can be easily modified to create designs of any required size. For example, a parameter called LPM_WIDTH can be set to a given value to make a counter or shift register from 1 to 256 bits wide, subject to the number of logic cells available in a given CPLD.

Figure 29.1 shows the symbol for the **lpm_shiftreg** component with its complete set of ports and parameters. The function of each port and permissible values of the parameters are listed in the Help for each component, accessible from the **Properties** dialog box or the Quartus II **Help** menu.

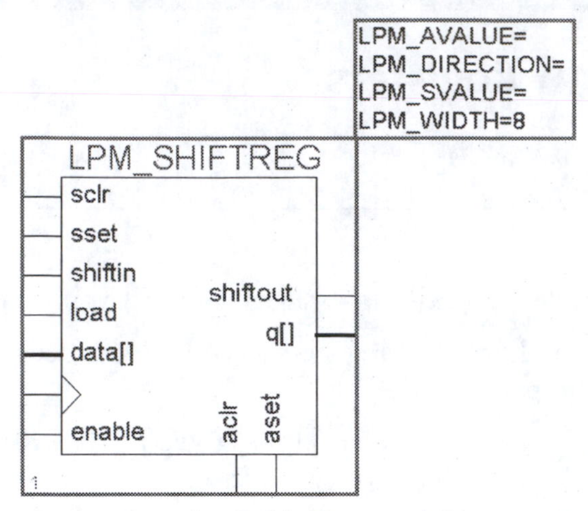

Figure 29.1 Available Port and Parameters for an LPM Shift Register

Procedure

LPM Shift Register

1. Refer to Altera's *LPM Quick Reference* or Quartus II Help to find the functions of the LPM shift register, **lpm_shiftreg.** Create a Block Diagram File for an 8-bit shift register having the following features:

   a. parallel load;

   b. active-LOW synchronous clear;

   c. serial input, set to 0;

   d. parallel outputs.

2. Create a simulation file for this shift register that fully demonstrates the functions listed. Show the simulation waveforms to your instructor.

   Instructor's Initials ____________

3. Add a switch debouncer as a component in the design file, using the Quartus II Graphic Editor. If you are using the Altera UP-2 or DeVry eSOC board, make sure to invert the outputs for the active-LOW LEDs. Assign pin numbers, compile the file, and download it to the CPLD test board. Demonstrate the operation of the shift register to your instructor.

   Instructor's Initials ____________

Lab 30

Sequential Project: Time Division Multiplexing

Name ______________________ Class ____________ Date ____________

Objectives Upon completion of this laboratory exercise, you should be able to:

- Design, program, and test a circuit that transmits and receives four 4-bit numbers in a multiplexed sequence along a single transmission path, using techniques and components from previous laboratory exercises.
- Use the Quartus II simulation tool to test the individual components of the 4-channel multiplexer and the whole circuit.
- Transmit serial data between two CPLD test boards.

Reference Ken Reid and Robert Dueck, *Introduction to Digital Electronics*

Chapter 9: Counters and Shift Registers
Lab 12: Binary and Seven-Segment Decoders
Lab 16: Multiplexer Applications
Lab 23: Latches and Flip-Flops
Lab 28: Shift Registers
Lab 26: Parameterized Counters
Lab 29: Parameterized Shift Registers

Equipment Required CPLD Trainer:

Altera UP-2 circuit board with ByteBlaster download cable, or
DeVry eSOC board with USB cable, or
RSR PLDT-2 circuit board with straight-through parallel port cable, or
equivalent CPLD trainer board with Altera EPM7128S CPLD

Quartus II Web Edition software
AC adapter, minimum output: 7 VDC, 250 mA DC
Anti-static wrist strap
#22 solid-core wire
Wire strippers

Note Hardware implementation of this project required 16 toggle switch inputs; the DeVry eSOC board has 8 toggle switch inputs. Your instructor may wish to modify this experiment for implementation using the DeVry eSOC board.

Experimental Notes

Time-division multiplexing is a method of improving the efficiency of a transmission system by sharing one transmission path among many signals. For example, if we wish to transmit four 4-bit numbers over a transmission line, we can send them one after the other, as shown in Figure 30.1.

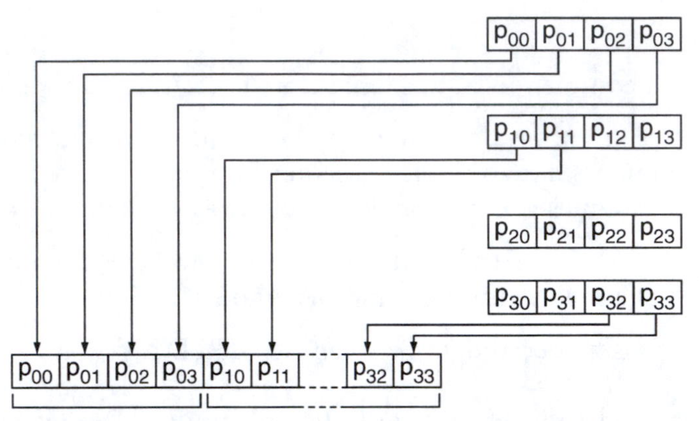

Figure 30.1 Time-Division Multiplexing of Four 4-Bit Words

Each bit is assigned a **time slot** in a sequence. During that time, it has sole access to the transmission line. When its time elapses, the next bit is sent and so on in sequence, until the channel assignment returns to the original location. In Figure 30.1, we see the 4-bit word p0 transmitted, LSB first, followed by p1, p2 (not shown), then p3.

This can be achieved by connecting four 4-bit serial shift registers to the inputs of a 4-channel multiplexer, as shown in the diagram for the 4-channel transmitter in Figure 30.2.

When a MUX channel is selected by a counter, its corresponding shift register has sole access to the transmission path. This lasts for four clock pulses, after which access to the transmission path passes to the next channel. Channel selection on the shift registers is controlled by a counter and decoder.

At the receive end of the circuit, shown in Figure 30.3, the incoming data stream is directed to one of four serial shift registers by a counter and decoder. Since only one shift register accepts data at any time, this has the effect of demultiplexing the data.

After a shift register receives its data, they are stored in a 4-bit latch. Transfer into the output latch happens only after the serial shifting has finished on that channel. During the shift period, old data are held in the latch so that the LED outputs do not flicker due to new data moving in.

Transmitter Implementation

The transmitter portion of this system consists of four main components: a multiplexer, four 4-bit shift registers, a 4-bit counter, and a decoder.

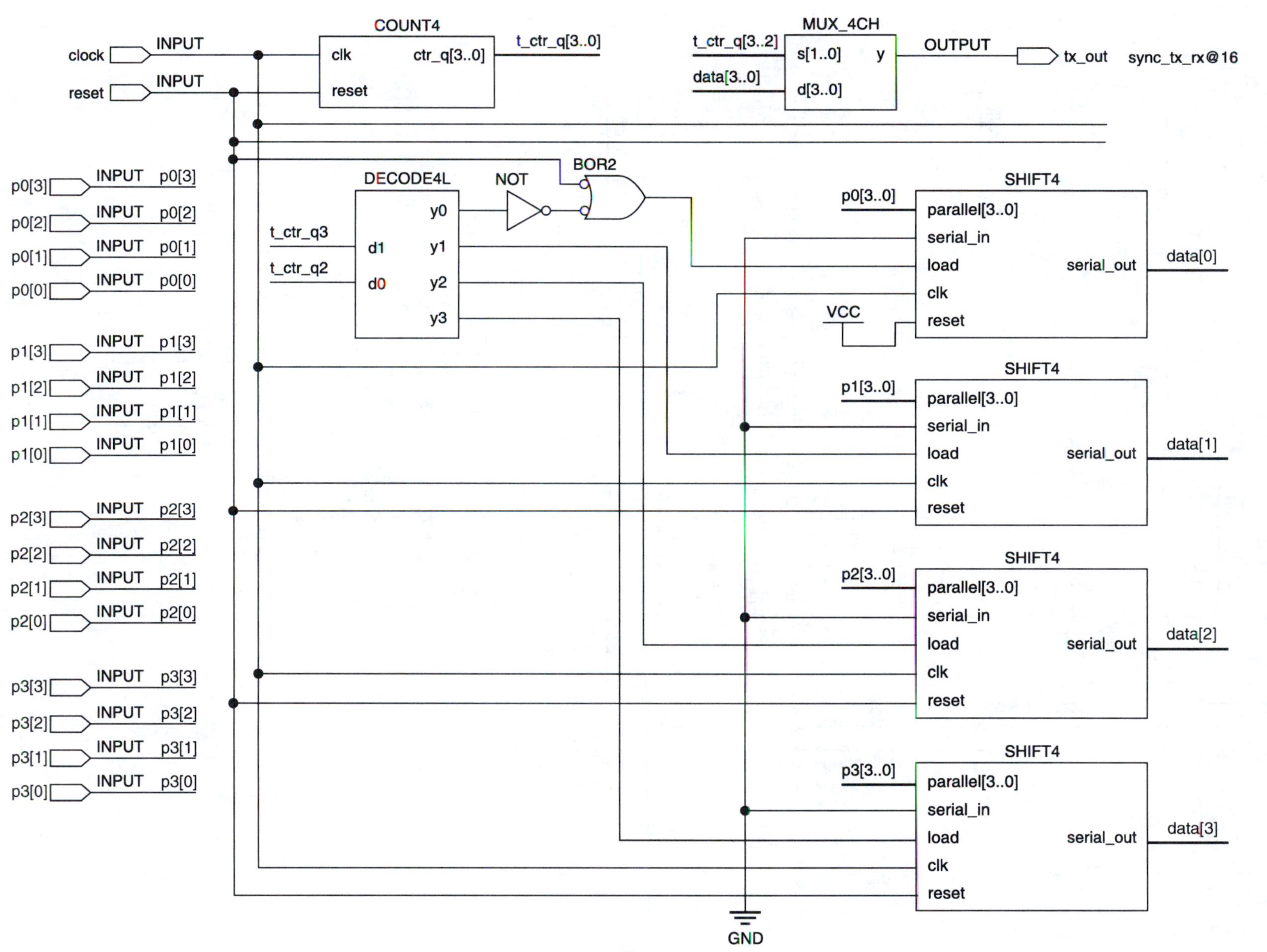

Figure 30.2 Multiplexing Transmitter

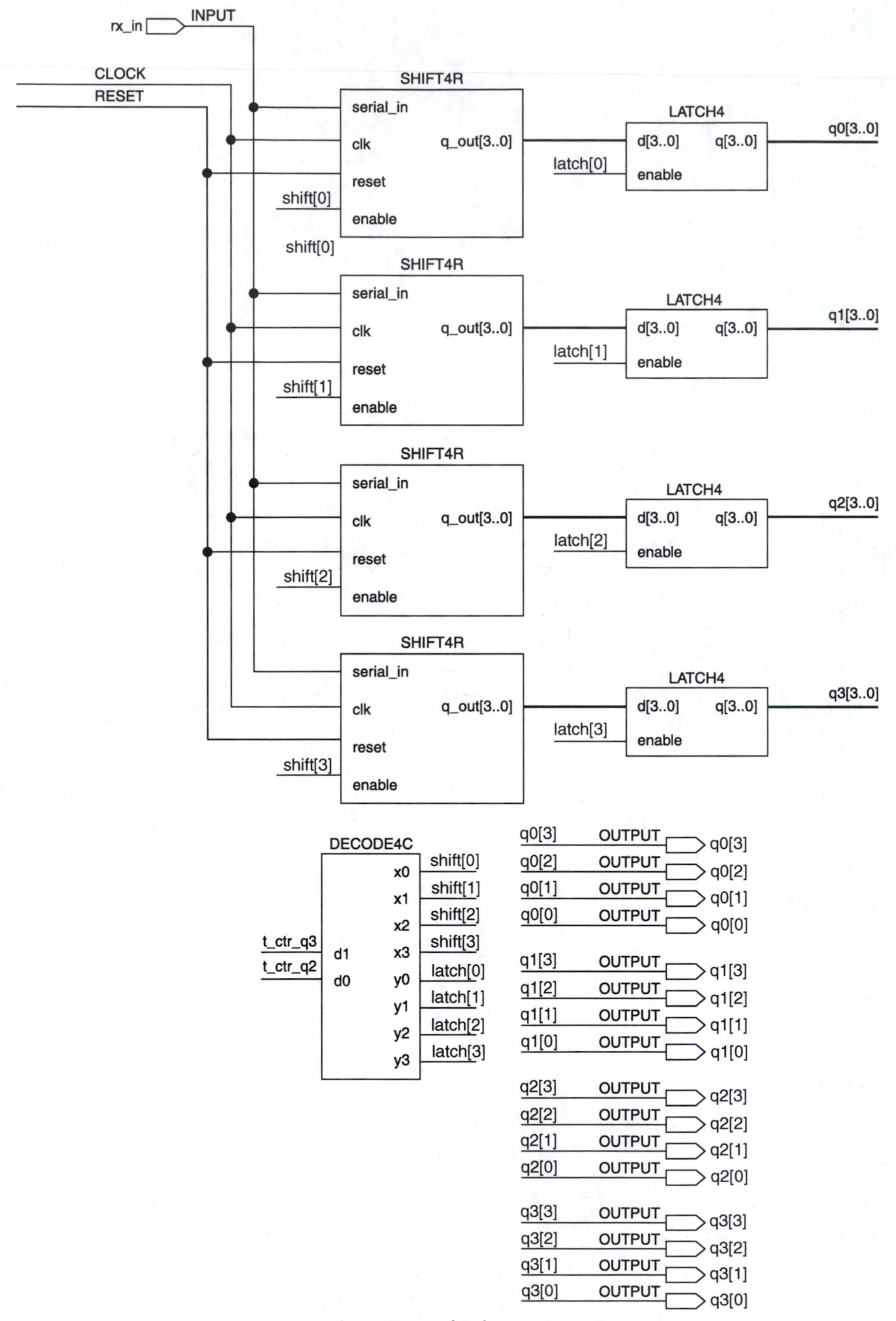

Figure 30.3 Demultiplexing Receiver

Multiplexer

Figure 30.2 shows the 4-channel multiplexer as a component named **mux_4ch**. The channel is selected by the upper two bits of a 4-bit counter. The data inputs are supplied by the serial outputs of the shift registers. The output, **tx_out**, provides the transmission path to the receiver portion of the circuit.

Shift Register

Figure 30.2 shows the symbol for the transmit shift registers as **shift4**. Each of these 4-bit shift registers has parallel inputs and a serial output. The serial output connects to one channel of the transmit multiplexer and sends four bits of data to the MUX when its channel is selected. Four bits of data require four clock pulses to serially transfer all data, so each shift register is selected for four clock pulses.

When the shift register is not selected, its LOAD input is HIGH, causing parallel data from four DIP switches to be transferred into the shift register. The parallel data are shifted out next time the channel is selected.

Each shift register, except for the first one, has an asynchronous reset. The first shift register (for channel 0) has a logic circuit that will synchronously load its data when the circuit is reset. Otherwise, if this register shared a reset line with the other registers, it would have no data during the first part of the transmit cycle, as it would not yet have loaded its parallel inputs.

Counter and Decoder

Figure 30.2 shows how the 4-bit counter and decoder control the sequence of selected channels in the transmitter circuit.

The decoder examines the two most significant bits of the counter and generates a LOW value on the output that corresponds to the binary value of these two bits. For example, when the two counter MSBs are 00, then Y0 goes LOW and Y1, Y2, and Y3 are all HIGH.

The LOW on Y0 sets the Channel 0 shift register into serial shift mode and keeps it there for four clock pulses. At the same time, the counter's two MSBs select Channel 0 on the multiplexer. (When channel 0 is selected, the counter goes through output values 0000, 0001, 0010, and 0011.) The transmitter serially shifts the four bits of the Channel 0 shift register to the transmission line output.

In a similar way, the next four counter states (0100, 0101, 0110, and 0111) select Channel 1 for serial transmission. Channel 2 transmits on the next four states, and then Channel 3. Following this, the transmit channel cycle repeats.

Receiver Implementation

The Block Diagram File of a 4 × 4-bit receiver is shown in Figure 30.3. The 4-channel receiver consists of the following components: 4-bit counter, four 4-bit shift registers and latches, and a 4-bit decoder with complementary outputs.

Counter

As in the transmitter, the counter acts as a sequencing device. The upper two bits are decoded to select an active channel, each channel for four clock periods. The counter is the same as the one used for the transmitter.

Shift Register

Each shift register accepts serial input data whenever its ENABLE input is HIGH. At all other times, it holds its previous value, even though the input data are applied to its serial input. Shift register outputs are parallel (**q[3..0]**). The register has an asynchronous reset.

Latch

The latch on each channel has four D inputs and four Q outputs. When *ENABLE* = 1, *Q* follows *D* (transparent mode). When *ENABLE* = 0, the latch holds its previous value (store mode). The latch is in store mode when the shift register is receiving serial data. This prevents the outputs from flickering while new data are being shifted in. When the shifting is complete, the latch becomes transparent, displaying the data in the shift register.

Decoder

The decoder, shown in Figure 30.3 as **decode4c**, selects the active shift register and latch channels. When a channel is active, the corresponding decoder **shift** output goes HIGH and its **latch** output goes LOW. For example, when the counter selects channel 2 (d_1d_0 = 10), **shift2** is HIGH and **latch2** is LOW. In this case, data will shift into receiver shift register 2, since it is enabled by **shift2**. The associated latch will be deactivated (i.e., in store mode) until a new channel is selected, at which time it will be allowed to go transparent and display the new data.

Transmitter Design Requirements

1. Design each of the transmitter modules and create a simulation for each one.
2. As you are building up the circuit from components, think of some ways to determine if the assembled portion of the circuit is working. For example, create a simulation for the counter. If that works, connect a switch debouncer or clock divider to the counter, assign it some pins and test it on your CPLD board to see if it produces a binary output on LEDs. Connect the counter outputs to the decoder and see if you can make the decoder pattern appear on LEDs, and so on. Show your instructor a simulation of the multiplexing transmitter.

Instructor's Initials ____________

3. Assign pins to the transmitter module as indicated in Table 30.1 or Table 30.2 at the end of this lab. If you are using the Altera UP-2 or De Vry eSOC board, invert the MUX output to make it compatible with the active-LOW LEDs on the board. Parallel inputs correspond to the input DIP switches. Clock is **PB1**. Reset is **PB2. Tx_out** is pin 44 for the UP-2 or PLDT-2 board or pin 4 for the eSOC board. (This is the pin for LED1. We will use this connection for testing only.)
4. Test the MUX transmitter by adding a clock divider or switch debouncer. Your should be able to see serial data appearing on LED1 of the CPLD test board. You should be able to change the data pattern (lights blinking in a repeating sequence) by changing the pattern of DIP switches. Show this to your instructor.

Instructor's Initials ____________

Receiver Design Requirements

1. Create each of the receiver components and create a simulation for each one.
2. Add the receiver components to the same drawing as the transmitter and interconnect the receiver components. Keep the receiver input and transmitter output on separate pins. Do not use a clock divider or switch debouncer as part of the design.
3. Create a simulation for the complete system.

Hint Run a simulation for the transmitter. Copy the resulting waveform from **Tx_out** to **Rx_in.** Run the simulation again.

Show this to your instructor.

Instructor's Initials ____________

4. Assign pin numbers to the receiver, according to Table 30.1 at the end of the handout. **Clock** and **reset** lines should be connected to the transmitter **clock** and **reset** lines. **Change the output of the transmitter to pin 16** (UP-2 or PLDT-2). For DeVry eSOC, consult your instructor. Compile and download the combined transmitter and receiver.
5. Connect a wire jumper from **tx_out** to **rx_in.** The LED outputs should follow the changes in the DIP switch inputs. Demonstrate the operation of the circuit to your instructor.

Instructor's Initials ____________

Communication Between Two Boards

Set up two boards so that the switches on one board change the LEDs on the other via a serial channel. Since communication will be possible in both directions, the connection is **full duplex.**

In order to communicate between two boards, it is necessary to designate one board as the **Master** and one as the **Slave,** as shown in Figure 30.4. The Master provides the clock to both boards and controls the reset function. (Pin numbers are for UP-2 or PLDT-2 boards. For eSOC boards, consult your instructor.)

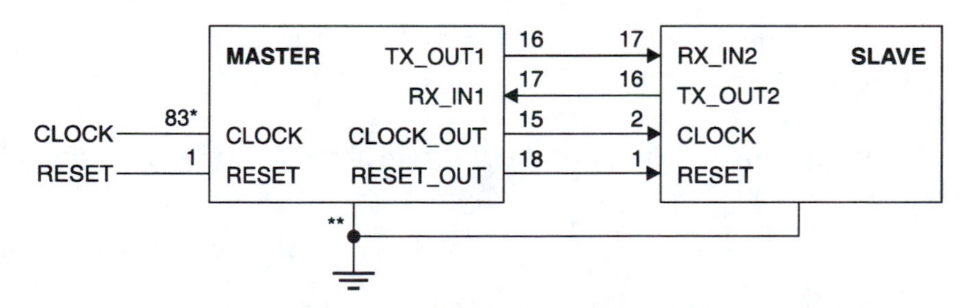

* Pin 83 (CLOCK) is connected internally. No external wire needed.

** This connection provides a common ground between the two boards. See text for details.

Figure 30.4 Communicating Between Two Boards

You will need to make some small changes to your pin assignments to make this configuration work. These changes are as follows.

Master

1. Since the Master must supply **clock** and **reset** functions to the Slave, you must add two "pass-through" pins for this function, as shown in Figure 30.5. Assign these pass-through pins to pins 15 (clock_out) and 18 (reset_out) (UP-2 or PLDT-2). Recompile and program the Master board.

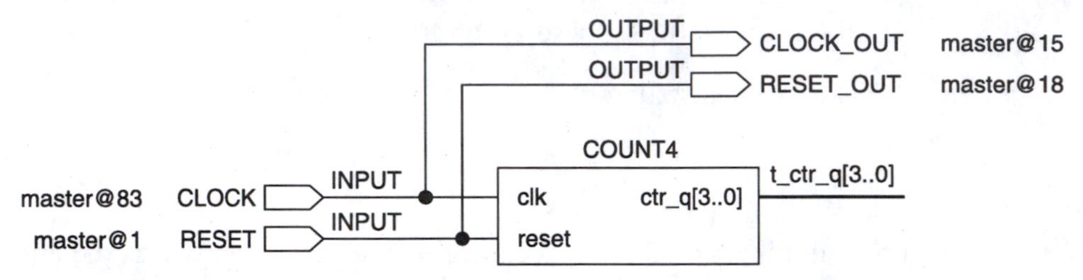

Figure 30.5 Configuring Clock and Reset for a Two-Board System

Slave

1. Change the clock pin assignment for the Slave clock to pin 2 (UP-2 or PLDT-2).
2. Recompile and program the Slave board.
3. Connect the two boards as shown in Figure 30.4. Note that the Master clock is hardwired from the on-board oscillator.
4. The ground pins must be connected between boards to ensure a common reference. There are a couple of options.
   - On the RSR PLDT-2, connect the (–) screw terminals for the board power supplies to each other;
   - On the Altera UP-2 board, solder a pin jack into the hole for the (–) terminal of the external power supply on each board and connect a wire between the board grounds, or;
   - Use pin 2 on the JTAG-OUT connector as a board ground. This is the bottom right pin when the board is oriented with the DIP switches on the side closest to you. Connect a wire between these points on both boards.
5. The switches on either board should control the LEDs on the other. Demonstrate this function to your instructor.

Instructor's Initials ____________________

Table 30.1 EPM7128LC84-7 Pin Assignments Altera UP-2 Board and RSR PLDT-2 Board

Pushbuttons and Clock			
Function	**Pin**	**Function**	**Pin**
PB1	11	Reset	1
Clock			

DIP Switches			
Function	**Pin**	**Function**	**Pin**
p0[3]	34	p2[3]	28
p0[2]	33	p2[2]	29
p0[1]	36	p2[1]	30
p0[0]	35	p2[0]	31
p1[3]	37	p3[3]	57
p1[2]	40	p3[2]	55
p1[1]	39	p3[1]	56
p1[0]	41	p3[0]	54

LED Outputs			
Function	**Pin**	**Function**	**Pin**
q0[3]	44	q2[3]	80
q0[2]	45	q2[2]	81
q0[1]	46	q2[1]	4
q0[0]	48	q2[0]	5
q1[3]	49	q3[3]	6
q1[2]	50	q3[2]	8
q1[1]	51	q3[1]	9
q1[0]	52	q3[0]	10

Transmit and Receive			
Function	**Pin**	**Function**	**Pin**
tx_out	16	rx_in	17

Note DeVry eSOC users should refer to Table 10.2 and consult your instructor for pin assignments.

Lab 31

State Machines

Name ______________________ Class ____________ Date ____________

Objectives Upon completion of this laboratory exercise, you should be able to:

- Draw the state diagram of a state machine from a verbal description.
- Enter the state machine design.
- Create a simulation that verifies the correctness of the state machine design.
- Combine the state machine with a switch debouncer and clock divider in a Quartus II Block Diagram File.
- Test the design on a CPLD test board.

Reference Ken Reid and Robert Dueck, *Introduction to Digital Electronics*

Chapter 10: State Machine Design

Equipment Required CPLD Trainer:

Altera UP-2 circuit board with ByteBlaster download cable, or
DeVry eSOC board with USB cable, or
RSR PLDT-2 circuit board with straight-through parallel port cable, or
equivalent CPLD trainer board with Altera EPM7128S CPLD

Quartus II Web Edition software
AC adapter, minimum output: 7 VDC, 250 mA DC
Anti-static wrist strap
#22 solid-core wire
Wire strippers

Experimental Notes

In this lab exercise you will design a state machine that will wait for a pushbutton to be pressed, wait for it to be released, then generate a sequence of four active LED outputs, one after the other. The LED sequence starts only after the pushbutton has been released. The pushbutton can be held for any length of time, but the sequence of outputs is always the same length of time.

For an Altera UP-2 board, an active LED is LOW. For an RSR PLDT-2 or DeVry eSOC board, an active LED is HIGH.

The state machine interprets a logic HIGH as a pressed push-button and a logic LOW as a pushbutton that is not being pressed. This accords with the logic levels generated by the switch debouncer component used with this circuit.

Procedure

State Diagram, Design Entry, and Simulation

1. Draw a state diagram of the state machine described in the Experimental Notes section of this lab. The detailed behavior of the machine is as follows:
   - The machine has six states, labeled s0 to s5. There is an input, called **input,** and four outputs, called **output[1..4]**.
   - When the machine is in **s0,** it monitors the input and stays in **s0** if **input** = 0. If **input** = 1, the machine transitions to **s1**. All outputs are OFF for both conditions.
   - When the machine is in **s1,** it monitors the input and stays in **s1,** with all outputs OFF, if **input** = 1. If **input** = 0, the machine makes a transition to **s2** and **output[1]** goes ON.
   - The machine makes an unconditional transition to s3. **Output** [1] goes OFF and **output[2]** goes ON.
   - The machine makes an unconditional transition to **s4**. **Output[2]** goes OFF and **output[3]** goes ON.
   - The machine makes an unconditional transition to **s5**. **Output[3]** goes OFF and **output[4]** goes ON.
   - The machine makes an unconditional transition to **s0**. All outputs are OFF.
2. Use the graphical editor to design the state machine just defined. Save the file as ***drive:*\qdesigns\labs\lab31\pulser\pulser.bdf** and use it to make a new project in Quartus II. Compile the project.
3. Write a set of simulation criteria to verify the correctness of your design. Use the criteria to create a simulation in Quartus II.

Simulation Criteria

Show the criteria and the simulation to your instructor.

Instructor's Initials ________________

Test Circuit

Figure 31.1 and Figure 31.2 show test circuits for the state machine designed in the previous section. Figure 31.1 is for the Altera UP-2 board. Figure 31.2 is for the RSR PLDT-2 or DeVry eSOC board.

1. Make a new folder for the block diagram file shown in Figure 31.1 or Figure 31.2. Use the New Project Wizard to create a project in the new folder.

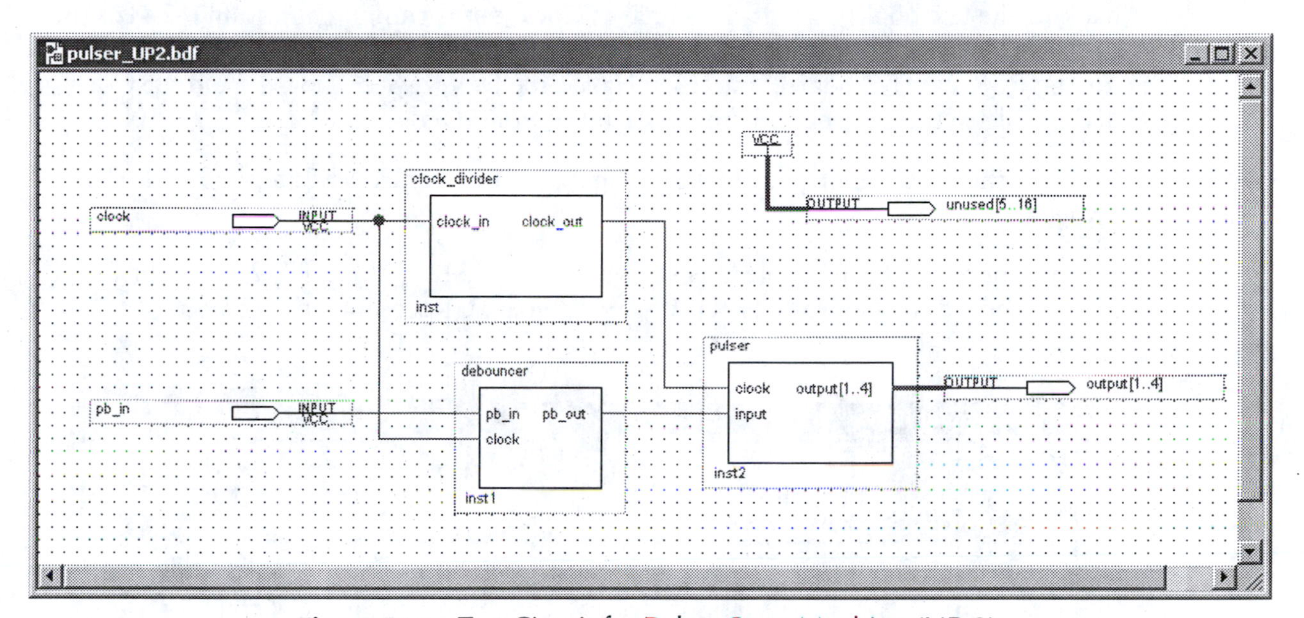

Figure 31.1 Test Circuit for Pulser State Machine (UP-2)

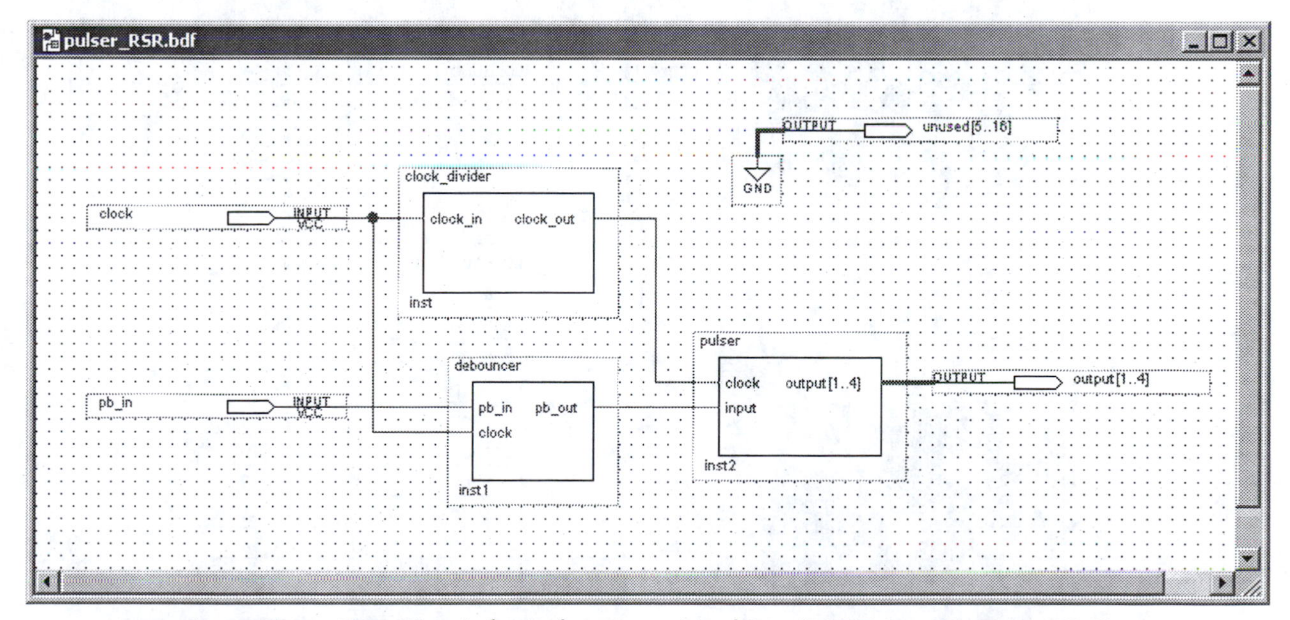

Figure 31.2 Test Circuit for Pulser State Machine (PLDT-2 or DeVry eSOC)

2. Copy the .bdf file for the state machine you entered in the previous section to the new folder. Save the file and add it to the new project. Create a symbol file for the state machine.

3. Copy the design file for the debouncer from the **Student_Lab_Files** folder on the CD placed in the back of the textbook to the working folder for this lab. Add the debouncer file to the project and create a symbol file.

4. Make a new file for the clock divider (see files for Lab 14 on the CD placed in the back of the textbook) so that the signal at **clock_out** is running at about 1–2 Hz. The signal at **clock_in** has a frequency of 25.175 MHz for an Altera UP-2 board, 4 MHz for the RSR PLDT-2 board, and 24 MHz for the DeVry eSOC board. Add the clock divider file to the project and create a symbol from the design file.

5. Connect the components, as shown in Figure 31.1 or Figure 31.2, and compile the project.

6. Add pin numbers to the project as shown in Figure 31.3 (Altera UP-2 board or RSR PLDT-2 board) or Figure 31.4 (DeVry eSOC board). Recompile the project.

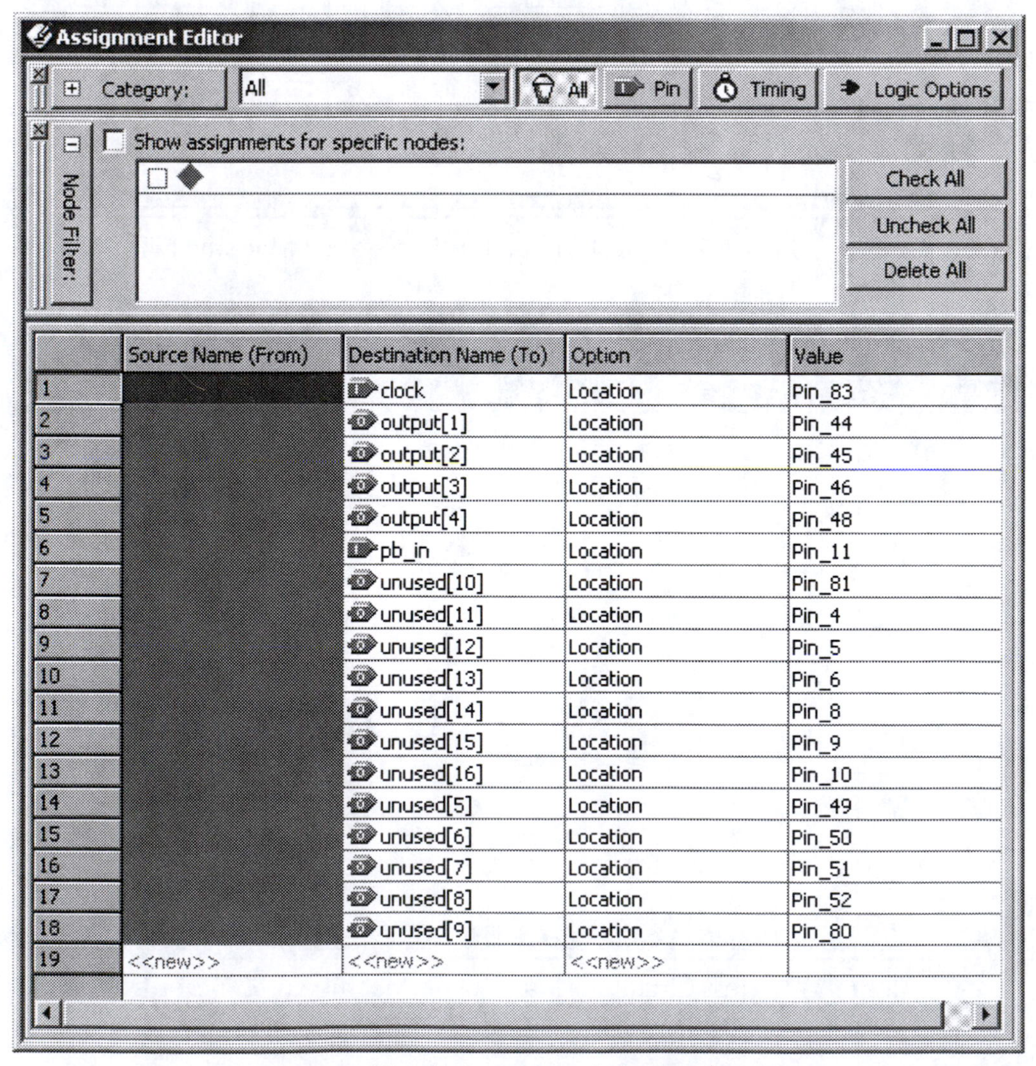

	Source Name (From)	Destination Name (To)	Option	Value
1		clock	Location	Pin_83
2		output[1]	Location	Pin_44
3		output[2]	Location	Pin_45
4		output[3]	Location	Pin_46
5		output[4]	Location	Pin_48
6		pb_in	Location	Pin_11
7		unused[10]	Location	Pin_81
8		unused[11]	Location	Pin_4
9		unused[12]	Location	Pin_5
10		unused[13]	Location	Pin_6
11		unused[14]	Location	Pin_8
12		unused[15]	Location	Pin_9
13		unused[16]	Location	Pin_10
14		unused[5]	Location	Pin_49
15		unused[6]	Location	Pin_50
16		unused[7]	Location	Pin_51
17		unused[8]	Location	Pin_52
18		unused[9]	Location	Pin_80
19	<<new>>	<<new>>	<<new>>	

Figure 31.3 Pin Assignments (UP-2 and PLDT-2)

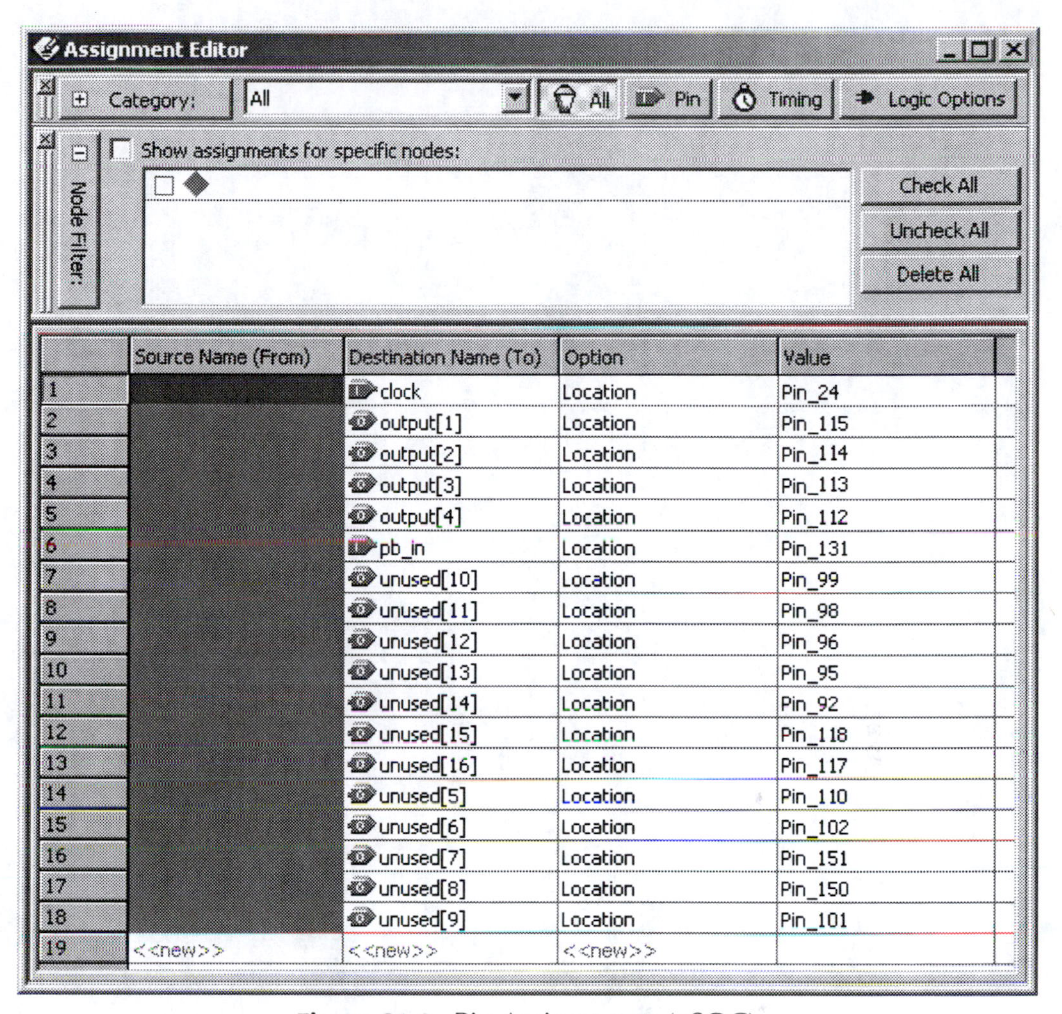

	Source Name (From)	Destination Name (To)	Option	Value
1		clock	Location	Pin_24
2		output[1]	Location	Pin_115
3		output[2]	Location	Pin_114
4		output[3]	Location	Pin_113
5		output[4]	Location	Pin_112
6		pb_in	Location	Pin_131
7		unused[10]	Location	Pin_99
8		unused[11]	Location	Pin_98
9		unused[12]	Location	Pin_96
10		unused[13]	Location	Pin_95
11		unused[14]	Location	Pin_92
12		unused[15]	Location	Pin_118
13		unused[16]	Location	Pin_117
14		unused[5]	Location	Pin_110
15		unused[6]	Location	Pin_102
16		unused[7]	Location	Pin_151
17		unused[8]	Location	Pin_150
18		unused[9]	Location	Pin_101
19	<<new>>	<<new>>	<<new>>	

Figure 31.4 Pin Assignments (eSOC)

7. Program your CPLD test board. Press the pushbutton assigned to **pb_in**. What do you observe?

8. Experiment with holding the pushbutton for different lengths of time. What is the shortest time that allows the machine to operate correctly? What happens if you hold down the pushbutton for a long time? Demonstrate the operation of the machine to your instructor.

Instructor's Initials ____________

Lab 32

Sequential Project: Traffic Light Controller

Name ______________________ Class ____________ Date ____________

Objectives Upon completion of this laboratory exercise, you should be able to:

- Design, simulate, and program a traffic-light controller.

Reference Ken Reid and Robert Dueck, *Introduction to Digital Electronics*
Chapter 10: State Machine Design

Equipment Required CPLD Trainer:
Altera UP-2 circuit board with ByteBlaster download cable, or
DeVry eSOC board with parallel port cable, or
RSR PLDT-2 circuit board with USB cable, or
equivalent CPLD trainer board with Altera EPM7128S CPLD
Quartus II Web Edition software
AC adapter, minimum output: 7 VDC, 250 mA DC
Anti-static wrist strap
#22 solid-core wire
Wire strippers

Experimental Notes

A **state machine** is a synchronous sequential circuit whose states progress according to the inherent design of the machine and possibly according to the states of one or more control inputs.

Procedure

Traffic Light Controller

A simple traffic light controller can be implemented by a state machine that has a state diagram such as the one shown in Figure 32.1.

The circuit has control over a North-South road and an East-West road. The North-South lights are controlled by outputs called **nsr, nsy,** and **nsg** (North-South red, yellow, green). The East-West road is controlled by similar outputs called **ewr, ewy, ewg.**

The cycle is controlled by an input called TIMER which controls the length of the two green-light cycles (s0 represents the EW green; s2 represents the NS green.) When TIMER = 1, a transition from s0 to s1 or from s2 to s3 is possible. This accompanies a change from green to yellow on the active road. The light on the other road stays red. An unconditional transition follows, changing the yellow light to red on one road and the red light to green on the other.

The outputs in the state diagram of Figure 32.1 are indicated as active-LOW, which is suitable for the Altera UP-2 or DeVry eSOC board. For the RSR PLDT-2 board, you must use active-HIGH outputs instead.

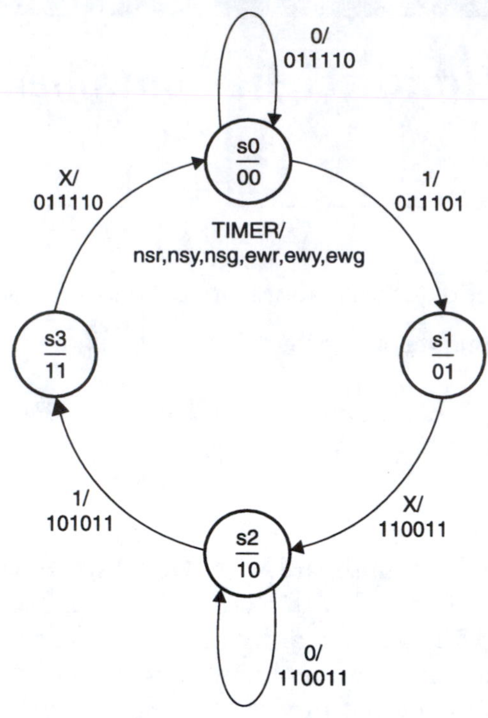

Figure 32.1 State Diagram for a Traffic Light Controller

The cycle can be set to any length by changing the signal on the TIMER input. (The yellow light will always be on for one clock pulse.) For ease of observation, we will use a cycle of ten clock pulses for any one road: 4 clocks GREEN, 1 clock YELLOW, 5 clocks RED. This can be generated by a mod-5 counter, as shown in Figure 32.2.

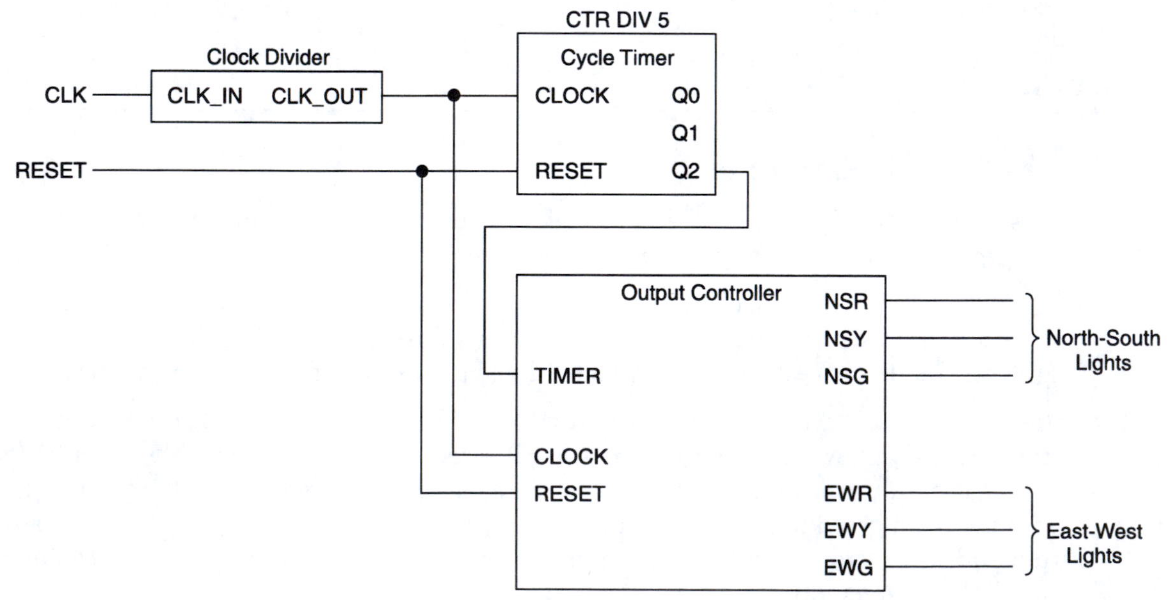

Figure 32.2 Logic Diagram for a Traffic Light Controller

The clock divider brings the on-board oscillator frequency down to the range of visible observation for our CPLD board. A 25-bit counter is used for the Altera UP-2 board, which has an on-board oscillator frequency of 25.175 MHz or a DeVry eSOC board, which has a 24 MHz clock. A 22-bit counter is suitable for the RSR PLDT-2 board, each of which has an on-board oscillator with a frequency of 4 MHz. Calculate the frequency of the state machine clock for your CPLD board.

f = ________

Draw the timing diagram of the mod-5 counter in the space provided:

CLK

Q0

Q1

Q2

How does the counter set the green-light cycle length to four clock pulses?

____

____

Creating a Traffic Light Controller in VHDL

1. Create a graphical file to implement the traffic controller state diagram shown in Figure 32.1, combined with the mod-5 cycle timer, but not the clock divider.

2. Create a simulation that shows the combined operation of the output controller and cycle timer. Show the waveforms to your instructor.

   Instructor's Initials ________________

3. Add a clock divider for the traffic controller, as shown on the logic diagram of Figure 32.2. Set the clock divider width to 22 (RSR PLDT-2) or 25 (Altera UP-2 or DeVry eSOC). Assign pins to the design so that the North-South lights correspond to LED1–LED3 on the CPLD board and the East-West lights correspond to LED9–LED11. Assign a pin to the state machine clock (so that it can be observed directly) and make it correspond to LED16. **Disable all other LEDs.** Make sure that the controller outputs are at the correct active level for the LEDs on your CPLD board.

4. Download the file to the CPLD board and demonstrate the operation to your instructor.

   Instructor's Initials ________________

Traffic Controller with Walk Signal

1. Modify the files of the previous section to implement a traffic light controller with a walk signal, as shown in the logic diagram of Figure 32.3.

   The walk signal goes on for one green cycle of the direction related to the switch. For example, when you press the NS switch, the North-South walk signal goes on for the next North-South green cycle. On the next NS green cycle, the walk signal is off unless the NS switch has been pressed again.

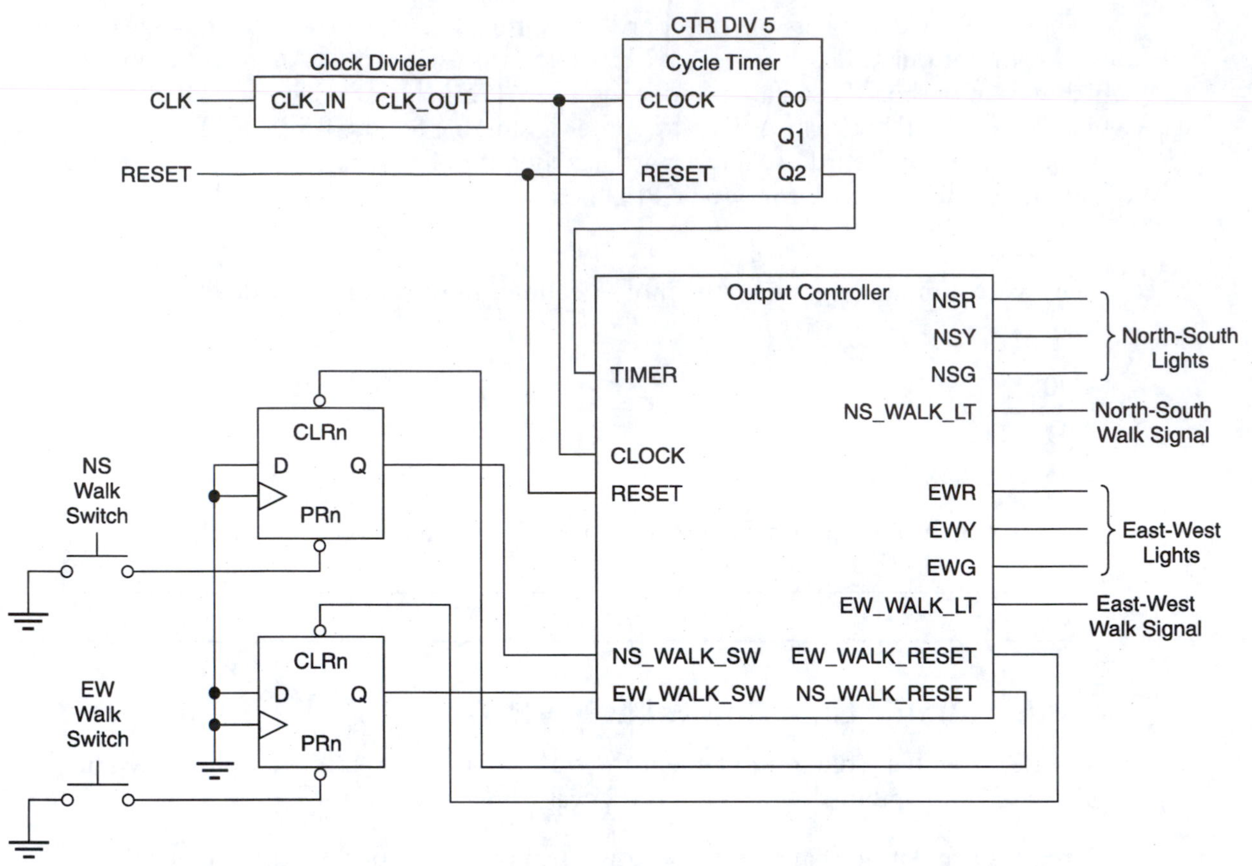

Figure 32.3 Logic Diagram for a Traffic Light Controller with a Walk Signal

The switches can be pressed at any time, as their states are stored in a pair of flip-flops. The status of the flip-flop determines the transition to a state for which the **walk** signal is (or is not) active. The flip-flop is asynchronously reset by the state machine at the end of its active cycle. The state diagram of Figure 32.1 forms the core of the new state diagram. The machine operates as follows:

- The machine resets to North-South red and East-West green (s0) and remains in this state until the TIMER input goes HIGH. Both walk signals are off.
- The machine transitions to s1 when TIMER = 1. East-West goes yellow.
- If the machine is in s1 and the North-South walk switch has not been pressed, the machine goes to s2. North-South is green for four clocks, until Timer = 1. East-West is red and both walk signals are off. The machine goes to s3 when TIMER = 1. Outputs are NS yellow and EW red. Walk signals are off.
- If the machine is in s1 and the North-South walk switch has been pressed, the machine goes to a new state, s4, which behaves exactly the same as s2, except that the NS walk signal is now on. As long as TIMER = 0, the machine remains in s4, with North-South light green and East-West light red. When TIMER = 1, s4 makes a transition to s3. In this transition, the walk signal turns off and the latch reset output (NS_WALK_RESET) goes LOW for one clock pulse. Outputs are NS yellow and EW red.
- If the machine is in s3 and the EW walk switch has not been pressed, the machine makes a transition to s0. Outputs are NS red and EW green. Walk signals are off.

- If the machine is in s3 and the EW walk switch has been pressed, this is stored as the HIGH state of a latch output. This is sensed at input EW_WALK_SW on the state machine and there is a transition to s5. This behaves the same as s0, except that the EW walk signal is on. As long as TIMER = 0, the machine stays in s5, with outputs North-South red and East-West green. When TIMER = 1, the machine makes a transition to s1. The walk signal turns off and a LOW pulse on EW_WALK_RESET resets the EW latch. The outputs are now NS red and EW yellow.

2. Draw the modified state diagram for the controller, as previously described, in the space provided on this page.
3. Implement the modified design.
4. Assign additional pins for the walk signals: NS walk on LED4 and EW walk on LED12. Compile and download the file to the CPLD test board. Demonstrate the operation of the circuit to your instructor.

Instructor's Initials ____________

Drawing of the modified state diagram for the controller:

Lab 33

Tristate Bussing in Altera CPLDs

Name ______________________ Class ____________ Date ____________

Objectives Upon completion of this laboratory exercise, you should be able to:

- Use Quartus II to create a data transfer system that uses a tristate bus to transfer data between two sources and two destinations.
- Create a simulation of the tristate bus system.
- Test the tristate bus system on a CPLD test board with an Altera MAX 7000S CPLD.

Reference Ken Reid and Robert Dueck, *Introduction to Digital Electronics*
Chapter 12: Memory Devices, Systems and Microprocessors

Equipment Required
CPLD Trainer:
Altera UP-2 circuit board with ByteBlaster download cable, or
DeVry eSOC board with USB cable, or
RSR PLDT-2 circuit board with straight-through parallel port cable, or
Equivalent CPLD trainer board with Altera EPM7128S CPLD
Quartus II Web Edition software
AC adapter, minimum output: 7 VDC, 250 mA DC
Anti-static wrist strap
#22 solid-core wire
Wire strippers

Procedure

Tristate Busses

1. Read Section 12.9 in Chapter 12 of *Introduction to Digital Electronics*.
2. Create a tristate bus system with two source registers and two destination registers, as shown in Figure 33.1. The components should have the following characteristics:

   Source1 and Source2:

   - 4-bit register.
   - Loads 4-bit data (**d_inx[3..0]**) from input pins when **srcx_ld** = 1 and clock is applied.
   - *Q* outputs, labeled **srcx_q[3..0]**, are brought to output pins for on-board monitoring.
   - *Q* outputs are applied to a 4-bit input of a quad 2-to-1 multiplexer.

   Destination1 and Destination2:

   - 4-bit register.
   - Loads 4-bit data from tristate data bus when **destx_ld** = 1 and clock is applied.
   - *Q* outputs, labeled **destx_q[3..0]** are brought to output pins for on-board monitoring.

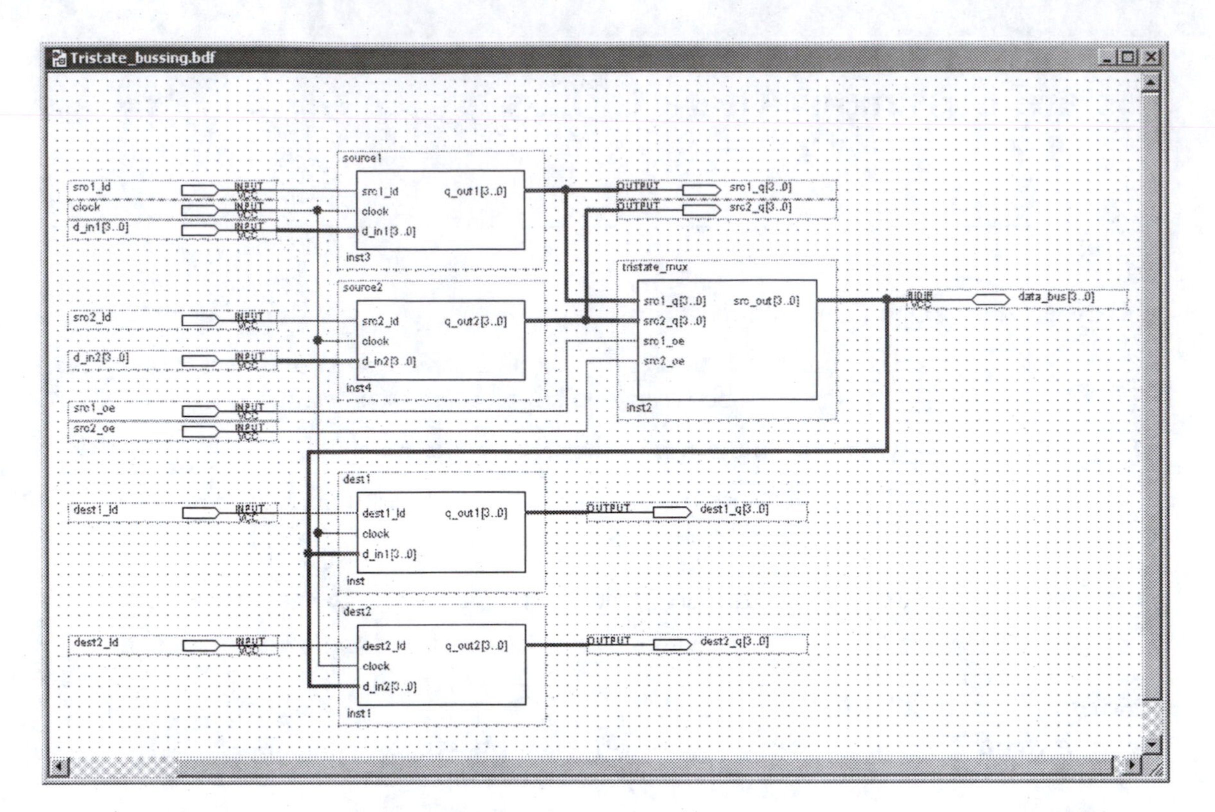

Figure 33.1 Tristate Bus System

Tristate_MUX:

- Quad 2-to-1 MUX.
- Data inputs connect to the outputs of the source registers.
- There are two select inputs, one for each source register.
- When **src1_oe** = 1 and **src2_oe** = 0, data from Source1 is on the data bus.
- When **src1_oe** = 0 and **src2_oe** = 1, data from Source2 is on the data bus.
- Otherwise, the bus is in the high-impedance state.

Note The pin on the output of the tristate MUX must be a component called **bidir.** A bidirectional pin is required on a tristate bus on an Altera CPLD.

3. Save and compile the file. Make a simulation that shows the following data transfers:
   - Load Source1 with 5H and Source2 with CH.
   - Transfer the contents of Source1 to Destination2.
   - Transfer the contents of Source2 to Destination1.

Instructor's Initials ________

4. Add a switch debouncer and pin numbers to the design. Recompile the design. Download the design to the target CPLD and demonstrate the operation of the circuit to your instructor, using the same sequence of data transfers as in the simulation.

Instructor's Initials ________

Table 33.1 EPM7128LC84-7 Pin Assignments Altera UP-2 Board and PLDT-2 Board

Pushbuttons and Clock			
Function	**Pin**	**Function**	**Pin**
PB1(pb_in)	11	PB2	1
clock	83		

DIP Switches			
Function	**Pin**	**Function**	**Pin**
src1_ld	34	src2_ld	28
src1_oe	33	src2_oe	29
dest1_ld	36	dest2_ld	30
	35		31
d_in1[3]	37	d_in2[3]	57
d_in1[2]	40	d_in2[2]	55
d_in1[1]	39	d_in2[1]	56
d_in1[0]	41	d_in2[0]	54

LED Outputs			
Function	**Pin**	**Function**	**Pin**
src1_q[3]	44	src2_q[3]	80
src1_q[2]	45	src2_q[2]	81
src1_q[1]	46	src2_q[1]	4
src1_q[0]	48	src2_q[0]	5
dest1_q[3]	49	dest2_q[3]	6
dest1_q[2]	50	dest2_q[2]	8
dest1_q[1]	51	dest2_q[1]	9
dest1_q[0]	52	dest2_q[0]	10

Table 33.2 EP2C8Q208C8N Pin Assignments DeVry eSOC Board

Pushbuttons and Clock			
Function	**Pin**	**Function**	**Pin**
PB1(pb_in)	130		
clock	24		

DIP Switches			
Function	**Pin**	**Function**	**Pin**
src1_ld	131	src2_ld	146
src1_oe	145	src2_oe	147
dest1_ld	163	dest2_ld	144
d_in1[3]	160	d_in2[3]	168
d_in1[2]	161	d_in2[2]	141
d_in1[1]	162	d_in2[1]	142
d_in1[0]*		d_in2[0]*	

LED Outputs			
Function	**Pin**	**Function**	**Pin**
src1_q[3]	115	src2_q[3]	101
src1_q[2]	114	src2_q[2]	99
src1_q[1]	113	src2_q[1]	97
src1_q[0]	112	src2_q[0]	96
dest1_q[3]	110	dest2_q[3]	95
dest1_q[2]	102	dest2_q[2]	92
dest1_q[1]	151	dest2_q[1]	118
dest1_q[0]	150	dest2_q[0]	117

**Note:* The eSOC has 8 toggle switch inputs. Wire inputs d_in1[0] and d_in2[0] to ground to allow these switches for other functions.

Lab 34

RAM and ROM in Altera CPLDs

Name ______________________ Class ____________ Date ____________

Objectives Upon completion of this laboratory exercise, you should be able to:

- Instantiate LPM components for RAM and ROM in a Quartus II Block Diagram File.
- Create simulations for the instantiated RAM and ROM components.
- Test the RAM and ROM on a FLEX 10K device on an Altera UP-1 or UP-2 CPLD test board.

Reference Ken Reid and Robert Dueck, *Introduction to Digital Electronics*
Chapter 12: Memory Devices, Systems and Microprocessors

Equipment Required
Altera LPM Quick Reference Guide
Altera UP-2 circuit board with ByteBlaster download cable
Quartus II Web Edition software
AC adapter, minimum output: 7 VDC, 250 mA DC
Anti-static wrist strap

Experimental Notes

Memory devices such as RAM and ROM are available as components in the Altera Library of Parameterized Modules (LPM). These components are implemented in the Embedded Array Blocks of a FLEX 10K device and in other higher-level Altera components. They are not supported in the MAX 7000S family.

We will be examining some very basic memory concepts, such as address and data, using an LPM RAM and an LPM ROM. The RAM that we will use (LPM_RAM_DQ) has separate data inputs and outputs. We use this component to avoid damaging the FLEX 10K device due to bus contention. We connect logic switches to the address and data inputs and monitor the output in the form of a hexadecimal digit. The 4-bit address is also monitored as a hex digit.

The ROM is configured in a similar way. Since the device is read-only, it must be initialized before it can be used. This can be done using a Memory Initialization File (MIF), a text file containing the data to be stored in ROM.

Procedure

RAM in an Altera CPLD

1. Create a Block Diagram File in Quartus II and enter the symbol for the component. LPM_RAM_DQ. Set the ports and parameters of the RAM to those shown in Figure 34.1. Make sure to invert the line for **we (write_enable)**.

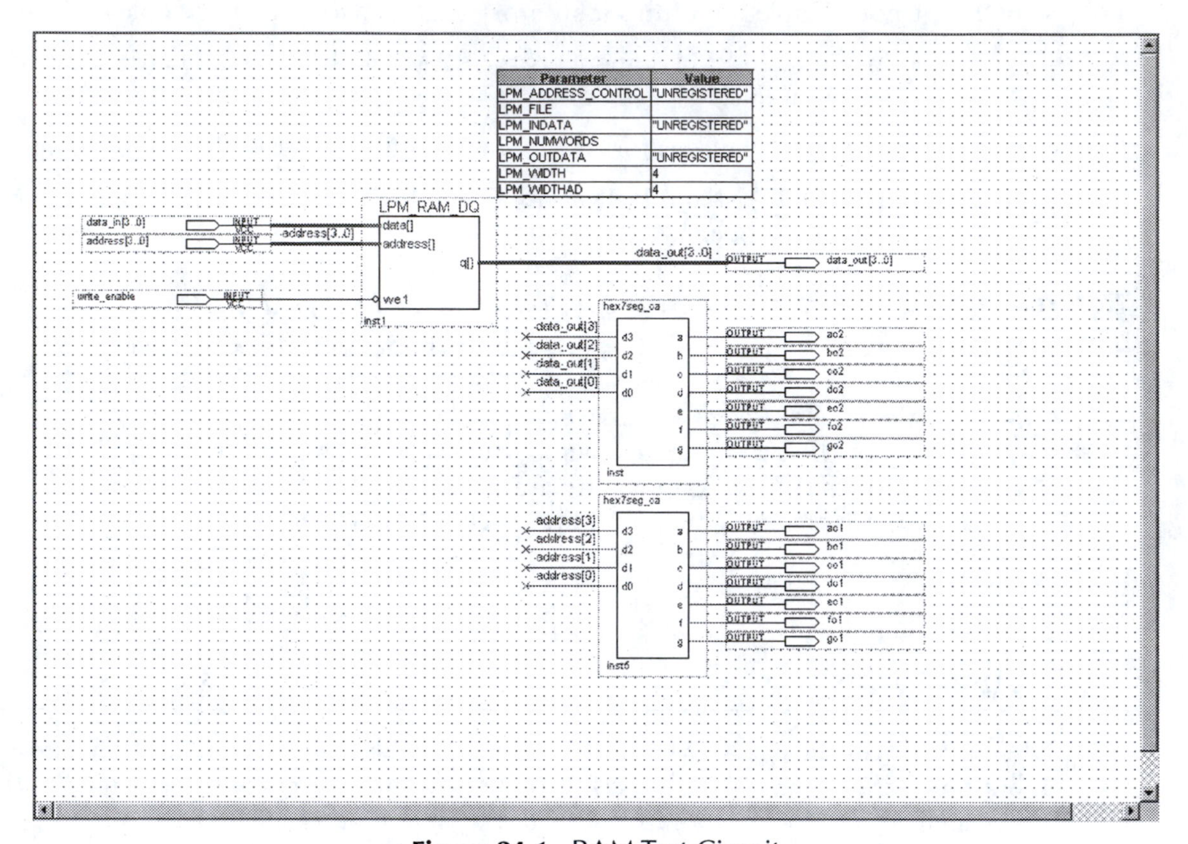

Figure 34.1 RAM Test Circuit

2. Save the file and use it to create a project in Quartus II. The device family for the project is FLEX10K. The device is EPF10K70RC240-4 for the Altera UP-2 board.

3. Copy the file for a hexadecimal-to-seven-segment converter to the working folder for the project. Save the file and include it in the new project. Use the file to create a symbol for the decoder. (Note that the seven-segment decoder must have active-low outputs.)

4. Connect the components as shown in Figure 34.1. Save and compile the project.

5. Write a set of simulation criteria that can be used to test the correctness of the design. Use the criteria to create a simulation in Quartus II. Show the simulation to your instructor.

Simulation Criteria

Instructor's Initials ________

6. Assign pin numbers to the design, as shown in Figure 34.2. Compile the project again.

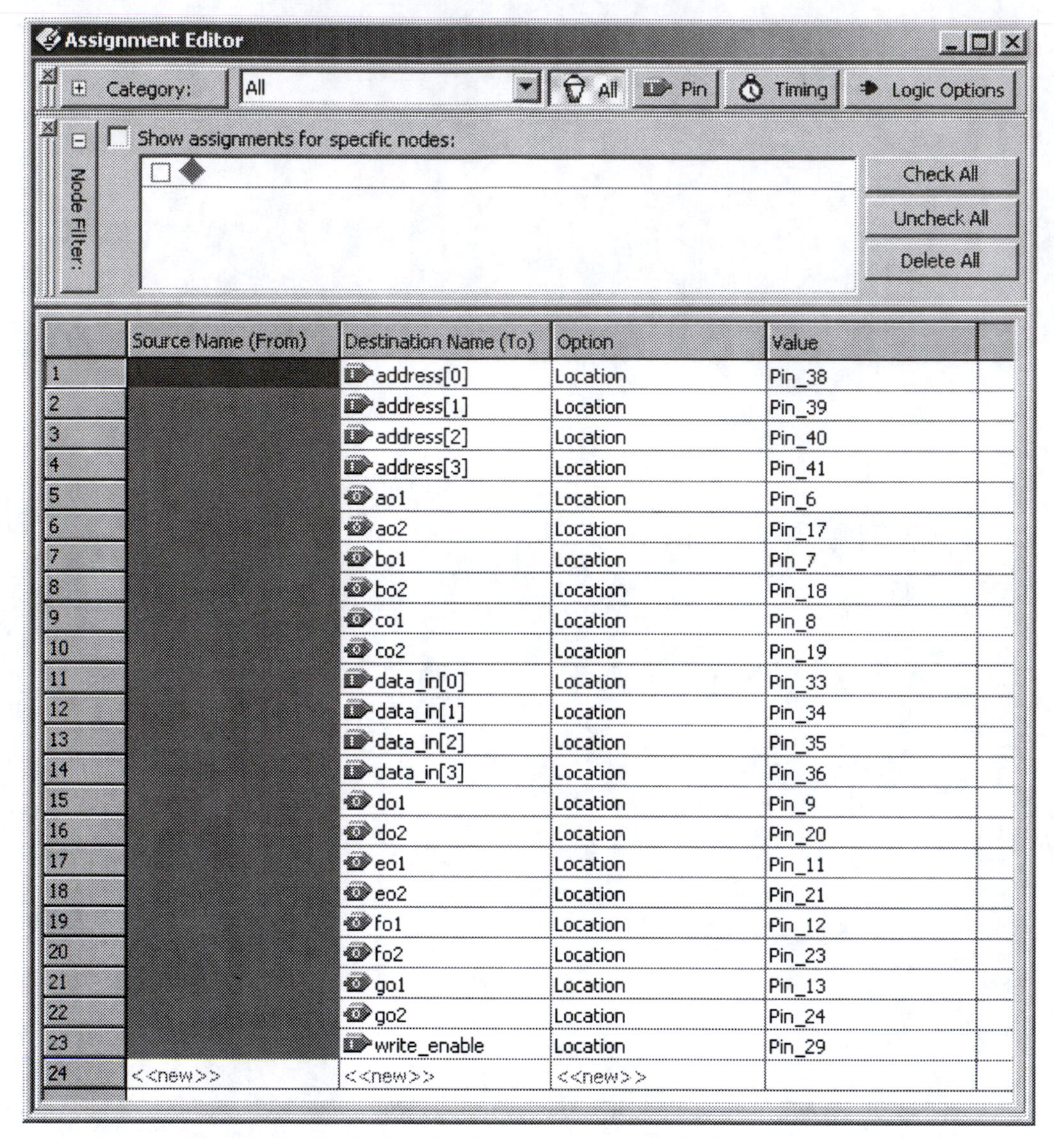

	Source Name (From)	Destination Name (To)	Option	Value
1		address[0]	Location	Pin_38
2		address[1]	Location	Pin_39
3		address[2]	Location	Pin_40
4		address[3]	Location	Pin_41
5		ao1	Location	Pin_6
6		ao2	Location	Pin_17
7		bo1	Location	Pin_7
8		bo2	Location	Pin_18
9		co1	Location	Pin_8
10		co2	Location	Pin_19
11		data_in[0]	Location	Pin_33
12		data_in[1]	Location	Pin_34
13		data_in[2]	Location	Pin_35
14		data_in[3]	Location	Pin_36
15		do1	Location	Pin_9
16		do2	Location	Pin_20
17		eo1	Location	Pin_11
18		eo2	Location	Pin_21
19		fo1	Location	Pin_12
20		fo2	Location	Pin_23
21		go1	Location	Pin_13
22		go2	Location	Pin_24
23		write_enable	Location	Pin_29
24	<<new>>	<<new>>	<<new>>	

Figure 34.2 RAM Pin Assignments (FLEX 10K Device)

7. The Altera UP-1/UP-2 board has four programming jumpers, located between the connector for the ByteBlaster cable (JTAG_IN) and the EPM7128S chip. These jumpers must be set to determine which chip on the board is to be programmed by the Quartus II software. Figure 34.3 shows the positions of the jumpers for each of the chips on the UP-1/UP-2 board.

   Change the jumpering on the Altera UP-1 or UP-2 board, as shown in Figure 34.3, to configure the EPF10K20 or EPF10K70 device.

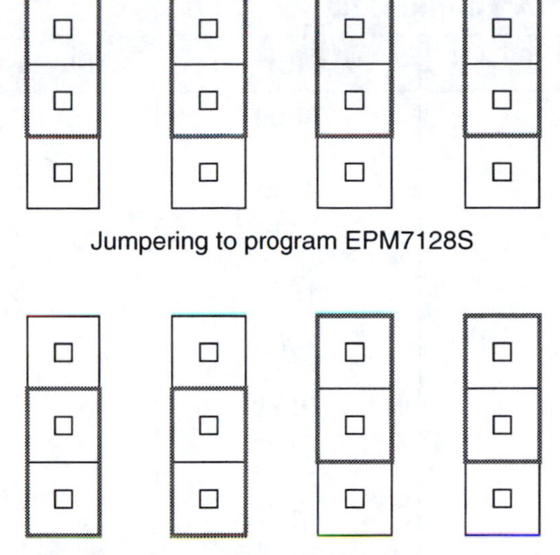

Figure 34.3 Jumpering for the Altera UP-1 or UP-2 Board

8. Open the Quartus II programmer and program the board as usual.
9. The **address** lines are the four leftmost DIP switches on the 8-position DIP switch next to the FLEX 10K device. The **data_in** lines are the rightmost four of those DIP switches. The **write_enable** line is connected to the pushbutton labeled **FLEX_PB2**. These connections are all hardwired.

   To enter a value into RAM, set the **address** switches with the location at which the data will be entered and the **data_in** switches with the data. Press the **write_enable** pushbutton. Data is now stored at the locations selected by the **address** inputs.

   The seven-segment display on the left side shows the selected address. The right-hand display shows the data stored at that address. Once data has been stored, it can be read on the seven-segment display by selecting the corresponding address.

   Fill the RAM with data shown in Table 34.1. Data are shown in binary to indicate switch positions and hex to show how the output display should look.

Table 34.1 RAM Input Data

Address (Binary)	Data (Binary)	Address (Hex)	Data (Hex)
0000	0000	0	0
0001	0011	1	3
0010	0110	2	6
0011	1001	3	9
0100	1100	4	C
0101	1111	5	F
0110	0010	6	2
0111	0101	7	5
1000	1000	8	8
1001	1011	9	B
1010	1110	A	E
1011	0001	B	1
1100	0100	C	4
1101	0111	D	7
1110	1010	E	A
1111	1101	F	D

10. Show the contents of the RAM to your instructor.

Instructor's Initials ________

ROM in an Altera CPLD

1. Create a new Block Diagram File and use it to make a new project in Quartus II. The device family for the project is FLEX10K. The device is EPF10K70RC240-4 for the Altera UP-2 board.

2. Copy the memory initialization file (**rom_init.mif**) from the "**Lab Manual Files**" folder on the CD at the back of *Introduction to Digital Electronics* to the working folder on your computer.

3. Enter the LPM_ROM component in the Block Diagram File and set its ports and parameters to those shown in Figure 34.4. Make sure that the parameter LPM_FILE is set to "rom_init.mif", with the filename in quotes.

4. Enter the symbols for the hexadecimal-to-seven-segment decoder and connect all the components, as shown in Figure 34.4. Compile the project.

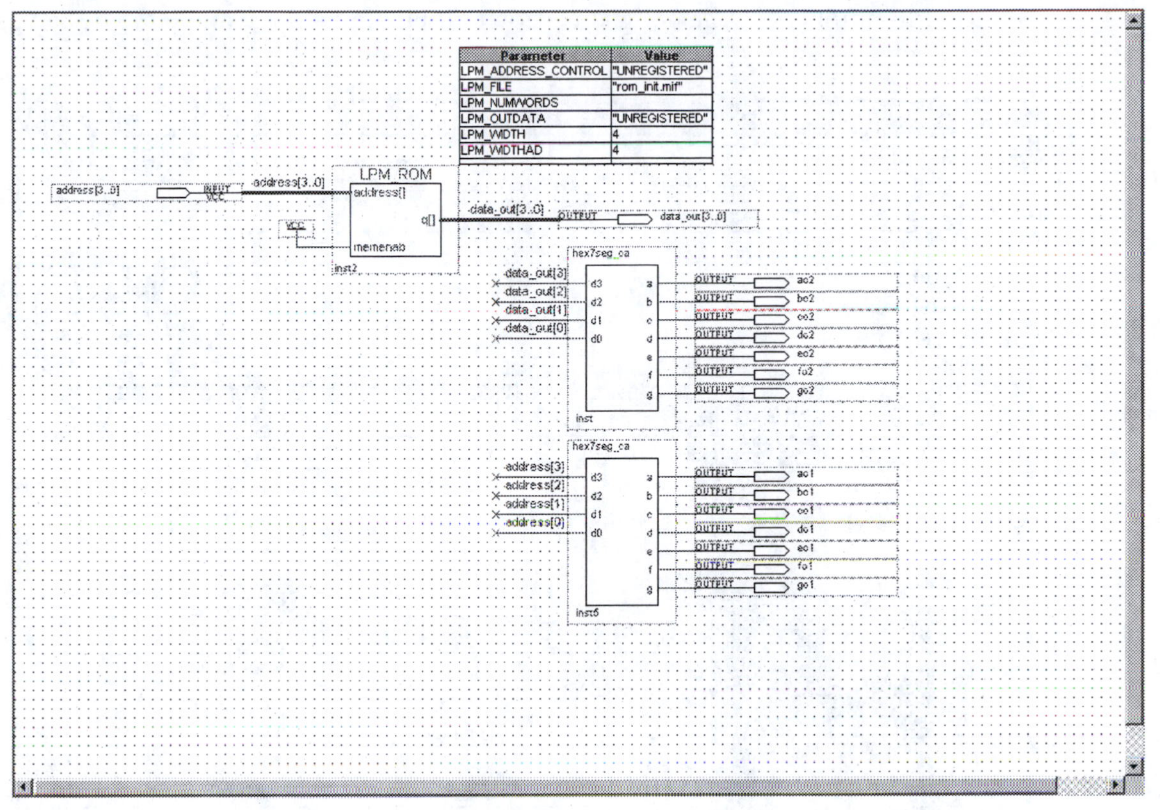

Figure 34.4 ROM Test Circuit

5. Open the file **rom_init.mif** by selecting **Open** from the **File** menu, and choosing the file type **Memory files (*.hex, *.mif)**. Fill in Table 34.2 with the data stored in the file.

Table 34.2 Data Stored in Memory Initialization File

Address	Data

6. Add pin numbers to the design, as shown in Figure 34.5. Compile the design again.

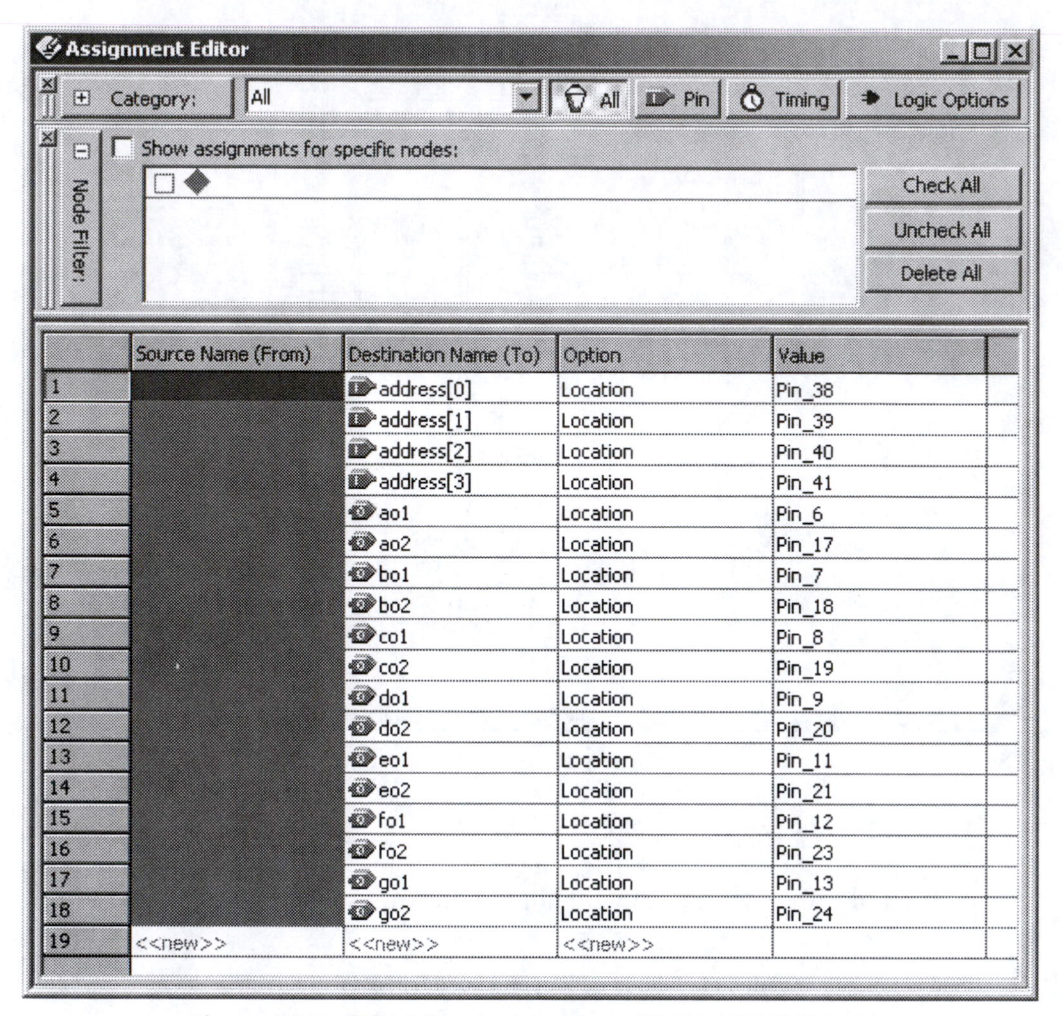

Figure 34.5 ROM Pin Assignments (FLEX 10K Device)

7. Program the CPLD test board. Address lines are the four leftmost DIP switches next to the FLEX 10K device. Apply all address combinations and read back data from the ROM. Compare the output values to those shown in Table 34.2. Demonstrate the operation of the ROM to your instructor.

Instructor's Initials ________

Lab 35

CPLD-Based Microprocessor Core

Name ______________________ Class __________ Date __________

Objectives Upon completion of this laboratory exercise, you should be able to:

- Instantiate and write machine language instructions for an intellectual property (IP) core for a simple microcomputer system (with Load, Add, Output, and Halt instructions).
- Create a simulation of the IP core showing the operation of the required program.
- Download the MCU core into an Altera FLEX 10K device to perform simple functions.
- Modify the IP core to create additional instructions (Increment, Decrement, Complement, Subtract, AND, OR, and XOR) for the MPU.
- Simulate the modified MCU to show operation of the new instructions.
- Download the modified MCU core to a FLEX 10K device and run simple programs.
- Modify the MCU to include a branch instruction. Write a machine language program that uses the branch instruction, create a simulation file, and download the modified MCU core to an Altera FLEX 10K device to demonstrate its operation.

Reference Ken Reid and Robert Dueck, *Introduction to Digital Electronics*

Chapter 12: Memory Devices, Systems, and Microprocessors

Equipment Required
Altera UP-2 circuit board with ByteBlaster download cable
Quartus II Web Edition software
AC adapter, minimum output: 7 VDC, 250 mA DC
Anti-static wrist strap

Experimental Notes

This lab exercise examines some of the functions of the CPLD-based microprocessor described in Chapter 12 of *Introduction to Digital Electronics*. Since the MAX 7000S family does not support the use of memory components such as LPM_ROM, the design must be implemented in a FLEX 10K device, such as found on the Altera UP-2 board. If the previous version, UP-1, board is used, the device is the EPF10K20RC240-4. If the UP-2 board is used, the device is the EPF10K70RC240-4. Pin numbers are the same in each case.

Procedure

Testing an IP Core

1. Copy the files for the 8-bit RISC machine from the CD at the back of *Introduction to Digital Electronics* to a working folder on your computer. This version of the machine has four instructions: Load, Add, Output, and Halt.

2. Open the project **risc8v1** and the file **risc8v1.bdf**, shown in Figure 35.1.

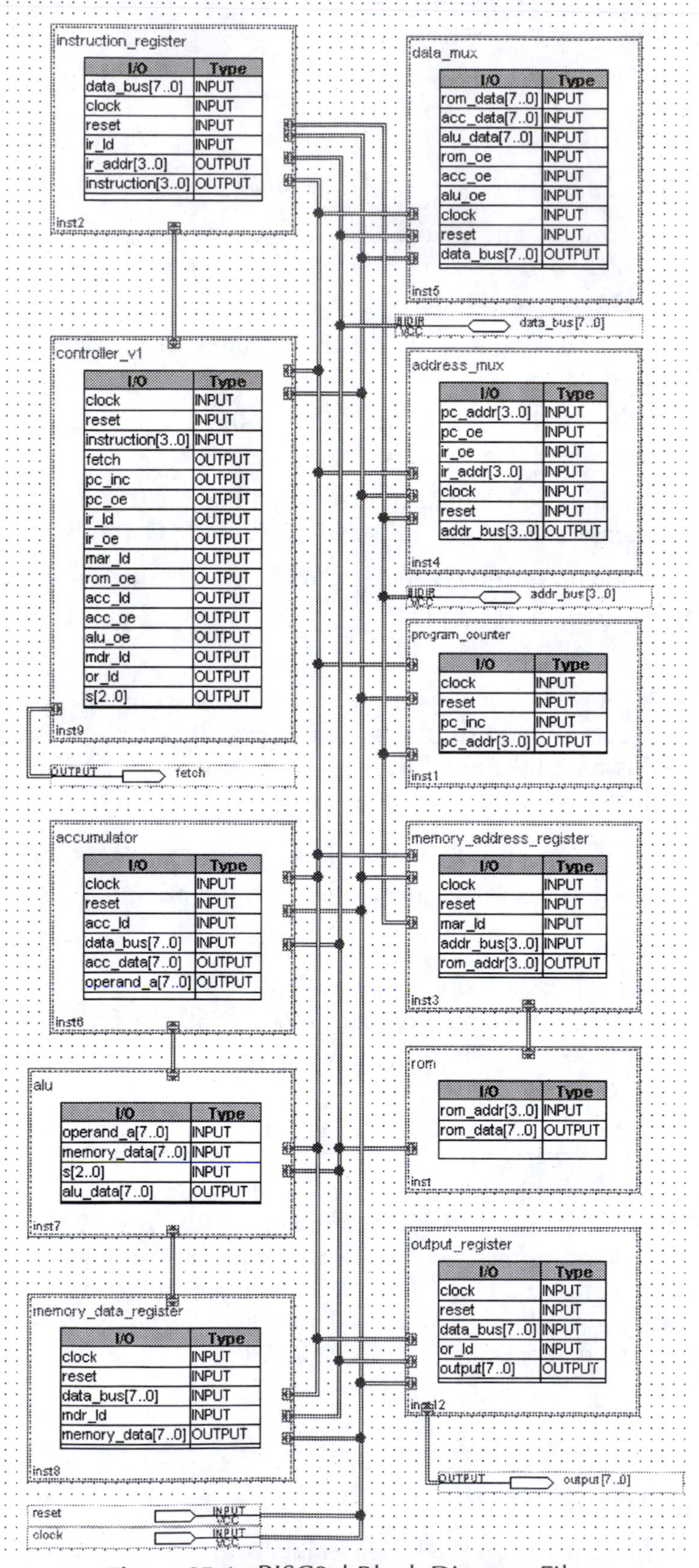

Figure 35.1 RISC8vl Block Diagram File

3. From the **File** menu, select **Open** and select **Memory files** (***.hex,*.mif**). Open the Memory Initialization File for the ROM (**program.mif**) and examine its contents. State what the program will do, step by step:

4. Compile the top-level (**risc8v1.gdf**). Device family: FLEX 10K. Target device: EPF10K70RC240-4 (Altera UP-2 board). Create a simulation that will examine the operation of the RISC machine. The only inputs required are **clock, reset, data_bus**[7..0], and **addr_bus**[3..0]. Set the clock period to 100ns. Set the ***input*** portions of the data and address bussess to high-impedance (group values of ZZ and Z, respectively).

5. Run the simulation and delete all redundant lines on the simulation result. Point out to your instructor the following features on the simulation:
   - Program counter contents.
   - Address bus contents:
     - Instruction address.
     - Operand address.
   - Data bus contents:
     - Instruction code/operand address.
     - Operand data.
   - Instruction register contents and components:
     - Instruction op code.
     - Operand address.
   - ROM addresses:
     - Instruction address.
     - Operand address.

- Accumulator contents.
- Memory data register contents.
- Output register contents.
- Controller state.

Instructor's Initials _________

6. Create a new file that contains a component for the RISC8v1 MCU. Add a 20-bit clock divider and hex-to-seven-segment decoders, as shown in Figure 35.2. Save the file as ***drive:*\qdesigns\labs\lab35\risc_on_board\risc_on_board.bdf.**

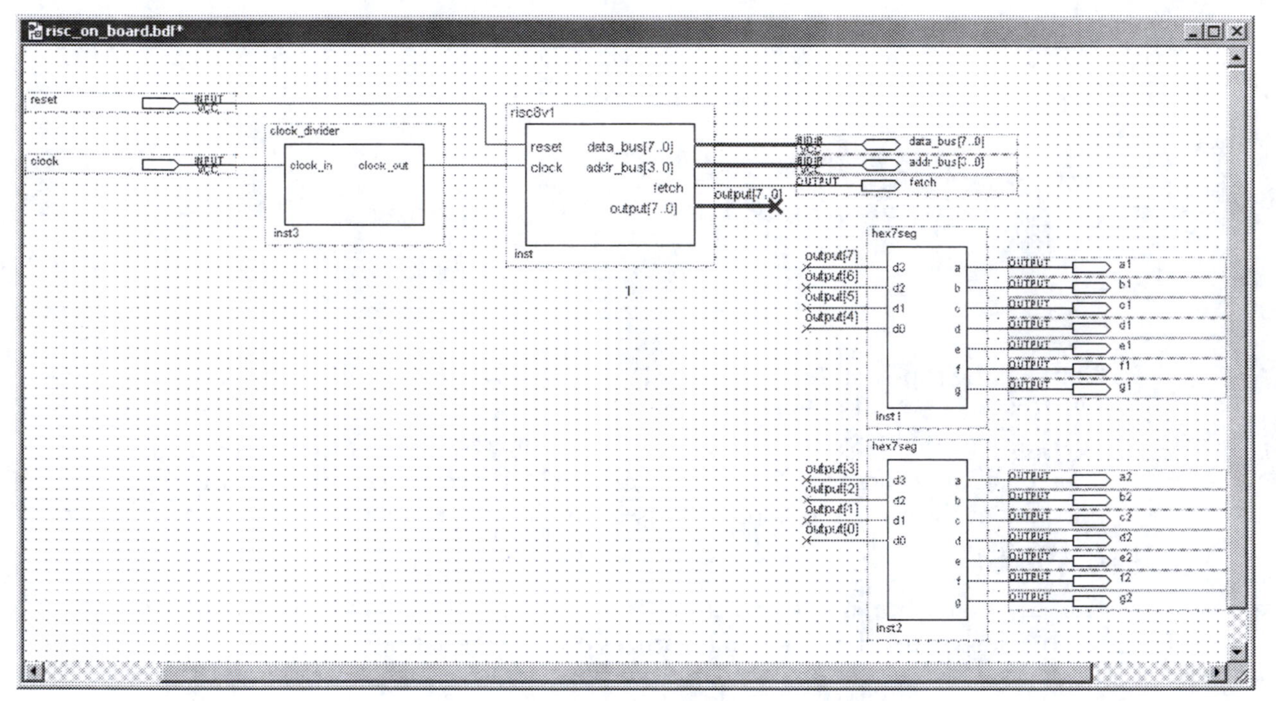

Figure 35.2 RISC Test Circuit

7. Assign pin numbers as follows: clock = pin 91, reset = pin 29, address and data busses as indicated in the following table and seven-segment displays as shown in the table in Figure 35.3.

addr_bus[3]	149	data_bus[7]	159
addr_bus[2]	148	data_bus[6]	158
addr_bus[1]	147	data_bus[5]	157
addr_bus[0]	146	data_bus[4]	156
		data_bus[3]	154
		data_bus[2]	153
		data_bus[1]	152
		data_bus[0]	151

FLEX_DIGIT Segment I/O Connections

Display Segment	Pin for Digit 1	Pin for Digit 2
a	6	17
b	7	18
c	8	19
d	9	20
e	11	21
f	12	23
g	13	24
Decimal point	14	25

Figure 35.3 Pin Assignments for Seven-Segment Digits

8. Recompile the design. Follow the remaining steps to configure the FLEX 10K chip on the Altera UP-2 board.

9. The Altera UP-2 has four programming jumpers, located between the connector for the ByteBlaster cable (JTAG_IN) and the EPM7128S chip. These jumpers must be set to determine which chip on the board is to be programmed by the Quartus II software. Figure 35.4 shows the positions of the jumpers for each of the chips on the UP-2 board.

   Change the jumpering on the Altera UP-2 board, as shown in Figure 35.4, to configure the EPF10K20 or EPF10K70 device.

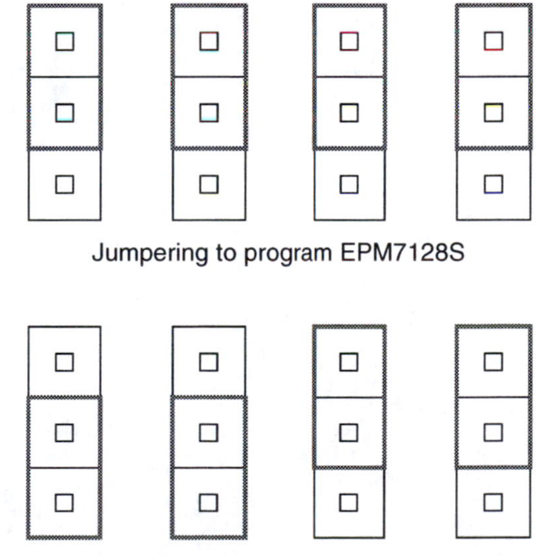

Figure 35.4 Jumpering for the Altera UP-1 or UP-2 Board

10. Open the Quartus II programmer and program as usual.

11. Demonstrate the operation of the RISC8v1 MCU to your instructor.

Instructor's Initials ________

Lab 36

Digital-to-Analog Conversion

Name ____________________ Class __________ Date __________

Objectives Upon completion of this laboratory exercise, you should be able to:

- Interface an integrated circuit digital-to-analog converter to an Altera CPLD.
- Program an Altera CPLD to control the DAC for manual input.

Reference Ken Reid and Robert Dueck, *Introduction to Digital Electronics*

Chapter 13: Interfacing Analog and Digital Circuits

Equipment Required

CPLD Trainer:

Altera UP-2 circuit board with ByteBlaster download cable, or
DeVry eSOC board with USB cable, or
RSR PLDT-2 circuit board with straight-through parallel port cable, or
equivalent CPLD trainer board with Altera EPM7128S CPLD

Quartus II Web Edition software
AC adapter, minimum output: 7 VDC, 250 mA DC
Anti-static wrist strap
Wire strippers
#22 solid-core wire
Solderless breadboard
±12-volt power supply
DAC0808 or MC1408 digital-to-analog converter
High-speed op amp (MC34071 or TL071)
0.1 μF capacitor
0.75 pF capacitor
2.7 kΩ resistor
4.7 kΩ resistor
6.8 kΩ resistor
5 kΩ potentiometers (2)

Experimental Notes

A monolithic (one-chip) digital-to-analog converter (DAC), such as the DAC0808 or MC1408, can be easily interfaced to a CPLD for a static or dynamic output.

The CPLD can simply pass through the values of a set of DIP switches, or other digital inputs, so that the DC output can be adjusted to a desired value within its range. A procedure for calibrating such a circuit is detailed in Chapter 13 of *Introduction to Digital Electronics*.

The DAC can also be configured to generate a periodic time-varying output.

Procedure

DAC/CPLD Interface

1. Wire the circuit for a DAC0808 digital-to-analog converter onto a solderless breadboard, as shown in Figure 36.1. The pin numbers for the device are indicated in parentheses.

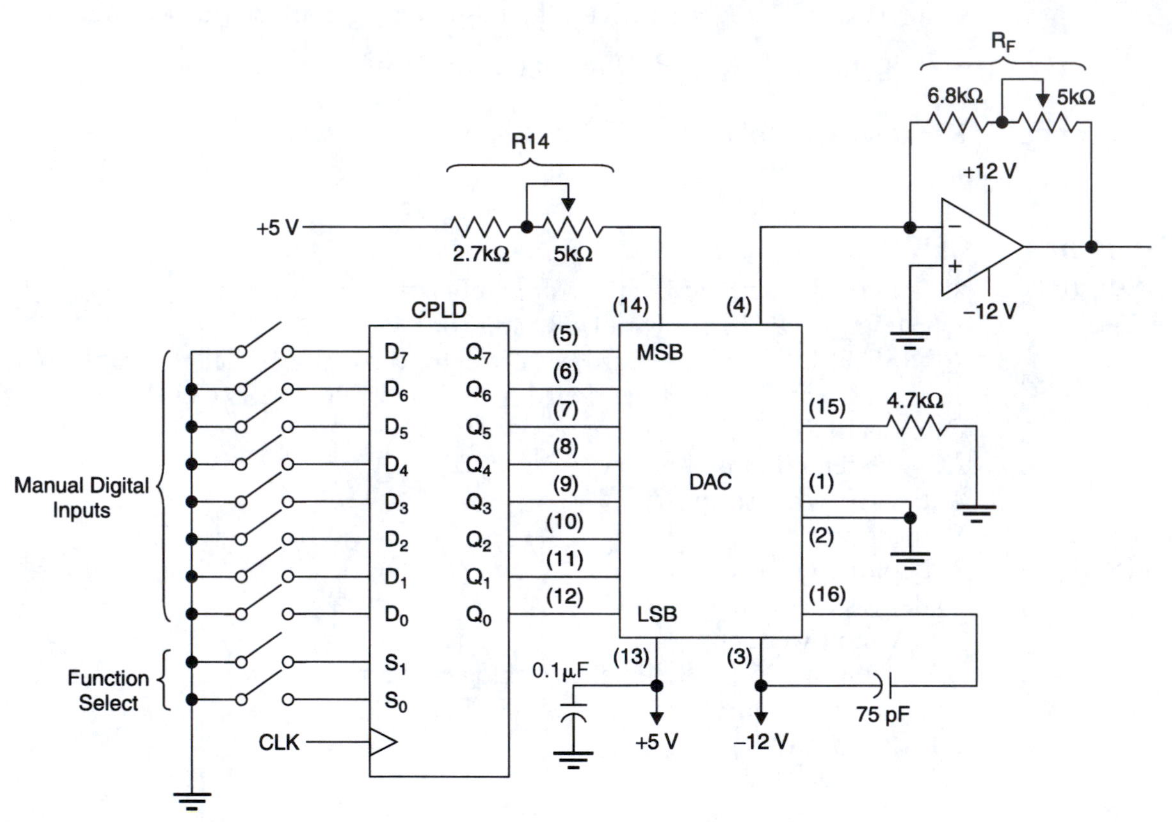

Figure 36.1 DAC Function Generator

2. Create a Quartus II file that will read the values of eight DIP switches at a set of CPLD input pins and pass them through the CPLD to a set of output pins without modification (i.e., an internal pin-to-pin connection). Although the DAC can be directly connected to the DIP switches, it is efficient to pass the signals through the CPLD so that we can keep these same pin connections, without rewiring, for later circuits explored in Lab 37.

3. Use the pin assignments shown in Table 36.1. Disconnect any LED connections for LED9 through LED16. Use the pins normally assigned to these LEDs in other projects to connect from the CPLD to the DAC inputs.

Table 36.1 Pin Assignments for DAC/CPLD Interface

		Pin Numbers		
Function	**Device**	**UP-2**	**PLDT-2**	**eSOC**
D[7]	SW1-1	34	34	160
D[6]	SW1-2	33	33	161
D[5]	SW1-3	36	36	162
D[4]	SW1-4	35	35	163
D[3]	SW1-5	37	37	168
D[2]	SW1-6	40	40	141
D[1]	SW1-7	39	39	142
D[0]	SW1-8	41	41	144
Q[7]	DAC pin 5 (MSB)	80	80	101
Q[6]	DAC pin 6	81	81	99
Q[5]	DAC pin 7	4	4	97
Q[4]	DAC pin 8	5	5	96
Q[3]	DAC pin 9	6	6	95
Q[2]	DAC pin 10	8	8	92
Q[1]	DAC pin 11	9	9	118
Q[0]	DAC pin 12 (MSB)	10	10	117
S[1]	SW2-7	56	56	130
S[0]	SW2-8	54	54	131
CLK	clock	83	83	24

4. Compile and download the file to the CPLD board.

5. The resistor networks shown in Figure 36.2 allow us to set our input reference current and output gain to values within a specified range.

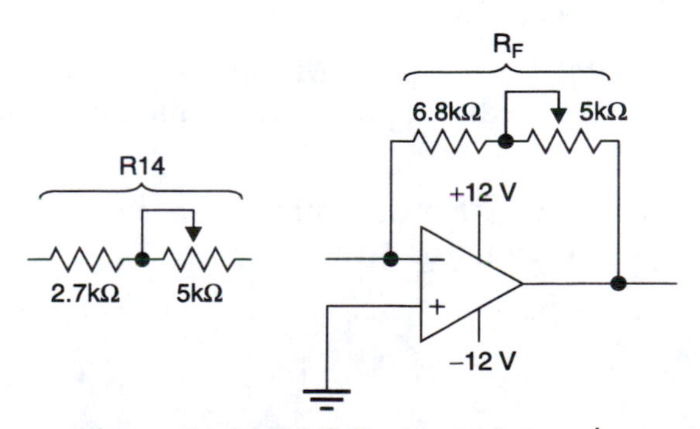

Figure 36.2 DAC Resistor Networks

Using the values shown in Figure 36.2, fill in Table 36.2 for the cases when V_o is at minimum and maximum, and when the pots are at their midpoint values. (R_{14} is the total value of the input reference resistance. R_F is the total resistance of the op amp feedback network.) Assume that the DAC input is set to 1111 1111. Show calculations in the space provided.

Table 36.2 DAC Output Range

	$R_{14}(\Omega)$	$R_F(\Omega)$	I_{ref}(mA)	I_o(mA)	V_o(V)
Minimum V_o					
Maximum V_o					
Pots at midpoint					

Calculations:

6. Calibrate the DAC as follows:

Connect a digital multimeter (DMM) to the op amp output. Apply power to the circuit and adjust the feedback pot for the minimum value of voltage. Set the input code to 10000000.

Adjust the R_{14} pot so that the output voltage of the op amp is 3.4 volts. What value of I_{ref} does this correspond to? I_{ref} = ______.

Set the feedback pot so that the output voltage is 5 volts. What value of R_F does this correspond to? R_F = ______

Measure the output voltage of the DAC circuit for the digital input values in Table 36.3.

Table 36.3 DAC Output Voltages for Manual Inputs

Input Code	Output Voltage (Calculated)	Output Voltage (Measured)	Error (volts)	Error (LSB)
00000000				
00000001				
00000011				
00000111				
00001111				
10000000				
11000000				
11100000				
11111111				

State the calculated value of resolution (i.e., the value of 1 LSB) for this DAC.

Resolution = ______ volts

Instructor's Initials ________

Do not disassemble the DAC circuit. It will be reused in Lab 37.

Lab 37

DAC-Based Function Generator

Name ______________________ Class ____________ Date ____________

Objectives Upon completion of this laboratory exercise, you should be able to:

- Interface an integrated circuit digital-to-analog converter to an EPM7128S CPLD.
- Program an Altera CPLD to make the DAC generate sawtooth, square, and triangle waves, with function selectable by DIP switch input.

Reference Ken Reid and Robert Dueck, *Introduction to Digital Electronics*
Chapter 13: Interfacing Analog and Digital Circuits

Equipment Required
CPLD Trainer:
Altera UP-2 circuit board with ByteBlaster download cable, or
DeVry eSOC board with USB cable, or
RSR PLDT-2 circuit board with straight-through parallel port cable, or
equivalent CPLD trainer board with Altera EPM7128S CPLD
Quartus II Web Edition software
AC adapter, minimum output: 7 VDC, 250 mA DC
Anti-static wrist strap
Wire strippers
#22 solid-core wire
Solderless breadboard
±12-volt power supply
DAC0808 or MC1408 digital-to-analog converter
High-speed op amp (MC34071 or TL071)
0.1 μF capacitor
0.75 pF capacitor
2.7 kΩ resistor
4.7 kΩ resistor
6.8 kΩ resistor
5 kΩ potentiometers (2)

Experimental Notes

A DAC can be configured to generate a periodic time-varying output. To do so, the changing set of values is fed to the DAC inputs. If the output of a binary counter is connected to the DAC input, the result is a linearly increasing output with an (ideally) instantaneous return to zero at the end of each cycle. This function is also known as a sawtooth or ramp waveform. An example of a DAC-based ramp generator is shown in example 13.14 of *Introduction to Digital Electronics*.

Other waveforms can be generated by changing the output sequence of the counter that drives the DAC inputs. The output behavior of such a counter can be described by a series of VHDL statements.

Procedure

DAC Ramp Generator

1. Refer to the DAC-based ramp generator in example 13.14 from *Introduction to Digital Electronics*. Create a circuit similar to the ramp generator in Figure 13.20 in the textbook by programming an 8-bit counter into the EPM7128S CPLD. The DAC-to-CPLD interface is the same as the one used in Lab 36 (Figure 36.1). The counter driving the ramp generator should be clocked at about 1.57–2 MHz by the output of a clock divider. (The on-board oscillator of the Altera UP-2 board runs at 25.175 MHz. The RSR PLDT-2 board has an on-board oscillator with a frequency of 4 MHz. The DeVry eSOC board has a 24 MHz oscillator.)

2. Assign pins to the counter design file so that the counter outputs are the same as the pins connected to the DAC inputs. (Refer to Table 37.1) Compile and download the Quartus II counter file.

Table 37.1 Pin Assignments for DAC/CPLD Interface

		Pin Numbers		
Function	**Device**	**UP-2**	**PLDT-2**	**eSOC**
D[7]	SW1-1	34	34	50
D[6]	SW1-2	33	33	51
D[5]	SW1-3	36	36	52
D[4]	SW1-4	35	35	54
D[3]	SW1-5	37	37	55
D[2]	SW1-6	40	40	56
D[1]	SW1-7	39	39	57
D[0]	SW1-8	41	41	58
Q[7]	DAC pin 5 (MSB)	80	80	16
Q[6]	DAC pin 6	81	81	17
Q[5]	DAC pin 7	4	4	18
Q[4]	DAC pin 8	5	5	20
Q[3]	DAC pin 9	6	6	21
Q[2]	DAC pin 10	8	8	22
Q[1]	DAC pin 11	9	9	24
Q[0]	DAC pin 12 (MSB)	10	10	25
S[1]	SW2-7	56	56	68
S[0]	SW2-8	54	54	69
CLK	clock	83	83	83

3. Connect an oscilloscope to the DAC op amp output. Draw the sawtooth waveform generated by the DAC. Measure its period and calculate the sawtooth frequency.

   T = __________; = f__________.

   Sketch of a sawtooth waveform:

4. Divide the counter clock frequency (f_c) by the DAC output frequency (f_{DAC}) to get an estimate of the number of clock pulses per sawtooth cycle. Since the clock signal from the on-board oscillator is divided by a clock divider inside the CPLD, the value of f_c cannot be measured directly unless the divided clock is brought out on a CPLD pin. Alternatively, you can measure the frequency at the LSB input of the DAC (pin 12) and multiply by 2.

   Compare the calculated ratio of f_c/f_{DAC} to the ideal value and determine the % error. Also state the % error of an oscilloscope measurement. (As long as you can measure to within the error of the oscilloscope, the measurement is reasonably accurate.) Estimate the oscilloscope error as follows:

   - Count the number of small divisions on the oscilloscope horizontal grid line.
   - Estimate what fraction of a small division it is possible to measure.
   - Divide the measurable fraction of a small division by the total number of small divisions and multiply by 100%.

f_c/f_{DAC} = _______ clock cycles (measured)

f_c/f_{DAC} = _______ clock cycles (ideal)

% error (frequency measurement) = _______

% error (oscilloscope screen) = _______

Instructor's Initials __________

Lab 38

Analog-to-Digital Conversion

Name ________________________ Class ____________ Date ____________

Objectives Upon completion of this laboratory exercise, you should be able to:

- Design, simulate, program, and test an interface between an Altera CPLD and an ADC0808 analog-to-digital converter.
- Determine the effect of sampling frequency on aliasing for an ADC0808 A/D converter.

Reference Ken Reid and Robert Dueck, *Introduction to Digital Electronics*

Chapter 13: Interfacing Analog and Digital Circuits

Equipment Required CPLD Trainer:

Altera UP-1 or UP-2 circuit board with ByteBlaster download cable, or
DeVry eSOC board with USB cable, or
RSR PLDT-2 circuit board with straight-through parallel port cable, or
equivalent CPLD trainer board with Altera EPM7128S CPLD

Quartus II Web Edition Software
AC adapter, minimum output: 7 VDC, 250 mA DC
Anti-static wrist
strap
#22 solid-core wire
Wire strippers
Solderless breadboard
ADC0808 analog-to-digital converter
1N4004 diodes or equivalent (4)

Experimental Notes

The ADC0808 analog-to-digital converter (ADC), shown in Figure 38.1, is a successive approximation ADC with eight multiplexed inputs. The ADC converts one channel at a time, as selected by the binary combination of inputs *ADDA*, *ADDB*, and *ADDC*, where *ADDC* is the most significant bit. The ADC requires a pulse on the address latch enable (*ALE*) line and on the *START* input to begin the conversion process.

An ADC output called *EOC* (End of Conversion) goes LOW near the beginning of the conversion process and then goes HIGH to indicate that the conversion is complete. Figure 38.2 shows the relative timing of the device. Further details are given in Chapter 13 of *Introduction to Digital Electronics*.

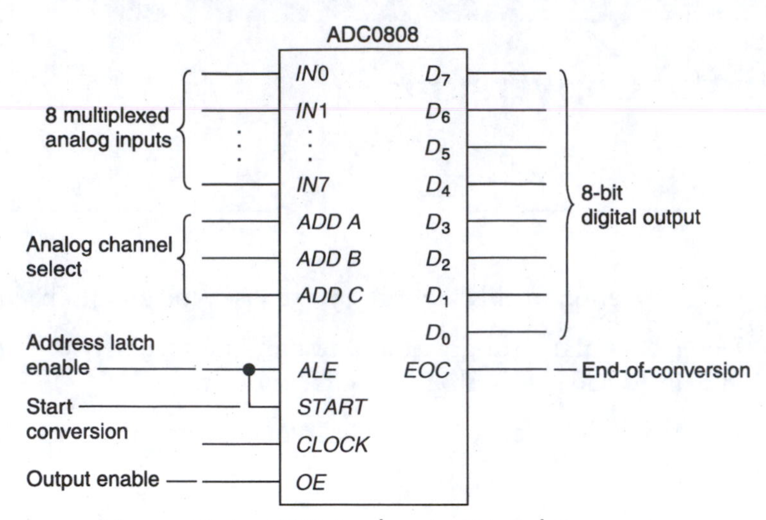

Figure 38.1 ADC Analog-to-Digital Converter

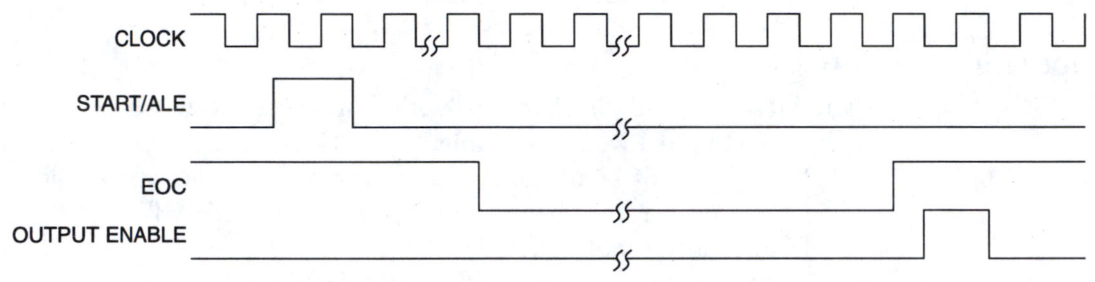

Figure 38.2 ADC0808 Timing

Procedure

CPLD-to-ADC Interface

1. Design a state machine controller for an ADC0808 converter, based on the state diagram of Figure 38.3. Create a Quartus II simulation for the controller. Show the controller waveforms to your instructor.

Instructor's Initials ________

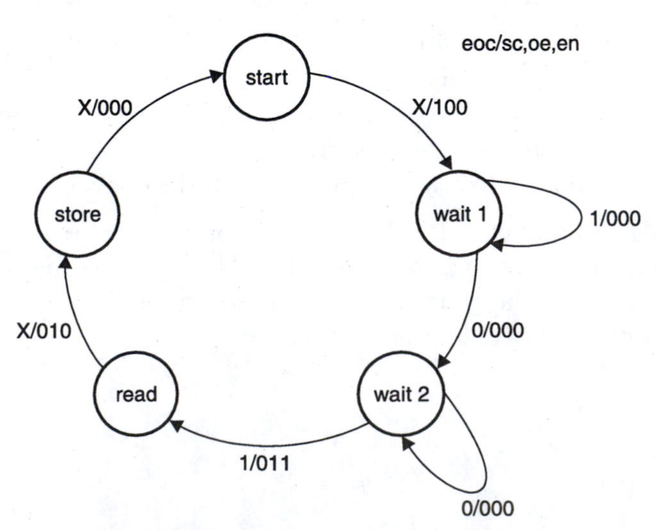

Figure 38.3 State Diagram for Continuous-Convert ADC Controller

2. Add a clock divider and an output latch to your controller, as shown in Figure 38.4. The clock frequency of the ADC0808 must be kept below 1280 kHz. Divide the clock of your CPLD board accordingly. (Altera UP-2 board: on-board oscillator is 25.175 MHz; RSR PLDT-2 on-board oscillator is 4 MHz; DeVry eSOC is 24 MHz.) Add a pair of seven-segment decoders to the latch outputs (Q_7 through Q_0).

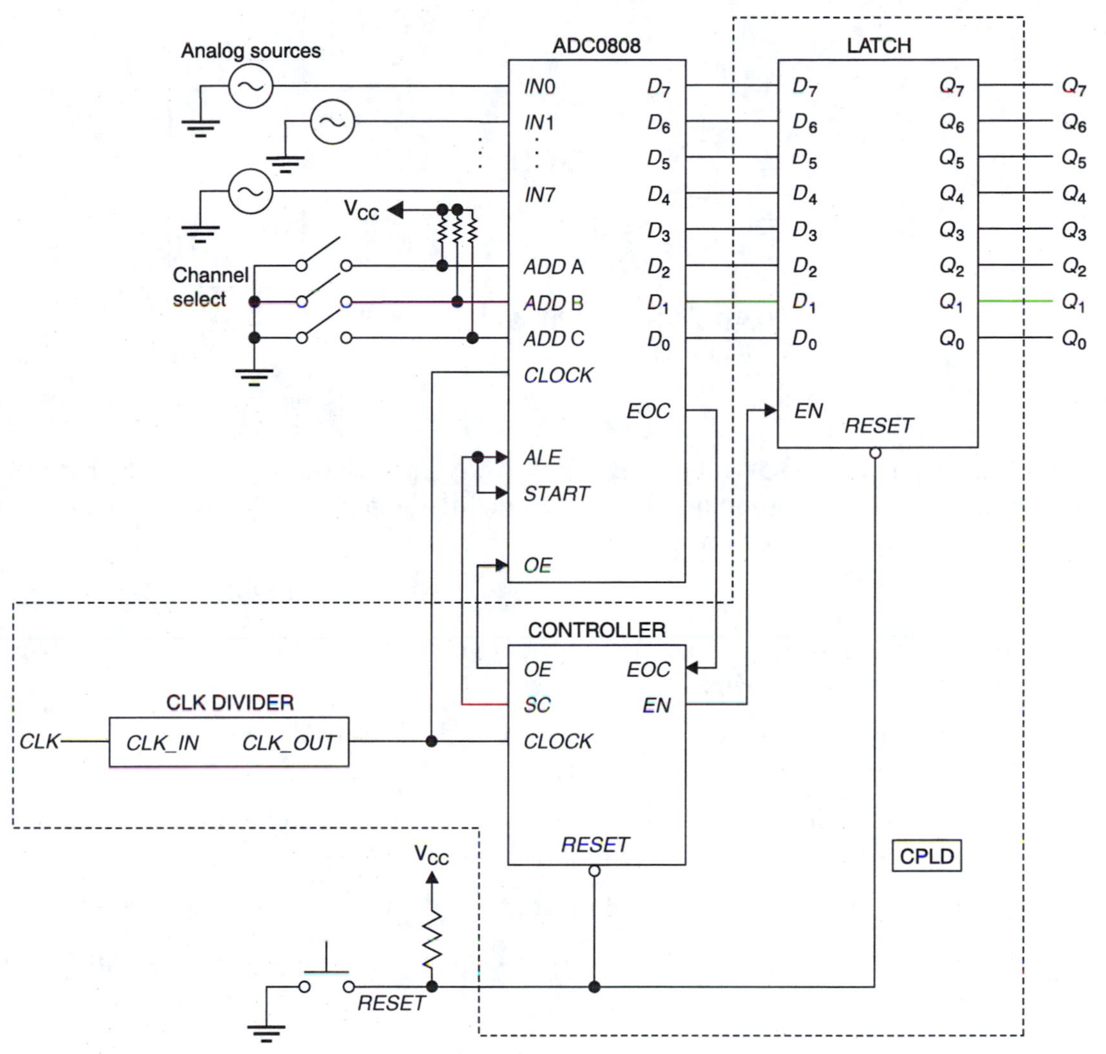

Figure 38.4 ADC Interface with One Output Channel and Manual Input Channel Selection

3. Assign pins to the CPLD for the design shown in Figure 38.4. Be sure to include pin assignments for the seven-segment displays. Compile and download the file to your CPLD board.

4. Refer to the datasheet for the ADC0808 (available on the Internet from National Semiconductor at **<http://www.national.com>**). Connect the ADC0808 to the CPLD board, as shown in Figure 38.4. Connect the current and voltage protection networks shown in Figure 38.5 to input channels *IN*0 and *IN*1.

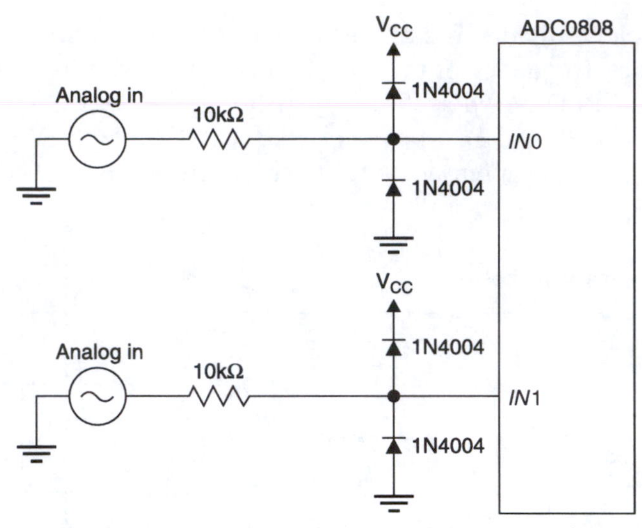

Figure 38.5 Current and Voltage Protection for ADC Inputs

5. Connect a variable 5-volt power supply to *IN*0 and select the channel by the position of the address DIP switches. Before turning on the power, show the wiring of your circuit to your instructor.

Instructor's Initials ________

6. Turn on the power and vary the analog input connected to *IN*0. Monitor the analog input voltage with a Digital Multimeter.
   - Note the input voltage at the point where the displayed digit changes from 00 to 01 and enter this value in Table 38.1.
   - For each of the remaining entries in the table, note the lowest voltage and highest voltage that will produce the specified two-digit code.
   - Find the average value of the lowest and highest voltages for each two-digit code.
   - Compare the average value with the calculated nominal input voltage for each code.

Table 38.1 Analog-to-Digital Converter Test Data (Channel 0)

Digital Output Code	Lowest Input Voltage	Highest Input Voltage	Average Input Voltage	Calculated Input Voltage
00				
01				
01				
80				
C0				
FF				

7. Repeat the measurements on Channel 1 and fill in Table 38.2.

Table 38.2 Analog-to-Digital Converter Test Data (Channel 1)

Digital Output Code	Lowest Input Voltage	Highest Input Voltage	Average Input Voltage	Calculated Input Voltage
00				
01				
01				
80				
C0				
FF				

8. Monitor the *START* pin of the ADC with an oscilloscope. What is the time between successive conversions, based on the spacing of pulses on the *START* line?

____

9. Demonstrate the operation of the circuit to your instructor.

Instructor's Initials ________